p. 384

WOLVES OF THE WORLD

NOYES SERIES
IN
ANIMAL BEHAVIOR, ECOLOGY,
CONSERVATION AND MANAGEMENT

A series of professional and reference books in ethology devoted to the better understanding of animal behavior, ecology, conservation, and management.

WOLVES OF THE WORLD

Perspectives of Behavior, Ecology, and Conservation

Edited by

FRED H. HARRINGTON

Mount Saint Vincent University
Halifax, Nova Scotia
Canada

and

PAUL C. PAQUET

Portland State University
Portland, Oregon
U.S.A.

np **NOYES PUBLICATIONS**
Park Ridge, New Jersey, U.S.A.

Published in the United States of America by
Noyes Publications
Mill Road, Park Ridge, New Jersey 07656

10 9 8 7 6 5 4 3 2 1

Library of Congress Cataloging in Publication Data
Main entry under title:

Wolves of the world.

Most of the papers included were originally
presented at the 1979 Portland International Wolf
Symposium.
Includes bibliographies and index.
1. Wolves--Congresses.   I. Harrington, Fred H.
II. Paquet, Paul C.   III. Portland International
Wolf Symposium (1979 : Or.)
QL737.C22W65   599.74'442   82-3397
ISBN 0-8155-0905-7   AACR2

Preface

"Only a mountain has lived long enough to
listen objectively to the howl of a wolf"
—Aldo Leopold, *Game Management*, 1933

Aldo Leopold was mistaken. We have lived and interacted with
wolves from prehistoric to modern times, certainly ample time to
comprehend and view them rationally. Exactly how long our ac-
quaintance has endured is uncertain, but our association probably began when our ancestors first immigrated to the northern hemisphere.
There, we encountered our ecological counterpart and perhaps most
direct competitor, the wolf. Our lives intertwined to such an extent
that we selected the wolf (or were selected by it) as our first domestic companion. Yet despite countless opportunities over untold generations, our perceptions of wolves reflected our prejudices more
than reality.

Our fancied notions about wolves finally began to erode in the
1940s when Adolf Murie, Rudolf Schenkel, Ian McTaggart Cowan
and Sigurd Olsen pioneered modern, scientific study into the animal's ways. During the following decade, insights continued to accumulate, although investigators were few and the research lacked a
consistent direction. Near the end of the 1950s, however, two pro-
jects were conceived which would soon provide the needed direction.

Durward Allen recognized that Isle Royale was the perfect wilderness laboratory—isolated, relatively small, yet teeming with wolves,
moose and beaver. For two decades he guided a line of graduate students through a series of ground-breaking studies. At the same time,
Douglas Pimlott saw the potential of Algonquin Provincial Park.
Pimlott's students, like Allen's, broke ground in a number of areas,
including the introduction of the radio-collar in wolf research.

By 1970, the stage was set for synthesis. David Mech, the first of
Allen's students on Isle Royale, gathered together what was then

known into *The Wolf*. The book provided an immediate focus for subsequent research efforts which, until then, had been somewhat diffused. At the same time, Mech brought the radio-collar to northern Minnesota and initiated a study which would become a model and standard for future work. The effects were staggering. Currently, field studies using radio-telemetry are being pursued from Alaska south to Minnesota and Montana, in Europe, the Middle East, and perhaps soon in Asia. At present, observations of wolves are outstripping our ability to synthesize and place them into perspective.

The 1980s should provide us with a more complete and refined picture of wolves, a picture with sufficient complexity to encompass the extensive variation now seen amongst them. This book is part of that process. Most of the chapters included were originally presented at the 1979 Portland Wolf Symposium, Portland, Oregon. That meeting brought together the majority of wolf researchers from around the world. Our original hope was that the conference would generate an updated synthesis of wolf ecology and behavior. We now realize that our hope was premature. As more data have accumulated, the less secure our previous generalizations about wolves seem. More data are required, data from diverse areas under varied environmental conditions. We must avoid hurrying our observations toward premature conclusions that so easily, and erroneously, imply "this is THE WOLF."

At present, there is an unspoken urgency as wolf researchers pursue their quarry. So often, intriguing observations are left dangling as wolves, packs, or even populations succumb to civilization's onslaught. Our studies are no longer esoteric exercises in biology and psychology. What we learn may, or rather must, be used eventually to ensure that wolves continue to exist. Thus, the papers in this volume sample areas which we feel are necessary if we are to secure this planet for the wolves.

The first two sections survey the areas of ecology and behavior in the wild. Because an animal's behavior is dependent on its ecology, we have made no effort to segregate these two areas. Rather, we have made our division between "disturbed" and "undisturbed" habitats. Studies of wolves in North America provide a view of wolves as classic predators of big game in relatively undisturbed wilderness habitat. Here we can patiently attempt to disentangle the complex web of interrelations that characterize "natural" communities. As we learn more about the workings of these systems, we can make wiser decisions on management necessitated by our society's appetite for minerals, lumber and energy. Studies of wolves in Eurasia can then provide an important and necessary contrast, and perhaps a glimpse into the future prospects of wolves in areas still pristine by 1980 standards. Eurasian wolves have managed to survive in habitats long altered by us. If we can understand how they have managed to adapt to the drastic alterations we have made in their environment, we may have an idea of how great a shock wolves can withstand, and what we can do to lessen that shock.

The third section provides a sampling of behavior studies conducted in captivity. Captive studies have often been maligned because they inevitably must confine a wide-ranging animal to a mere fraction of what it might normally traverse. But wide-ranging wolves are also elusive. Captive studies allow us to focus on those rarely seen aspects of wolf life and permit us deeper insights into other areas just glimpsed in the field.

Conservation is the subject of the fourth section. With wolf populations variously labeled as "threatened", "endangered" and "extinct", and further inroads being made into wolf habitat day by day, it may soon not be enough to merely set aside refuges for the remaining populations. As several papers in this section point out, we must use our knowledge of wolf behavior and ecology to balance our demands for the wolf's habitat with our goals for wolf preservation.

We conclude the book with two perspectives on wolves, and wolf study, from observers outside the typical biologist/psychologist frame of mind. As philosophers of science have pointed out for years, our observations as scientists are biased by our cultural, sociological and scientific backgrounds. A recognition and understanding of other viewpoints can only serve to open our eyes wider and allow us to "see" phenomena to which we might otherwise have been blind.

Although we have presented the major areas of present-day wolf research, our collection is by no means exhaustive. Some areas have been underrepresented and others, unfortunately, are absent. The areas which are included, however, should provide an overview indicating where the study of wolves is headed during the 1980s.

Halifax, Nova Scotia   Fred H. Harrington
Portland, Oregon   Paul C. Paquet
April 1982

Contributors

Warren B. Ballard
Alaska Department of Fish &
Game
Glennallen, Alaska

William E. Berg
Minnesota Department of
Natural Resources
Grand Rapids, Minnesota

Dmitri I. Bibikov
A.N. Severstov Institute of
Animal Evolution
Morphology and Ecology
USSR Academy of Science
Moscow, USSR

Anders Bjarvall
National Environmental
Protection Board
Solna, Sweden

Luigi Boitani
Istituto di Zoologia,
Universita di Roma
Rome, Italy

Susan Bragdon
Williams College
Williamstown, Massachusetts

Ludwig N. Carbyn
Canadian Wildlife Service
Edmonton, Alberta, Canada

Carl D. Cheney
Institute of Animal Behavior
Utah State University
Logan, Utah

Rick Davies
B.C. Fish and Wildlife Branch
Nanaimo, British Columbia,
Canada

John C. Fentress
Department of Psychology
Delhousie University
Halifax, Nova Scotia, Canada

Patrick M. Ghezzi
Utah State University
Logan, Utah

Fred H. Harrington
Psychology Department
Mt. St. Vincent University
Halifax, Nova Scotia, Canada

Daryll M. Hebert
B.C. Fish and Wildlife Branch
Nanaimo, British Columbia, Canada

Robert E. Henshaw
N.Y. Department of Environmental
Conservation
Albany, New York

Richard A. Hook
Department of Biology
Northern Michigan University
Marquette, Michigan

Erik Isakson
National Environmental Protection
Board
Solna, Sweden

David James
Alaska Department of Fish & Game
Fairbanks, Alaska

Doug Janz
B.C. Fish and Wildlife Branch
Nanaimo, British Columbia, Canada

Paul Joslin
Chicago Zoological Society
Brookfield, Illinois

David W. Kuehn
Minnesota Department of
  Natural Resources
Grand Rapids, Minnesota

Herb Langin
B.C. Fish and Wildlife Branch
Nanaimo, British Columbia,
  Canada

Charles A. Lyons
Department of Psychology
Utah State University
Logan, Utah

Ursula I. Mattson
University of Montana
Missoula, Montana

Stephen McCusker
Washington Park Zoo
Portland, Oregon

L. David Mech
U.S. Fish & Wildlife Service
North Central Forest
  Experimental Station
St. Paul, Minnesota

H. Mendelssohn
Department of Zoology
Tel Aviv University
Tel Aviv, Israel

Sebastian M. Oosenbrug
Canadian Wildlife Service
Edmonton, Alberta, Canada

Paul C. Paquet
Portland State University
Portland, Oregon

Sverre Pedersen
Alaska Department of Fish &
  Game
Fairbanks, Alaska

Rolf O. Peterson
Michigan Technological
  University
Houghton, Michigan

Errki Pulliainen
Department of Zoology
University of Oulu
Oulu, Finland

Robert R. Ream
University of Montana
Missoula, Montana

William L. Robinson
Department of Biology
Northern Michigan University
Marquette, Michigan

Jenny Ryon
Department of Psychology
Delhousie University
Halifax, Nova Scotia, Canada

Barbara M.V. Scott
Department of Animal Science
University of British Columbia
Vancouver, British Columbia, Canada

David M. Shackleton
Department of Animal Science
University of British Columbia
Vancouver, British Columbia, Canada

Henry S. Sharp
Department of Sociology &
  Anthropology
Simon Fraser University
Burnaby, British Columbia, Canada

Gordon W. Smith
B.C. Fish & Wildlife Branch
Nanaimo, British Columbia, Canada

Robert O. Stephenson
Alaska Department of Fish & Game
Fairbanks, Alaska

James D. Woolington
U.S. Fish and Wildlife Service
Kenai National Moose Range
Soldotna, Alaska

John Youds
B.C. Fish & Wildlife Branch
Nanaimo, British Columbia, Canada

Erik Zimen
Universitat des Saarlandes
Saarbrucken, West Germany

Acknowledgments

This book was made possible by the cooperation of numerous individuals and organizations. John O. Sullivan, C. Peter Nielsen, Sandra Gray-Thatcher and Paul C. Paquet organized the original symposium, obtained the necessary financing and attracted the many participants. Participants submitted their manuscripts promptly (51 in all), and we appreciate the care with which they prepared them and the patience displayed during the review process. We could not include all submitted papers because of space and subject limitations, and we thank those authors whose papers were not included for their understanding. We are especially grateful to all the individuals who refereed one or more manuscripts: Vic Van Ballenberghe, Lu Carbyn, Ian McTaggart Cowan, John Fentress, Steve Fritts, Jane Packard, Dave Mech, Rolf Peterson, Jenny Ryon, Henry Sharp, Margaret Skeele and Bob Stephenson.

Furthermore, we would like to thank all those who contributed to the success of the Portland International Wolf Symposium: the Audubon Society, Portland; the Collins Foundation, Portland; Friends of the Zoo, Portland; Greenpeace Oregon, Inc.; the Humane Society of the U.S.; Lewis and Clark College, Portland; the North American Wolf Society (NAWS); Northwest Trek, Washington; the Northwest Wolf Preservation Society; the National Wildlife Federation; Portland State University; the Sierra Club; Southern Oregon State College, Ashland, Oregon; the Evergreen Wolf Research Project, Evergreen College, Evergreen, Washington; Mrs. Natalie Friendly; William and Virginia Horn of Cascade Meat Products, White City, Oregon; Ms. Ann Mueller, Missoula, Montana; Mr. and Mrs. N.C. Nielsen; Dr. Joseph F. Paquet, Portland; Dr. Evelyn Aiello, Portland; Marlis Shaffer, Ashland, Oregon; the Washington Park Zoo, Portland; Ms. Sandi Shipley, Julie's Travel Desk, Lake Oswego, Oregon; and Mary Rosenblum, Portland.

A very special thanks to Jennifer W. Paquet for preparing the manuscripts and to the Department of Medicine, The Oregon Health Sciences University, for use of their word processor.

And lastly, our families—Nora, Ian and Justin, Jennifer, Jake, Sheba, Buk and Niarome—deserve a special thanks for tolerating our continual disappearance into the "office."

Contents

Contents

Part II
Behavior and Ecology of Wild Wolves in Eurasia

Part III
Behavior of Wolves in Captivity

Part IV
Conservation

Part I

Behavior and Ecology
of Wild Wolves in North America

Introduction

Since the early 1940s North America has been the focus for studies of free-ranging wolves. The reason is obvious. Much of Canada and most of Alaska support numerous viable, and sometimes thriving wolf populations. Although influenced to various degrees by human activities, these populations have retained much of what is nostalgically referred to as "real," "typical," or "natural" behavior. They court, mate and raise their young, test, capture and kill their prey, as they have done for countless generations. Environmental factors which helped mold the wolf in times past are still present today. Study of the North American wolf should provide insights into the development, causation, function and evolution of wolf behavior.

This first section considers the behavior and ecology of wolves in North America. Common to all these studies is a reliance on radio-telemetry, first employed in Ontario and Minnesota during the 1960s. Prior to the introduction of telemetry for wolf research, the inadequacies of methodology typically created more questions than answers: "Is this the same pack I observed yesterday? Is this one pack of eight wolves, or two packs of three and five? Do wolves occupy territories, home ranges with seasonal overlap, or completely overlapping ranges?" Wolves living in forest habitat essentially vanished by spring, leaving only howls and tracks as clues to their presence. Only a few areas with relatively small numbers of wolves in isolated conditions (such as Isle Royale), could produce reasonably reliable data. The questions were endless and the answers, by necessity, very, very tentative.

For most other areas, radio-telemetry provided a secure and somewhat continuous foothold (or neck lock) on at least a segment of the population. Wolves ceased being phantoms. They now had numbers, their packs were assigned names, and although one often could not distinguish the wolves from the trees, the constant monotonous beeping of the receiver was reassurance that forest wolves did not "pack it in" for the summer. The use of radio-telemetry exploded during the 1970s, and today virtually all studies employ it. Consequently, we are now collecting detailed and precise data on known individuals and packs. More and more accurate statements about the activities of particular wolves or packs under a variety of conditions are now possible.

We must hope that the proliferation of wolf studies following publication of *The Wolf: The Ecology and Behavior of an Endangered Species* in 1970, will continue. As Durward Allen (1979) emphasized, studies must span decades if we are to fully understand the wolf and its relation to its prey and environment.

In this section, the first five papers deal with classic topics of wolf ecology: movements, patterns of home range use, pack dynamics, and predator-prey relations. William Berg and David Kuehn describe territorial populations and document long-distance dispersal for young animals in Minnesota. Barbara Scott and David Shackleton describe similar phenomena for the recently recovered and largely unstudied Vancouver Island wolf. In fewer than ten years, this subspecies has progressed from endangered status to the densest population in North America, if not in the world. Robert Stephenson and David James describe an Alaskan wolf population which regularly migrates between summer and winter ranges in pursuit of caribou prey. Sebastian Oosenbrug and Lu Carbyn focus on wolf-bison interactions in The Northwest Territories. Due to the limited distribution and reduced numbers of bison in North America, these observations are of both ecological and historical importance. Daryl Hebert et al. discuss predator-prey dynamics on Vancouver Island, concentrating on wolf-blacktail deer relations. The chapter describes the impact of an emergent wolf population on a previously stable prey population.

Warren Ballard presents observations on the recently appreciated relationship between wolves and brown bears in Alaska, both in terms of their effects on one another and their influence on the prey they share. Next, Fred Harrington and L. David Mech focus on summer movements of individual wolves about homesites in Minnesota. Lu Carbyn concludes the section with a discussion of the often neglected role of disease in wolf population dynamics.

Ecology of Wolves
in North-Central Minnesota

William E. Berg and David W. Kuehn

INTRODUCTION

Minnesota's wolf (*Canis lupus*) range totals approximately 87,000 km², or 40% of the state's area. The primary range expanded from 31,000 km² in the early 1950s (Stenlund, 1955) to more than 37,000 km² in 1970 (Leirfallom 1970) (Figure 2.1). Wolf densities in this wild and relatively inaccessible area currently average about one per 42 km² (Minnesota Department of Natural Resources [MDNR], 1980). The peripheral range is a 50,000 km² mosaic of deciduous and coniferous forests with 17% of the area in agriculture, developed lakeshores, and rural and urban development (Bailey, 1978). All but the most isolated areas, which seldom exceed 260 km², are accessible by road.

Most wolf research in Minnesota has been in the wolves' primary range. Areas studied include the Superior National Forest (SNF) in the northeast (e.g. Olson, 1938; Stenlund, 1955; Van Ballenberghe et al., 1975; Mech and Frenzel, 1971; Mech 1977c) and the Beltrami Island State Forest (BISF) in the northwest (Fritts and Mech, 1981) (Figure 2.1).

The MDNR began research in the Hill City Study Area (HCSA) in 1970 (Figure 2.1). In 1975, work was expanded to the Chippewa National Forest (CNF) and, in 1977, to the Bearville Study Area (BSA) (Figure 2.1). All of these areas are within the peripheral wolf range, as identified by Bailey (1978). MDNR estimates of wolf numbers and delineation of areas occupied by wolves in 1969 (Nelson, 1971) and 1979 encompassed the state's entire wolf range.

This report summarizes the results of research by MDNR from 1970 to 1979 on movements and mortality of wolves in north-central Minnesota, and the results of the 1979 range-wide wolf presence and population survey.

4

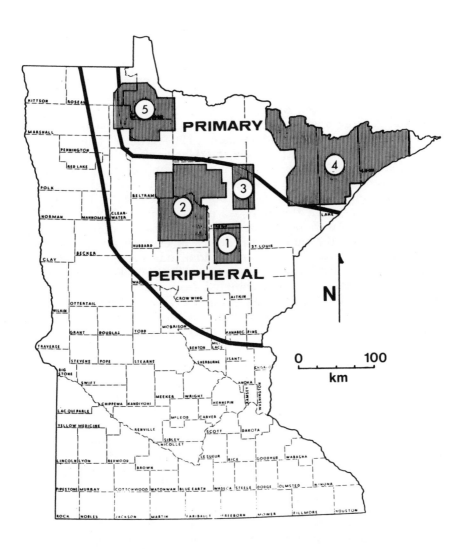

Figure 2.1: Location of Minnesota Department of Natural Resources timber wolf Hill City (1), Chippewa National Forest (2) and Bearville (3) study areas in relation to the primary and peripheral wolf ranges, and Superior National Forest (4) and Beltrami Island State Forest (5) study areas.

METHODS

Wolves were trapped with No. 4 or 14 double longspring leghold traps, and were immobilized with intramuscular injections of phencyclidine hydrochloride (SernylanTM, Bioceutics, St. Louis, MO) and promazine hydrochloride (SparineTM, Wyeth Laboratories, Philadelphia, PA) (Seal and Erickson, 1969). Each wolf was fitted with a radio-collar and ear tagged.

Radios transmitted on the 150 or 164 MHz bands and weighed 300-450 g. Wolves were usually located once or twice weekly by aerial telemetry (Kolenosky and Johnston, 1967; Gilmer et al., 1981).

Information on wolf presence and density was obtained in winter 1979-1980 from MDNR wildlife biologists and managers, conservation officers, foresters and parks personnel across the entire Minnesota wolf range, and from U.S. Forest Service (USFS) forestry and wildlife personnel on the SNF. All persons with field experience in their work units were requested to map the areas occupied by wolves and to estimate the number of wolves in each area based on their best knowledge. Replies from more than one person regarding the same area were combined to prevent duplication. Wolf densities for some remote areas were extrapolated from the surrounding region. Known densities from the HCSA, BSA, CNF and BISF study areas, and aerial observations of wolf signs on the CNF supplemented information from field personnel. Density estimates were then made on the basis of MDNR Deer Management Units (Figure 2.2).

RESULTS AND DISCUSSION

Pack Ranges

From 1970 to 1979, 1,144 locations were obtained for 29 wolves. Fifteen other wolves provided inadequate telemetry data. Ten wolves were radio-tracked on the HCSA, nine on the CNF, and 10 on the BSA. The mean radio-tracking period was 328 days, with the longest being 1,847 days.

Wolves in the HCSA pack were the most intensively studied, with at least one wolf radio-collared from September 1973 to January 1979. This pack numbered up to eight wolves and occupied a 260 km^2 area of wild land surrounded by agricultural development and heavily traveled roads. The HCSA pack was typical of much of north-central Minnesota, where only the scattered wilderness areas are inhabited by wolves, and these seldom abut the ranges of other packs.

Thirteen wolves from four packs containing five, five, seven and eight individuals were radio-tracked on the CNF and BSA between September 1977 and August 1979. Six other wolves not associated with packs were also monitored in these areas. Although packs appeared to abut in the northern portions of both the CNF and BSA, in

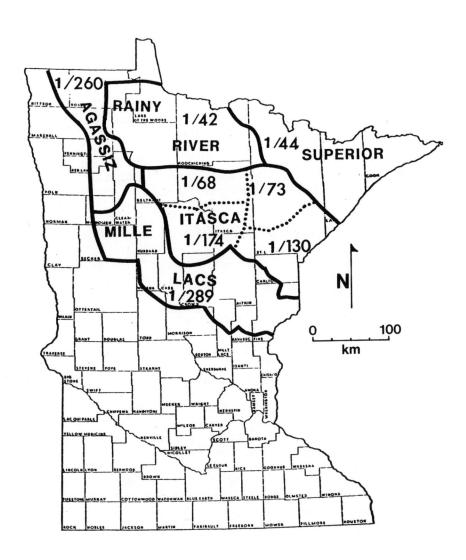

Figure 2.2: Wolf densities per km^2 in Minnesota's five northern Deer Management Units (DMU) in 1979. Dotted lines separate the subunits of the Itasca DMU, for which more complete data were available.

the southern portions they occupied "islands" of wild land surrounded by more developed areas. Pack ranges varied from 143 to
370 km², and averaged 230 km². The peak wolf density on these four
ranges averaged one per 37 km². The range sizes occupied by lone
wolves were not obtainable. Lone wolves did not overlap ranges of
radioed packs.

Dispersal

At least seven of 23 juvenile (<1 year old) and yearling (1-2
years) wolves dispersed from their pack ranges (Table 1). Two juvenile females dispersed in their first winter and five males dispersed
between one and two years of age. Most dispersal movements were to
areas of lower wolf densities (Figure 2.3). Straight-line dispersal distances varied from 37 to 432 km, and averaged 148 km. The previous
record wolf dispersal in Minnesota was 390 km from the BISF (Fritts
and Mech, 1981).

Table 2.1: Dispersal of Juvenile and Yearling Wolves

Study Area	Animal*	Month of Dispersal	Duration**	Distance Covered***
HCSA	004 (F70)	Unknown	max. 7 mo.	79 km
HCSA	256 (M74)	January 1976	8 mo.	195 km
HCSA	258 (M74)	Unknown	max. 6 mo.	146 km
HCSA	342 (M74)	December 1976	1 mo.	46 km
BSA	553 (F78)	April 1979	4 mo.	37 km
BSA	555 (M78)	June 1979	8 mo.	432 km
CNF	608 (M76)	January 1978	4 mo.	98 km

*Animal's identification number (sex and year born)
**From initiation to maximum distance traveled
***Straight-line distance beweeen initial and final location

In addition, a female live-trapped as a pup near BISF by U.S.
Fish and Wildlife Service (FWS) predator control personnel, and released in the SNF, was radio-collared as a yearling by MDNR in the
BSA. She did not establish a territory and her signal was lost after
391 days, 80 km west of the BSA trap site. She had traveled over a
2,540 km² area.

Two male wolves, having neither a clearly defined home range
nor pack affiliation, also dispersed from the areas in which they were
trapped. One, from the CNF, traveled over a 10,500 km² area prior
to pairing with a radio-collared female dispersing from the BISF
(Fritts and Mech, 1981).

Mortality

Fifty-seven percent (n = 25) of the wolves tagged by MDNR
were eventually recovered, compared to a 14% recovery rate in the
less accessible SNF (Van Ballenberghe et al., 1975). Excluding cap-

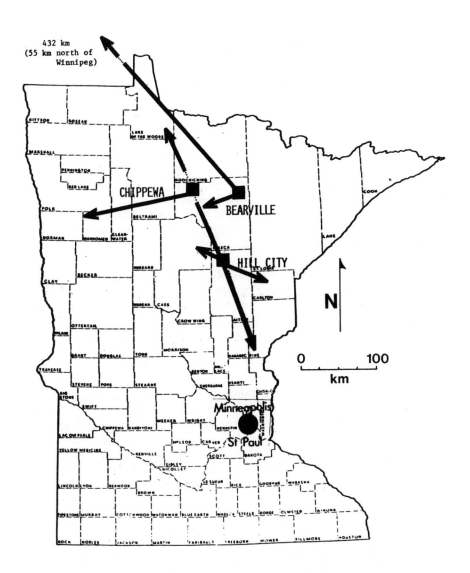

Figure 2.3: Dispersals of seven radio-collared juvenile and yearling wolves from three study areas in north-central Minnesota.

ture-related mortality (n = 4), 16 of 21 (76%) documented cases of wolf mortality were caused by humans, compared to 42% in SNF (Mech, 1977c). Five wolves were killed by cars and six were legally trapped or shot, including three taken by private individuals prior to 1973, one trapped in Manitoba, one killed in 1971 under the MDNR-directed predator control program, and one killed by a FWS predator control agent. The remaining five wolves were killed illegally; one of these was shot for cause by a farmer and another was shot while in a MDNR trap in the HCSA. Five wolves apparently died of natural causes. An adult male whose physical condition and blood parameters were subnormal, died within a few days after release. Two emaciated juveniles died of malnutrition as judged from weight loss, another juvenile died of malnutrition after injuring a leg (J.M. Higbee, pers. comm.), and a fourth died of unknown causes.

Juvenile and yearling dispersing wolves were more vulnerable than dispersing adults. Seven of eight dispersing wolves less than two years old were killed, whereas neither of the two dispersing adults is known to have died during the study period.

Wolf Presence and Population Survey

Radio-tracking (Mech, 1973, 1977a, 1977c; Van Ballenberghe et al., 1975; Fritts and Mech, 1981) and intensive aerial observation (Mech, 1966; Pimlott et al., 1969; Wolfe and Allen, 1973; Peterson, 1977) can provide wolf population estimates and approximate pack boundaries within intensively studied areas. Group observations by experienced personnel alone (Olson, 1938; E. Clem and J. Mathisen, pers. comm.; W. Berg, unpub. CNF Report), or in combination with aerial observations (Stenlund, 1955; Fritts and Mech, 1981) can also provide data on wolf densities and pack boundaries.

Stenlund (1955) used aerial and ground reconnaissance in combination with other reports to derive a population estimate of between 205 and 273 wolves on a portion of the SNF. Extrapolation to 18,200 km² of primary range in northeastern Minnesota estimated the population at not more than 400 wolves in the early 1950s. He did not estimate a population for the remainder of northern Minnesota, but believed that due to exploitation, wolf densities were much lower than in the SNF (M. Stenlund, pers. comm.).

In 1969, a survey of MDNR Enforcement and Wildlife personnel in 100,000 km² of northern Minnesota indicated a "minimal" population of 750 wolves (Nelson, 1971). Mech (in Bailey, 1978, Appendix C) estimated Minnesota's population at 1,000 to 1,200 wolves in the mid-1970s, based on densities in the SNF, CNF, BISF and HCSA.

The wolf presence and density survey in winter 1979 provided input from more than 120 MDNR and SNF field personnel. It identified 138 areas occupied by two or more wolves, and 22 areas — mainly near the edge of the peripheral range — that contained lone wolves.

This survey estimated a statewide population of 1,235 wolves (MDNR, 1980). When calculated on the basis of MDNR Deer Management Units, densities varied from one wolf per 43 km^2 in extreme northern Minnesota (including SNF and BISF) to approximately one per 270 km^2 on the southern and western fringes of the peripheral range (Figure 2.2).

Although the wolf population in a portion of northeastern Minnesota has declined in recent years (Mech 1977c; Mech and Karns, 1977), the 1979 statewide population estimate suggests an increase of 76% from 1969. The population estimate for the CNF, based on telemetry and aerial track reconnaissance, indicates a twofold increase from 10-15 wolves in 1975-76 to 30-39 wolves in 1978-79.

Wolf abundance, as indicated by scent station lines (Linhart and Knowlton, 1975), in the primary wolf range has also increased as evidenced by mean visitation indices of eight in 1976 (n = 4 lines) and 32 in 1980 (n = 9 lines) (Berg, 1981). Similarly, the proportion of scent station lines visited by wolves increased from 50% in 1976 to 100% in 1979. The mean visitation index was much lower in the peripheral range, but increased from zero in 1976 to four in 1980.

SUMMARY AND CONCLUSIONS

Research on wolves in north-central Minnesota indicated that: 1) pack territories varied from 143 to 310 km^2; 2) packs occupied "islands" of wild land in the southern part of the peripheral range, but abutted each other in the north; 3) seven dispersing wolves less than two years old traveled a mean distance of 148 km, generally to areas of lower wolf density; and 4) most (76%) wolf mortality was human-related. The statewide wolf population was estimated at 1,235 in 1979 and, with the exception of the northeast, was gradually increasing.

Acknowledgements

We acknowledge the assistance of biologists R.A. Chesness and T. Bremicker, both associated with the study until 1974, and trappers L. Jewett and H. Niskanen. The FWS provided telemetry equipment for the first four years of study. Research on the CNF was funded, in part, by the USFS under a USFS-MDNR cooperative agreement. Animal movement data were analyzed on the Wang 2200 Computer System by W. Snow.

A Preliminary Study of the
Social Organization of the
Vancouver Island Wolf
(*Canis lupus crassodon;* Hall, 1932)

Barbara M.V. Scott and David M. Shackleton

INTRODUCTION

The Vancouver Island wolf (*Canis lupus crassodon*) is one of 24 currently recognized subspecies of North American wolves (Mech, 1974a). It was first described by Hall (1932) on the basis of pelage color and carnassial teeth characteristics. Since this initial work, studies of this subspecies have been limited to morphometric studies with limited cranial samples (Jolicoeur, 1959; Lawrence and Bossert, 1967). Although no accurate data exist on wolf numbers on the Island, Cowan (pers. comm.) believes they have undergone major fluctuations and may have been "virtually extirpated on either two or three occasions since the early 1920s." Starting in the early 1970s, there have been reports of an apparent increase in wolf numbers and concomitant reductions in populations of black-tailed deer (*Odocoileus hemionus columbianus*) (Hebert et al., this volume). However, at the same time, hunter harvest patterns of black-tailed deer on the Island have changed, including hunter effort and success (Hebert, 1979; Kale, 1979), and the effects of forest harvest practices (Cowan, 1947; Gates, 1968; Harestad, 1979), past and present, must also be considered for deer population dynamics (Smith, 1968; Bunnell, 1979).

This study represents the first attempt to study the social organization of the Vancouver Island wolf. Food habits of the wolves outlined in this paper have been previously reported (Scott and Shackleton, 1980).

STUDY AREA

The study site covered approximately 530 km² on the northeast portion of Vancouver Island, British Columbia (Figures 3.1 and 3.3). The major watersheds are the Adam and Eve Rivers, which flow in a north-northwesterly and northeast-northwesterly direction, respectively, through the area.

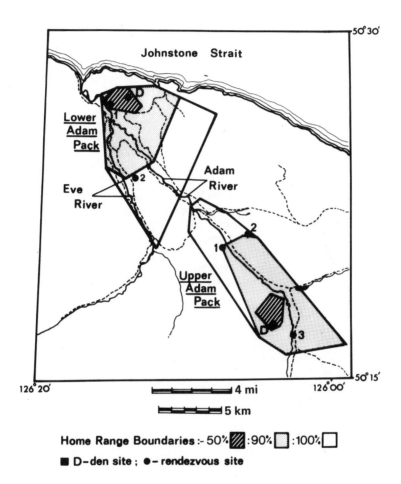

Figure 3.1: Study area showing home range boundaries of the two packs as determined from radio locations (Lower Adam — April to November 1978, Upper Adam — April to September 1978). Rendezvous site occupation periods were: Lower Adam, Site 1 — mid-July to late August 1978, Site 2 — September to November 1978; Upper Adam, Site 1 — mid-July to mid-August 1978, Site 2 — late August to mid-September 1978, Site 3 late December 1978 to May 1979. Broken lines indicate main logging roads.

The terrain is characterized by steep and narrow valley systems ranging in elevation from sea level to 2,159 m. Logging practices within the area have altered natural habitat types resulting in clear-cut and second growth zones running from valley bottoms up to sub-alpine. Pristine habitat remains at river headwaters and upper reaches of side hills within watersheds. Habitat types within these areas include old-growth forests, alpine and high-elevation wet meadows. Tree line occurs at 1,373 m. The study site includes three biogeoclimatic zones — subalpine mountain hemlock, coastal Douglas fir and coastal western hemlock zones (Krajina, 1965). Access to the site is afforded by numerous logging roads.

METHODS

Individuals were trapped, using fresh scats and urine as attractants, between April and November 1978. Captured animals were immobilized with a mixture of phencyclidine hydrochloride (Sernylan™, 100 mg/ml) and promazine hydrochloride (Sparine™, 50 mg/ml), administered intramuscularly, both at dosages of 1 mg/kg estimated body weight. Standard body measurements and weights were obtained, and animals fitted with radio-transmitters and ear tags. Further details are given in Scott (1979).

Locations of radio-collared wolves were obtained from both the ground and from the air during both daylight and nocturnal periods. Ground tracking involved three to five compass bearings taken at known map locations, while aerial locations were obtained from a Cessna 180 fixed-wing aircraft fitted with two four-element Yagi antennae on the wing struts. Activity switches in the collars allowed determinations of whether located animals were active or not.

Location bearings were plotted on 20 chain forestry maps and later converted to polar coordinates for computer analysis. Home range sizes were calculated from minimum convex polygons, and plotted by a program developed by Harestad (1979). Home range areas were calculated and plotted for 100, 90 and 50% of the nearest locations, and observation-area curves (Odum and Kuenzler, 1955) plotted for 100 and 90% of nearest locations.

Den and rendezvous sites (Joslin, 1967) were first identified from repeated radio locations from the same area. Both den sites were checked on the ground and, where possible, ground confirmation from the presence of scats, tracks and other signs were used for rendezvous sites.

RESULTS

Weights and Measures

Morphological data collected on the nine wolves captured in the

Table 3.1: Weights (kg) and Measurements (mm) of the Nine Vancouver Island Wolves (*Canis lupus crassodon*) Captured Between 10 April and 1 November 1978 Near Kelsey Bay, British Columbia

Pack and Wolf Identity	Month Captured	Sex	Age*	Total Wt.	Total Lgth.	Tail Lgth.	Hind Foot Length	Head Girth	Canine Length Upper	Canine Length Lower
Upper Adam (87)	April	M	A	38.6	1803	495	267	550	24.5	—
Lower Adam (86)	April	M	A	32.7	1800	483	241	787	28.3	25.1
Upper Adam**	May	M	A	36.7	1727	394	279	792	24.5	23.0
Lone wolf	November	M	A	36.6	1780	460	300	770	26.0	23.5
Average		M	A	36.2	1778	458	272	752	25.8	23.9
Standard Deviation				2.5	35	45	25	62	1.8	1.1
Lower Adam	July	M	P	10.0	1100	270	210	480	8.9	6.1
Lower Adam	October	F	P	21.8	1490	380	250	610	19.2	19.3
Lower Adam (97)	April	F	A	31.8	1638	432	254	660	26.8	22.0
Upper Adam (90)	April	F	A	31.4	1715	432	264	729	23.0	18.5
Upper Adam (4)	April	F	Y	28.2	1638	392	254	711	22.8	19.7
Average		F	A/Y	30.5	1664	419	257	700	24.2	20.1
Standard Deviation				2.0	45	23	6	36	2.3	1.8

*Age classes: A = adult, P = pup of year, Y = yearling
**Animal died shortly after handling

study are presented in Table 3.1. The respective weights of 36.3
and 31.3 kg for an adult male and an adult female from southern
Vancouver Island (Cowan and Guiget, 1965) are comparable to
weights in this study. Available weights from other parts of North
America (Table 3.2) suggest that animals in the Vancouver Island
sample are heavier than those from eastern North America, but
lighter than wolves from Alaska, the Northwest Territories, and
central and northern British Columbia.

Table 3.2: Average Body Weights (kg) of Adult *Canis lupus*, One Year of Age and Older, from Various North American Populations

...Weight (Sample Size)...				
Male		Female	Location	Source
45.0	(4)	37.2	(3)	Quesnel, B.C.	Cowan & Guiget, 1975
44.6	(18)	38.8	(21)	Wood Buffalo National Park, N.W.T.	Fuller & Novakowski, 1955
43.8	(80)	37.6	(66)	Great Slave Lake, N.W.T.	Kelsall, 1968
43.4	(9)	41.0	(4)	Horseranch Range, B.C.	B.C. Fish & Wildlife Branch (unpublished)
38.6*	(60)	32.2	(50)	Alaska	Rausch, 1967
36.2	(4)	30.5	(3)	Vancouver Island, B.C.	(This study)
35.5**	(84)	27.5	(60)	Northeastern Minnesota	Stenlund, 1955
34.5	(172)	28.2	(182)	Central-western Ontario	Kolenosky & Stanfield, 1975
30.6	(36)	26.3	(32)	Northeastern Minnesota	Van Ballenberghe, 1977
28.0	(17)	24.8	(26)	Central-western Quebec	Huot et al., 1978
27.7	(40)	24.5	(33)	Ontario	Pimlott et al., 1969
27.5	(129)	23.5	(112)	Southeastern Ontario	Kolenosky & Stanfield, 1975

*Skinned weights, 4.5-7 kg lighter than total weight
**Wolves less than one year old are probably included

In terms of body dimensions, both sexes from Vancouver Island
appear to have greater total body and tail lengths, and shorter or
equal hind foot and canine lengths compared to wolves from northern
Minnesota (Stenlund, 1955; Van Ballenberghe, 1977) and the Northwest Territories (Fuller and Novakowski, 1955; Kelsall, 1968).
Measurements from central British Columbia (Cowan and Guiget,
1965) appear similar except for shorter tails.

Social Organization

Pack Size and Composition: Estimates of pack size and composition are based upon limited direct observations, radio tracking,
tracks, and other signs. The two packs studied were immediately
adjacent to each other (Figure 3.1). The Upper Adam pack was
estimated to be originally composed of 10 animals: Two adult
males (one of which died shortly after handling), one adult female,
one yearling female, two unknown adults or yearlings, and four

pups of the year. The two radio-collared females in the pack disappeared on 4 September and the collared male on 11 September 1978. One collar was found shortly after their disappearance just off a road, but no other signs or locations were obtained despite repeated searches which extended well outside the study area. Our tentative conclusions were that they had been killed rather than moved elsewhere.

The lower Adam pack was estimated to contain only an adult male and an adult female, both collared, together with three pups (1 male, 1 female and one of undetermined sex). The radio-collar of the adult female began to malfunction in June and ceased transmitting by early July. The adult male was found shot in mid-November. The female pup was consistently located at the same place after being monitored for two days following capture. Inaccessibility of this area together with deep snow hampered attempts to determine if the collar had slipped off, or if the animal was dead.

Home Range Size: The concept of territory is not used here to describe space use because defense of an area was not investigated. Home range (*sensu* Burt, 1943) is used, instead, to denote the areas traversed by members of the packs. Estimates of home range size for individual wolves are shown in Table 3.3 and the locations of the packs in Figure 3.1.

The Lower Adam adult male occupied a larger area during the denning period than the adult female (Table 3.3). If, as estimated, this pack contained only two adult animals, the difference may have simply reflected the hunting activities of the male and the obligatory proximity of the female to her young pups at the den.

Of the three animals monitored in the Upper Adam pack, the yearling female occupied the larger home range area (Table 3.3) compared to either of the adults, particularly during the denning period. The yearling and adult female were located together 88% of the time for the whole study (61 out of 69 locations). During the post-denning period, they had essentially identical spatial and temporal distributions (Scott, 1979). The adult male from this pack was infrequently located with either of the females (45% of time for whole study) although, again in the post-denning period, locations for all three animals were similar (Scott, 1979). This similarity in post-denning distributions and home range areas occupied (Table 3.3) may indicate an increase in pack cohesion after the pups left the den site and became more mobile.

For both packs, location concentrations corresponded to den and rendezvous sites in the denning and post-denning period, respectively, and fell within the 50% boundaries. The plots shown in Figure 3.1 demonstrate that not only were the two packs occupying non-overlapping areas, but their homesites were maximally separated within their respective home ranges. The Lower Adam male occupied a larger home range in the whole period and during post-denning than other monitored animals (Table 3.3). However, locations for this

Table 3.3: Individual and Pack Home Range Sizes (km²) for Denning (11 April to 11 July), Post-Denning (Lower Adam – 12 July to 20 November, Upper Adam – 12 July to 10 September), and Total Study Periods for Two Wolf Packs on Vancouver Island

Wolf and Pack Identity	Denning Period:				Post-Denning Period:				Total Period:			
	Number of Locations	Home Range Sizes 100%	90%	50%	Number of Locations	Home Range Sizes 100%	90%	50%	Number of Locations	Home Range Sizes 100%	90%	50%
Lower Adam:												
Adult Male (86)	40	43.1	20.9	1.1	60	56.3	32.0	3.2	100	75.0	33.8	3.3
Adult Female (90)	25	16.4	15.1	0.3	10	10.5	2.5	0.1	35	18.5	15.1	1.7
Pack	65	43.2	25.3	0.7	70	56.3	29.0	3.3	135	75.0	33.8	3.7
Upper Adam:												
Adult Male (87)	43	42.1	20.9	1.6	29	20.7	15.5	0.9	72	43.8	24.8	3.0
Adult Female (97)	37	32.9	25.0	1.8	32	26.6	17.6	1.4	69	47.8	26.3	2.8
Yearling Female(4)	39	57.4	40.0	2.4	33	27.0	17.6	1.4	72	63.0	45.0	3.9
Pack	119	60.9	41.8	2.5	94	28.1	22.7	1.7	213	64.0	43.6	3.9

male extended over a longer period than for other animals, which may, in part, explain the differences in post-denning home ranges between members of the two packs.

Observation area curves shown in Figure 3.2 indicate that during the denning period, asymptotes were reached for all individuals for their 100% home range boundaries. Following denning, further increases in home range size and corresponding changes in the asymptote levels occurred. These changes coincided for most animals, with shifts in core-area use (away from dens to rendezvous sites). They may also reflect a general increase in movements between rendezvous sites as the pups became more mobile.

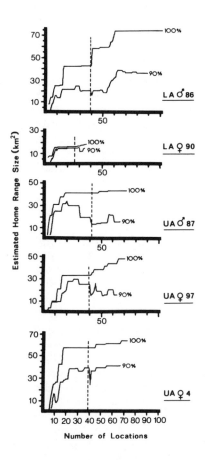

Figure 3.2: Observation area curves for radio-collared wolves' 100 and 90% home ranges. Vertical broken lines separate denning and post-denning periods. Pack affiliation, sex and identity number (see Table 3.3) are shown at right.

The observation area curves for 90% home range boundaries show a different pattern (Figure 3.2), with a decrease in area at or around the time the wolves vacated the den sites. This decrease in 90% home range area probably indicates that during this shift away from dens toward rendezvous sites, the wolves limited their movements within the previous 90% areas rather than travelling througout the home range or moving into previously unoccupied areas.

The data in Figure 3.2 further suggest that the cumulative home ranges of both packs, considered from the aspect of the annual cycle, may not have reached maximum size. Thus, home range sizes for the whole study period (Table 3.3) are most likely underestimates of the annual home range areas used by each pack. Other factors affecting these estimates are considered later.

Wolf Activities: The activity switches in the radio-collars permitted determination of whether located wolves were active or not. Although we obtained fewer nocturnal locations (14.7% nocturnal, n = 348), differences in activity between diurnal and nocturnal locations were significant for the combined data of both packs (p<0.05). At night, 90.2% of all locations (n = 51) were of active wolves, compared to 59.3% of all locations (n = 297) during daylight hours.

Home Sites: Den sites of both packs (Figure 3.1) were examined after the animals had vacated them. The general areas were initially located by radio. The Lower Adam pack had denned in the base of a red cedar *(Thuja plicata)* windfall, 24.5 m in length, which lay in a primary growth stand of red cedar and western hemlock *(Tsuga heterophyla)* 150 m above sea level. This site was immediately adjacent to a clearcut and a second growth hemlock area. Various water sources, standing and running, were also in the immediate vicinity.

The denning area of the Upper Adam pack was found at 750 m elevation on a north-facing ridge slope in a mature stand of red cedar and mountain hemlock *(T. mertensiana)*. No major den was located, but a number of excavated caves were found at the bases of large trees (cedar and hemlock). Several wet meadows and small streams were located in the immediate vicinity of the dens.

Four areas, classed as rendezvous sites (Joslin, 1967), were identified from concentrations of radio locations, and a fifth site was indicated during the preliminary stages of the study from signs and direct observations (Figure 3.1). Two sites, identified from radio locations, were examined on the ground. Abundant evidence of scats, tracks, scratchings, beds, trails and other signs were found. The sites were open areas: Site 1 of the Lower Adam pack was located on a river gravel bar; Site 2 of the Upper Adam pack was in a wet meadow surrounded by mature forest.

Lone Wolf Movements: One lone wolf, an adult male, was captured on 1 November 1978 and monitored until 15 June 1979 (Figure 3.3). The straight line distance between point of capture and last location, 7.5 months later, was 209 km. The minimum distance travelled, determined from straight line projections between the 14 locations obtained, was 346 km.

On 21 February 1979, the animal was observed from the air in the middle of a small lake feeding on a yearling female deer. Signs from air and later on the ground showed he had killed the deer after chasing it onto the frozen lake. Tracks indicated that while the deer was breaking through the snow, the wolf was able to run over the frozen surface crust. Other observations of this animal's tracks from the air indicated that he was travelling alone throughout the monitoring period. It was also apparent that while moving through steep mountainous terrain, the animal travelled along the "easiest" routes, through mountain passes and along watersheds. He was also observed to travel along cleared right-of ways for major power lines.

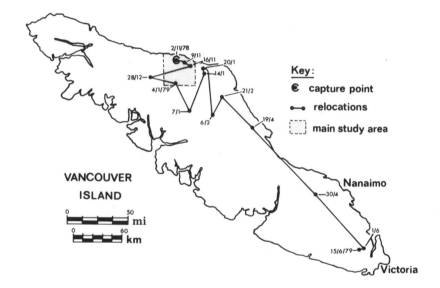

Figure 3.3: Movements of lone adult male from point of capture on 1 November 1978 until 15 June 1979 when observations were terminated. The straight line distances between location points total 346 km.

DISCUSSION

The physical characteristics of the captured wolves were discussed earlier in relation to other published data. With a larger sample size of weights and measurements, it will be interesting to compare how, in relation to wolves from other populations in North America, this subspecies fits with respect to Rosenzweig's (1968) relationships of carnivore body size with latitude-temperature and primary productivity.

Published reports of wolf pack sizes vary from pairs to groups of up to 21 animals (Mech, 1970), indicating that the Upper Adam pack of 10 individuals is typical, and the Lower Adam pack with only two adults may represent a newly formed pack. However, even allowing for some degree of underestimation in home range sizes, densities of the wolves in the two study packs (6.4 km² per Upper Adam wolf and 15.0 km² per Lower Adam pack) are higher than previously reported maximum densities of 26 km² per wolf in Algonquin Provincial Park, Ontario (Pimlott et al., 1969).

The high wolf densities in our study area would suggest that major prey was plentiful. Food habits of the two packs, determined from scat analyses (Scott and Shackleton, 1980), demonstrated that black-tailed deer were the major prey, with Roosevelt elk *(Cervus elaphus roosevelti)* and beaver *(Castor canadensis leucodontus)* as secondary prey. Young ungulates were apparently heavily used in spring, while beaver were taken in both winter and spring. Use of beaver in winter and spring, in contrast to their use primarily in summer by wolves in other studies (Mech, 1966; Pimlott et al., 1969; Peterson, 1977), may have reflected the milder winters experienced on the Island which probably increased availability of beaver during these seasons (Scott and Shackleton, 1980). Scat contents collected at den and rendezvous sites were found to differ significantly ($p < 0.05$) from those collected from other areas within the home ranges during corresponding periods (Scott and Shackleton, 1980). Similar variation in the contents of scats from rendezvous sites and from other areas within the home range were found by Theberge et al. (1978) in Algonquin Provincial Park, Ontario.

If the estimates of home range size obtained here are representative of wolves throughout Vancouver Island, it appears that these animals occupy considerably smaller ranges than those reported by other authors in North America (see Mech, 1970). Although a degree of underestimation of annual home range size may have been made, data indicated that home range areas were reaching an asymptote by the end of the study. From the relationships between home range, body weight and habitat productivity presented by Harestad and Bunnell (1979), the relatively small home range sizes on Vancouver Island suggest that their habitat supported high densities of their major prey. Indeed, high densities of black-tailed deer apparently do occur on Vancouver Island (Hebert, 1979; Hebert et al., this volume).

Seasonal changes in home range use were observed. Shifts in core areas (50% location areas; see Scott, 1979) were apparent as animals moved from den to rendezvous sites, and between rendezvous sites. The shifts between den and rendezvous sites are indicated in the observation area curves (Figure 3.2).

Only limited conclusions may be drawn from the data on pack

cohesiveness. In the denning period while her transmitter was still functioning, the Lower Adam female was restricted to the vicinity of the den site while the adult male moved extensively throughout the home range (Table 3.3). In the post-denning period, a number of direct observations of this pack with mobile pups were made (Scott, 1979), indicating that they were more cohesive during this period. Common locations (spatio-temporal) of the three collared Upper Adam wolves showed a strong association between the two females, especially during the post-denning period. The male in this pack was more frequently located away from these two animals, although again during the post-denning period, the data indicated an increase in the number of common locations and, hence, pack cohesiveness. Such temporary separation of pack members appears typical for wolves in general (Mech, 1966; Jordan et al., 1967; Van Ballenberghe et al., 1975; Peterson, 1977).

Activity data obtained for the wolves showed greater nocturnal activity and, from the continuous monitoring of one individual, longer travel distances at night. Such a diurnal pattern of activity has been reported for other wolf packs in North America (for example, Mech, 1966, 1970; Mech and Frenzel, 1971; Kolenosky, 1972). However, it is equally apparent from the extensive daylight observations of Mech (1966) and Haber (1977) that wolves are active for significant periods during daylight hours (see Harrington and Mech, this volume). While recognizing the inherent difficulties frequently encountered in obtaining nocturnal locations and activity measures, we would suggest that greater emphasis is required in this area and that it will probably lead to a certain re-evaluation of home range estimates which, up until now, have been based primarily upon data collected during the daytime.

Den sites of both packs were associated with mature forest stands, but at different elevations. Rendezvous sites, on the other hand, were typically located in small open meadows bordered by mature forest. They also tended to be at higher elevations relative to the rest of the home range, and this may reflect seasonal, altitudinal movements which their major prey, black-tailed deer, are observed to make (Harestad, 1979). The general characteristics of the dens and rendez-vous sites, together with their associated habitat types, are comparable to other wolves inhabiting forested areas (Mech, 1966; Joslin, 1967; Pimlott et al., 1969; Carbyn, 1974; Van Ballenberghe et al., 1975; Peterson, 1977).

Perhaps one of the most interesting aspects of this study was the behavior of the lone wolf. Over a period of 7.5 months he travelled at least 346 km, ranging from the northeast corner of the Island to its southern tip, apparently alone for the entire period. Similar straight line distances of lone wolf movements have been reported in other studies; a yearling female was found to have travelled 111 km between capture points over 14 months (Van Ballenberghe, 1972), a lone dispersing individual 206 km in two months (Mech and Frenzel,

1971), and a migrating individual moved 360 km (Kuyt, 1972) (also see Berg and Kuehn, this volume p. 8). Other reports of lone wolves suggest that they may either be outcasts from established packs (Peterson, 1977) or aged and subordinate individuals (Mech, 1966; Jordan et al., 1967) which may avoid established packs (Jordan et al., 1967; Mech and Frenzel, 1971; Van Ballenberghe et al., 1975). Little appears to be known of the significance of such behavior as, by its very nature, it is difficult to observe. Recent studies, however, have begun to concentrate on lone wolves and pairs (Rothman and Mech, 1979), so the significance of their behavior may become more apparent (Mech, 1970).

The immediate concerns for future work on the Vancouver Island wolf are to study a larger number of packs, study in areas representing a broader spectrum of habitat types and histories of forest exploitation, compare packs from northern and southern regions of the Island where prey conditions may vary (Harestad, 1979), and obtain satisfactory estimates of prey densities and availabilities. British Columbia, including Vancouver Island, relies heavily on forest harvesting with resulting changes in habitats and, as the number of truly pristine forested areas is reduced, it is imperative that studies of wolves inhabiting forested regions in this province be continued and expanded to permit the development of suitable management strategies.

SUMMARY

Two packs of Vancouver Island wolves were studied for 10 months. Weights and measurements suggest they are intermediate in size between wolves from eastern and northern North America. Home ranges determined from radio locations ranged from 64 to 75 km^2 and, as such, are smaller than those reported elsewhere in North America. Wolf densities were also higher and, along with the small home ranges, appear to be the result of high prey densities. Wolves were more active at night and shifted their core areas from around den sites to rendezvous sites later in the season. One lone adult male made extensive movements during a period of 7.5 months, during which he showed no fidelity to any given area.

Acknowledgements

Funds for this study were generously provided by the following organizations: Natural Sciences and Engineering Research Council of Canada, University of British Columbia, British Columbia Fish and Wildlife Branch, British Columbia Parks Branch, British Columbia Resource Analysis Branch, World Wildlife Fund (Canada), Wild Canid Research and Survival Center, Mittlenatch Society and the MacMillan Bloedel Company.

Among the many people who gave extensively of their time and help in this study, we would particularly like to thank Dr. D. Bowen, Dr. D. Eastman, Dr. A. Harestad, Dr. D. Hebert, Mr. D. Janz, Dr. L.D. Mech and Mr. D. Williamson. Ms. Susan Shaw was an invaluable assistant throughout the study, and Mr. Keith McKillican, the pilot, together with other Island Air employees, made the logistics of aerial locations essentially trouble-free. To all of the above and to others space does not allow us to name, we extend our sincere thanks.

Wolf Movements and Food Habits in Northwest Alaska

Robert O. Stephenson and David James

INTRODUCTION

In the large expanse of arctic tundra habitat in northern Alaska, caribou (*Rangifer tarandus*) have, in recent history, been the most abundant large mammal. Wolves (*Canis lupus*) and, to a lesser extent, red foxes (*Vulpes vulpes*), wolverines (*Gulo gulo*), grizzly bears (*Ursus arctos*), and several other species of small mammalian and avian predators and scavengers are dependent upon these caribou for food. Numerous studies of North American wolves show that in most areas, a wolf pack tends to remain in a home range that, although large, remains relatively stable throughout the year. This coincides with the relatively constant availability of prey such as deer (*Odocoileus* spp.) elk (*Cervus elaphus*), moose (*Alces alces*) and sheep (*Ovis* spp.) which often undergo seasonal movements, but over relatively short distances. Wolves that depend primarily upon caribou for prey, however, are often faced with drastic seasonal fluctuations in food availability. In the Northwest Territories, Kuyt (1972) found that ear-tagged wolves migrated as much as 360 km during winter in response to the southward movement of caribou to their winter range. Parker (1973) substantiated Kuyt's conclusions and both authors showed that extremely high densities of wolves may occur during the maximum compression of wintering caribou populations. Studies by Miller and Broughton (1974) and Miller (1975) on the Kaminuriak caribou herd suggest a similar relationship between wolves and caribou. On Baffin Island, Clark (1971) found that wolves relied almost totally on caribou during summer.

A number of authors (Kelsall, 1968; Parker, 1972; Bergerud, 1974; Miller and Broughton, 1974) have discussed the important interrelationships between wolves and caribou, including the important effects of wolf predation on caribou population dynamics. In our study, we sought primarily to elucidate the relationship

26

between the movements and behavior of wolves and the movements of the migratory Western Arctic caribou herd in northwest Alaska. The following is a general review of our findings during the first 24 months of study.

METHODS

The study area (Figure 4.1) includes portions of the Noatak and Utukok drainages and is located in the western Brooks Range about 720 km northwest of Fairbanks. The southern portion of the area consists mainly of rugged, glaciated ridges with local relief of 500–900 m. The northern portion is characterized by irregular buttes, mesas, east trending ridges and intervening tundra plains with elevations up to 760 m. Tussock meadows and dry upland meadows are the predominant plant communities; willows (*Salix* spp.) and alders (*Alnus* spp.) that reach 5 m in height predominate along stream borders and river floodplains.

Wolves were immobilized with intramuscular injections of phencyclidine hydrochloride (Sernylan™) administered with Cap-chur™ darts fired from a helicopter. Radio-collaring began in April 1977 and, by early July, seven wolves had been radio-collared. Ten additional wolves were radio-collared in April and June 1978. After immobilization, wolves were weighed, measured and examined for infirmities, and the age of each wolf was estimated on the basis of tooth eruption and wear. Blood samples were taken for disease studies. Radio-collars, manufactured by AVM Instrument Company or Telonics Electronic Consultants, were attached and instrumented wolves were radio-located with a PA-18-160 Super Cub equipped with two four-element Yagi antennae, or occasionally with Helio Courier or Cessna 180 aircraft equipped with one four-element Yagi. Data recorded for each radio-location included date, time, location, weather conditions and activity of radio-collared wolves and their associates. When wolves were observed in close proximity to prey animals or in the act of pursuing prey, the area was circled widely and observations were made until the outcome of the interaction was evident. When possible, the carcasses of wolf-killed ungulates were examined on the ground. During May, August and October 1978, we attempted to obtain accurate information regarding the rate of predation by the Iligluruk pack by locating all pack members twice daily for periods of 10 days.

In addition to studies of the movements of packs and individuals, we conducted aerial surveys in various parts of the area during April 1977 in order to estimate wolf population levels. Active wolf dens located during the study were inspected on the ground after being abandoned by wolves. The area around each den was searched and all scats were collected and food remains identified. Scats were later autoclaved and analyzed for the frequency of occurrence of various prey remains. A reference collection of skins, skeletal material and hair samples was used to help identify scat contents. Impressions of hair scale patterns and medulla structure (Williamson, 1951; Adorjan and Kolenosky, 1969; Moore et al., 1974) were used to aid in identifying prey remains that were not classifiable by gross examination.

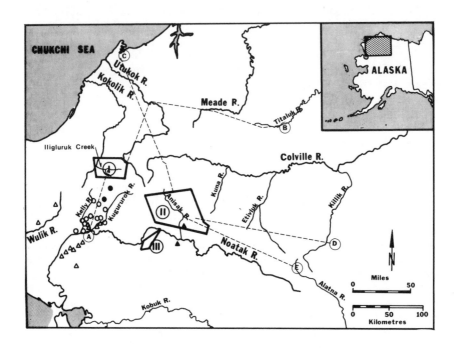

Figure 4.1: Summer range boundaries, winter movements and dispersals of radio-marked wolves in northwest Alaska.

— Approximate boundary of summer ranges of:
    I. Iligluruk Pack
    II. Anisak Pack
    III. Noatak Pack

○ Winter 1977-78 locations of Iligluruk Pack
● Winter 1978-79 locations of Iligluruk Pack
△ Winter 1977-78 locations of Anisak Pack
▲ Winter 1978-79 locations of Anisak Pack

-- Dispersal movements of:
    A.   NW 13, yearling female
    B.   NW 5, yearling female
    C.   NW 9, yearling male
    D.   NW 2, yearling male
    E.   NW 1, yearling male

Because the hair of caribou and moose up to two to three months of age is distinguishable from that of adults, we categorized ungulate remains in scats as adult or calf whenever possible; the adult classification included all animals older than three months. The number of scats in which the remains of a prey item occurred was recorded as frequency of occurrence (FQ). The proportion of food items in each scat was visually estimated to the nearest 0.25, and a value of 0.05 was assigned to remains occurring in trace amounts. These proportions were called relative estimated bulk (REB) (Lockie, 1959). REB was used to estimate the actual proportions of arctic ground squirrels and larger prey actually consumed by wolves according to the method derived by Floyd et al. (1978).

In addition to comparing the FQ and the actual proportion of the species categories of prey listed in Table 4.2, some comparisons were based on combined FQ for ungulates and small mammals. For the analysis of FQ data, chi square values were derived from 2 x 2 contingency tables (df = 1) with Yates correction for continuity (Zar, 1974:62): Statistical significance was at the five percent level.

An objective analysis of ungulate species FQ was difficult because adult ungulate hair in a majority of scats could not be identified to species. Exclusion of the unidentified adult ungulate FQ from subsequent analyses would have resulted in a substantial underestimation of the actual FQ and REB of adult ungulates in the wolf diet. Therefore, unidentified adult ungulate remains were considered to be adult caribou. This seemed justifiable because moose constituted a very small proportion of the remains identified in scats. This procedure probably resulted in an underestimation of the occurrence of adult moose, but we believe the error was negligible. This procedure did not affect analyses involving comparisons of the two general categories of ungulates and small mammals.

RESULTS

The aerial surveys in April 1977 corroborated earlier indications that wolf numbers were low in the area with density along the northern foothills of the Brooks Range being about one wolf per 390 km^2. In the northern and western portions of the North Slope, wolf density was about one wolf per 520 km^2.

Wolf numbers were low in the immediate vicinity of the study area, although at least two other packs existed in the areas adjacent to the summer ranges of the radio-marked packs. In the lower Noatak area where radio-marked packs wintered on three occasions, wolf numbers may have been slightly higher than to the north; nevertheless, density was low.

Iligluruk Pack

In mid-April 1977, the tracks of two wolves were seen along the upper Kokolik River, and during early summer 1977, an adult male and an adult female were radio-collared in the same area. They successfully reared seven pups and remained in an area of about 1,000 km^2 (Figure 4.1) throughout the summer and early autumn, relying primarily upon caribou for food. This pack was last located in the Iligluruk drainage on 19 October, after which inclement weather precluded flying until 28 November, when an aerial search of the summer range as well as the upper Colville River drainage indicated that the wolves were no longer present in the area. On 2 December, the entire pack was relocated near the mouth of the Kelly River in the lower Noatak drainage 80 km south of the 1977 den (Figure 4.1). They remained in this area until early January, after which several extensive searches failed to locate them.

They were again located near the mouth of the Kelly River on 6 April. At this point, the pack included seven individuals, including both adults and five pups. A trapper at the mouth of the Kelly River reported trapping two gray pups in mid-winter, and it is likely that these were members of the Iligluruk pack. On 11 April 1978, a male and two female yearlings were radio-collared in the Kugururok River valley. However, the collars were chewed off both female wolves during the next few days, apparently because a less durable collar material was used in a new design. The pack then moved 29 km northeast to the upper Kelly River where they were located on 19 April. By 22 April, the pack had returned to the area used during the previous summer where they were located 42 km north of their location of 19 April, within 4.8 km of the den used in 1977.

The pack resumed its use of the 1977 summer range and, by 18 May, began using a den on Iligluruk Creek 16 km upstream from the 1977 den. The adult female remained near the den and was often inside it from 13 May through mid-June. During late June and early July, the two female yearlings that had lost their radio-collars were recaptured and again outfitted with transmitters, two male yearlings were radio-collared, and the collar on the adult female was replaced. During summer 1978, the seven adult and yearling wolves in the pack ranged over an area only slightly larger (approximately 1,300 km^2) than that used by the two adults during 1977. On 1 October, the pack, which included two adults, five yearlings and four pups born in 1978, was still located in the summer range although one female yearling (NW 13) had moved alone to the Kugururok drainage (see discussion on extraterritorial movements).

By 16 November, the pack had again moved south and 10 pack members were observed near the Noatak River in the same general area used during the previous winter. On 11 January and 13 February 1979, the pack was also observed in this area and apparently remained in this "winter range" for a period similar to that observed in the previous winter. On 4 May, two pack members were again observed near the Iligluruk den and, in succeeding weeks, a minimum of five pack members and six pups were observed near the den. It appears that some radios failed during the late winter. We also are aware that at least one radio-marked wolf was shot, but have been unable to determine its identity. We are unable to account for four pack members and know only that the original adult female and male, one other radio-marked wolf whose transmitter failed, and two unmarked wolves (probably pups born in 1978) were present in the traditional summer range during 1979. At least six pups were being reared by the pack.

Anisak Pack

In mid-April 1977, the tracks of six to eight wolves were noted in the Anisak area during aerial surveys, and four male yearling wolves were collared in May at different locations up to 70 km apart in the vicinity of the Anisak River. Three of these were later found to be members of a pack that, during summer 1977, included a minimum of seven adult or juvenile wolves and two pups. The remaining radio-collared male (NW 4) disappeared after being found 3.2 km from the Anisak den eight days after being collared. This observation initially suggested that NW 4 was a member of the Anisak pack, but subsequent events, recounted below, suggest that NW 4 was originally from a different area.

The Anisak wolves ranged over an area of about 2,600 km² during both summers, although an area of about 1,300 km² was used far more intensively than the remainder. All pack members remained in association with the 1977 den until late summer when two of the yearling males (NW 1 and 2) disappeared. The location of both wolves was unknown until 29 November when NW 2 was radiolocated near the Killik River (Figure 4.1) 220 km east of the Anisak den. On 27 December, this wolf was trapped in the same area. On 1 December, the other missing wolf, NW 1, was trapped at the headwaters of the Alatna River 170 km east of the Anisak den and 35 km southeast of the location of NW 2. The remaining yearling male (NW 3) stayed with the pack at least through winter 1977-78.

The Anisak pack was last located in its summer range on 22 September, after which inclement weather and other problems precluded radio-tracking until 28 November when a high altitude search of the summer range failed to locate the pack. On 2 December, NW 3 and six other pack members were located 170 km southwest of the 1977 den in the lower Noatak drainage 64 km south of the Iligluruk pack. On 15 December, the pack was located 53 km north of its earlier location, and on 3 January 1978 it was located in the same area. By 3 March, the pack had moved 54 km northwest to the headwaters of the Wulik River, and on 12 March it was found 32 km to the south. By 8 April, the pack had moved 74 km east. On the following day, it was located 18 km to the south where three wolves, including a black adult male, a gray male

male, a gray male pup and a gray female pup, were radio-collared. By 11 April, the pack had moved 51 km north to the Kugururok drainage but, by 19 April, it had returned to the lower Noatak where it was also found on 20 April.

On 27 April, the pack was again found in the summer range at the headwaters of the Anisak River, having travelled more than 170 km sometime during the previous week. During early summer, at least six adult wolves were associated with the den, which was the same as that used in 1977. In early July, the female parent was radio-collared but died the following day of unknown causes. Despite this, the pack reared seven pups, and at least five adults and yearlings were present in the pack in late September. The summer range was similar to that used during 1977, and the pack remained in the Anisak area until at least 16 November.

Due to severe weather, the pack was not again located until 13 February 1979 when it was found 45 km south of the Anisak den, about 20 km south of the area used during summer. It was not located again until 1 June when two pack members were observed at the den used in 1977 and 1978. Although we cannot rule out the possibility that the Anisak wolves left the vicinity of their summer range during the intervals between these observations, we consider it unlikely.

Noatak Pair

As mentioned earlier, one of the yearling male wolves collared in the Anisak area in 1977 was not located in that area after 22 May. We assumed that the wolf had either dispersed or that its transmitter had failed, but almost a year later on 19 April 1978 he was located again. On this and eight subsequent occasions, NW 4 was associated with one other gray wolf in a relatively small area 70 km southwest of the Anisak den, where they were last located on 24 July despite several high altitude searches. It is probable that this wolf's transmitter failed in late summer. In retrospect, it appears that NW 4 was collared by chance during an "extraterritorial" or dispersal movement while in the Anisak area. NW 4 either returned to a previously used range southwest of the Anisak area or, in the course of dispersing from some other area, acquired an associate and the two established a territory.

Dispersals

As noted above, the two packs consistently occupied the same summer areas. They also travelled as fairly cohesive social units during winter, although without establishing any apparent territories. There were, however, a number of instances in which individual pack members left usual pack ranges and dispersed for varying lengths of time. These movements are summarized in Table 4.1, and shown in Figure 4.1.

Table 4.1: Summary of "Extraterritorial" (Short-Term) and Dispersal (Long-Term) Movements by Radio-Marked Wolves

Wolf	Pack	Sex	Age at Time of Movement	Approx. Date of Movement	Known Minimum Straight Line Distance Moved	Direction	Date of Last Location	Comments
NW 5	Unknown	F	≈12 mo.	5/77	177 km	north, then east	13/6/77	Dispersal, but origin unknown
NW 1	Anisak	M	15 mo.	after 14/8/77	170 km	east	12/1/78 (trapped)	Dispersal
NW 2	Anisak	M	14 mo.	≈ 7/7/77	35 km	north	7/7/77	Extraterritorial movement lasting only a few days
NW 2	Anisak	M	16 mo.	after 22/9/77	220 km	east	27/12/77 (trapped)	Dispersal
NW 9	Anisak	M	13 mo.	after 9/6/78	225 km	north	29/1/79 (shot)	Dispersal
NW 11	Iligluruk, accompanied by 2 other 12 mo. old wolves	M	12 mo.	13/5/78	48 km	northeast	21/5/78	Extraterritorial movement, wolves returned by 28/5/78
NW 13	Iligluruk	F	15 mo.	16/8/78	50 km	south	18/8/78	Extraterritorial movement, wolf returned to pack territory 29/8/78
NW 13	Iligluruk	F	16 mo.	after 7/9/78	90 km	south	5/6/79	Dispersal that resulted in new pack association

We categorized these movements as either short-term "extra-territorial movements" or long-term and probably permanent dispersals. We observed three short-term extraterritorial movements and five long-term dispersals involving two female and four male radio-marked wolves ranging in age from 12 to 16 months (Table 4.1). The straight-line distances moved ranged from 35 to 225 km. Only one of the five dispersing wolves is known to have acquired new associates; three of the five wolves were eventually trapped or shot by Eskimo hunters. Contact was lost with the remaining wolf.

Prey Abundance

Caribou are the only large prey species that reach any significant level of abundance in the study area, but the availability of caribou varies drastically, sometimes from one week to the next. During most years, there are extended periods when caribou are totally absent from major portions of the area. The general movements of caribou in relation to the study area are briefly described below. The study area is located in the range of the Western Arctic caribou herd which numbered about 106,000 animals in 1978. The primary winter range of this herd lies south of the Brooks Range along the northern fringe of the boreal forest, 200–300 km south of the study area, although small portions of the herd often winter in the mountains not far south of the study area. In contrast to the pattern observed during most of this century (Hemming, 1971), however, a significant portion of the herd has wintered north of the Brooks Range on the coastal plain from 100–300 km north and east of the study area during several recent winters.

The movement from winter range toward calving grounds usually begins in late March and continues until pregnant cows reach the primary calving ground centered about 150 km northeast of the Iligluruk Creek area. Bulls and yearlings generally reach the North Slope later than cows and, during May and early June, are widely distributed over the area. The peak of calving is generally about 5 June. By mid-June, cows and calves begin leaving the calving area and are joined by bulls and yearlings as the post-calving aggregation begins. This movement usually takes a large portion of the herd to the southwest, after which they move east across the northern mountains and foothills of the Brooks Range with dispersal beginning in early July. During late summer, caribou tend to be widely distributed over the North Slope. By late August, a slow movement south is usually evident, and during September and October, steady movements to the south occur. During mid-winter, members of this herd are relatively sedentary.

The following brief review of relative prey abundance within the ranges of the Iligluruk and Anisak packs is synthesized from our own observations and those made by colleagues studying other wildlife species in the area. Because the terrain in the study area is open and sparsely vegetated, observations of prey made during aerial tracking flights and observations from the ground provide reasonably good estimates of relative prey abundance.

The availability of prey other than caribou is limited in both areas. Moose are extremely scarce within the Iligluruk summer range during all seasons of the year, with only 15 being observed during an aerial survey of riparian willow habitat along the Utukok and Kokolik Rivers in 1977. Dall sheep (*Ovis dalli*) occur in low numbers in the mountains between the pack's summer and winter ranges. Smaller prey include ground squirrels (*Spermophilus groenlandicus*), tundra voles (*Microtus oeconomus*), singing voles (*Microtus muirus*), ptarmigan (*Lagopus* sp.) and small numbers of a variety of waterfowl and small birds. The availability of prey in the Anisak area is similar except that moose are more common along the larger rivers in the area, as are snowshoe hares (*Lepus americanus*), which are cyclically abundant in these limited areas.

The general pattern of caribou availability was as follows. Beginning in May, caribou were usually abundant in both areas with several thousand animals passing through during the northward spring migration to the calving grounds and summer range. In the Iligluruk summer range, this high level of abundance continued until early July due to the presence of hundreds of adult bulls and yearlings that occupied areas peripheral to the calving ground and to the movement of as many as 50,000 caribou through the area during the post-calving migration. In the Anisak range, which was south of the usual caribou summer range, caribou were abundant for only a short period in spring and early summer. Between mid-June and early September, caribou were very scarce with only a few caribou present in the entire summer range. In the Iligluruk range, caribou were scarce after mid-July, although the absolute number of caribou available usually remained somewhat higher than that observed in the Anisak range during late summer. The availability of caribou increased dramatically during September in both areas as hundreds, and sometimes thousands, of caribou passed through during the fall migration. During October, however, the number of caribou inhabiting the Iligluruk and Anisak areas declined sharply, and observations during the winter months indicated that caribou were absent from the Iligluruk summer range from October through April. In the Anisak range, we obtained only a few indications of caribou abundance during winter; these suggested that caribou were rare during winter 1977–78, while at least a few caribou remained in the area during the following winter.

In the winter range used by the Iligluruk pack during 1977–78 and 1978–79, and by the Anisak pack during 1977–78, scattered bands of caribou, including as many as 150 animals, were present. Moose were more common and more widespread in this area compared to the summer range, the densities of about one moose per km^2 occurring in the best lowland habitat. In addition, the moun-

tains surrounding the lower Noatak area support a low-density population of about 300 Dall Sheep.

Food Habits

The food habits of radio-marked packs were assessed through
observations of large prey carcasses noted in the course of aerial
and ground-based monitoring, and through analysis of scats collected at dens and rendezvous sites. During both summer and winter,
caribou were the predominant prey of radio-marked wolves and
their associates. During the period 1 May to 1 October in 1977 and
1978, the members of radio-marked packs were observed at 37 kills,
all of which were caribou. Twenty-nine of these carcasses were
found during monitoring of the Iligluruk pack, while the remaining
eight were killed by Anisak wolves. The two Noatak wolves were
not observed at any kills. Thirty-one of the kills observed during
summer were adult or yearling caribou, two were known to be
calves and four were of unknown age; 19 kills of known sex comprised nine bulls and 10 cows.

Identifiable prey items were detected in 920 of 1,023 scats. The
material in the 103 remaining scats appeared to consist largely of
residue from the digestion of soft tissue. Also not included in the
analysis were trace amounts of wolf hair and plant material, the
amounts of which did not suggest use as food.

General Characteristics of the Diet: The data in Tables 4.2 and
4.3 demonstrate some general aspects of wolf food habits in the
study area. Caribou constituted 97 and 96 percent of the biomass in
the Iligluruk and Anisak diets, respectively. In terms of the number
of caribou eaten, calves made up 20 percent of the Iligluruk diet and
six percent of the Anisak diet.

Moose constituted only one and four percent of the biomass
in the Iligluruk and Anisak diets, respectively. Moose calves occurred 2.5 times more frequently than adult moose in terms of
number of individuals in the packs' diets.

The diet of the Iligluruk pack included nearly four times as
many arctic ground squirrels as individual adult caribou, but squirrel
biomass was only three percent that of the caribou (Table 4.3).
The Anisak pack consumed a smaller proportion of squirrels relative to adult caribou, although the number of squirrels consumed
exceeded the combined number of caribou calves and moose adults
and calves eaten. Squirrels' biomass, however, was only 0.2 percent
of the biomass of caribou adults. The remaining categories of prey
contributed relatively little to the diet of the packs.

Selectivity: The FQ data in Table 4.2 indicate some differences
and similarities between feeding patterns of the Iligluruk and Anisak
packs. Caribou FQ was not significantly different (p>0.05) between
the two packs. Calf caribou FQ was significantly higher (p<0.01)
in the Iligluruk diet, and adult caribou FQ was significantly higher
(p<0.001) in the Anisak diet. Moose FQ appeared to be greater in
the Anisak diet, but statistical verification of this was prevented by

small sample size. Arctic ground squirrel FQ was significantly higher (p<0.001) in the Iligluruk scat collection. No significant difference was apparent between the microtine/shrew FQ in the Iligluruk and Anisak diets. Eight occurrences of ptarmigan remains were found in scats from the Iligluruk pack, whereas no ptarmigan remains were found in Anisak scats.

Table 4.2: Frequency of Occurrence (FQ) and Relative-Estimated Bulk (REB) of Prey Remains in Scats Collected at Summer Home-sites of the Iligluruk and Anisak Wolf Packs, 1977 and 1978

Food Item	Iligluruk Pack		Anisak Pack	
	FQ	REB	FQ	REB
Ungulate				
Unidentified adult ungulate*	294	263.05	—	—
Caribou adult	157	141.05	345**	339.50**
Caribou calf	70	63.10	—	—
Moose calf	2	2.00	6	5.90
Moose adult	1	0.95	3	3.00
Total	501	470.15	354	348.40
Small mammal				
Arctic ground squirrel	124	65.30	11	4.00
Microtine/shrew	57	10.60	41	5.85
Unidentified rodent	5	0.25	1	0.05
Snowshoe hare	1	0.05	0	0.00
Total	177	76.20	52	9.90
Other				
Ptarmigan	8	0.60	0	0.00
Unidentified carnivore	6	0.30	0	0.00
Unidentified bird	5	0.25	4	0.20
Arthropod	3	0.15	4	0.20
Mollusc	1	0.05	0	0.00
Eggshell	0	0.00	4	0.20
Unidentified fish	0	0.00	1	0.05
Total	23	1.35	13	0.65
Total scats	551		369	

*Assumed to be adult caribou

**The 1977 Anisak pack FQ and REB ratios are unknown. The 1978 Anisak pack FQ and REB ratios were 70:12:3 and 68.35:11.85: 3.00, respectively, for unidentified adult ungulate:caribou adult: caribou calf (n = 85).

Table 4.3: Relative Weights and Numbers of Three Key Species Which the Iligluruk and Anisak Wolf Packs Consumed During May–August 1977 and 1978*

Prey Type	Assumed Wt. of Prey (kg)	Relative Estimated Bulk		kg of Prey Eaten		No. Individual Prey Eaten	
		Iligluruk	Anisak	Iligluruk	Anisak	Iligluruk	Anisak
Caribou, adult	104	404.10	327.26**	994.09	805.06	9.56	7.74
Caribou, calf	22	63.10	12.24**	51.74	10.04	2.35	0.47
Moose, calf	72	2.00	5.90	3.64	10.74	0.05	0.15
Moose, adult	404	0.95	3.00	8.04	25.38	0.02	0.06
Arctic ground squirrel	0.7	65.30	4.00	25.73	1.58	36.76	2.26

*Estimation technique as described by Floyd et al. (1978)

**The relative proportions of adult and calf caribou in the Anisak pack's diet were estimated (see text)

The combined data for the two packs indicate that caribou were the most important prey of wolves. Moose were of far less importance, and we saw no indication of predation on Dall sheep. Although the Anisak wolves had a greater opportunity to take sheep because the northern portion of their range encompassed a low density Dall sheep habitat, members of this pack were rarely located in these areas.

In both 1977 and 1978, the proportion of calves among caribou on or near the Western Arctic herd calving ground during early June was about 39 percent, and calves further comprised about 29 percent of the post-calving aggregations (Alaska Department of Fish and Game files, Fairbanks, Alaska). The greater proportion of calves in the Iligluruk diet was probably due to the greater proximity of their summer range to the calving grounds. Additionally, most of the post-calving movements occurred north of the Anisak range and resulted in periodic surges of from several hundred to several thousand caribou, including calves, through the Iligluruk summer range. For several reasons, including the limited time during which calves were available to the wolves in any numbers, the proportion of caribou calves in the summer diets of the Iligluruk and Anisak packs did not exceed, and was probably less than, the proportion of calves in the population.

During the winter months of both years (1 October to 1 May), we located 11 kills during infrequent monitoring. These included six caribou and one moose killed by the Iligluruk wolves and four caribou killed by the Anisak wolves. Three of these were identified as adults, and the remainder were of unknown sex and age. In addition to the 11 verified wolf kills, there were an additional 14 indications that kills had been made, but we were unable to locate the carcasses.

Characteristics of Predation on Caribou

Observations of wolf-killed caribou and wolf-prey interactions provide a good indication of the rate at which the radio-marked wolves took caribou, as well as other aspects of predation. Below, we offer preliminary comments about the nature of interactions between wolves and caribou. Our data originate mainly from observations of the Iligluruk wolves from early summer to early winter in 1977 and 1978.

Our preliminary estimates of the rate at which the packs killed caribou are given in Table 4.4. We estimated rate of kill using two methods. In Method I, we selected periods during which contact was relatively frequent and calculated an average period between kills. We also calculated kill rates by expressing the number of kills found as a percent of the number of days on which each pack was located, and projecting this to a one year period (Method II). Both derived rates should be considered maximums and are representative of a situation in which wolves were heavily reliant upon caribou for food and where caribou were available at moderate to high levels.

Table 4.4: Predation Rates Calculated for Two Radio-Marked Wolf Packs in Northwest Alaska

Pack	No. Days Located	Total No. of Kills Observed	Average No. Days Between Kills	
			Method I	Method II
Iligluruk	105	36	2.5 (1.9)*	2.9 (2.2)*
Anisak	55	12	no data	4.6 (3.9)*

*Calculated with probable kills included in total

As shown, the estimated average number of days between kills ranged from 1.9 to 4.6 days, with the lowest rates being estimated for the Anisak wolves. This may be due to the less intensive monitoring of this pack, but also reflects a somewhat lower average availability of caribou in the area. The limited winter data do not permit the calculation of a specific predation rate for that season, but it is worth noting that 10 kills were located on 25 occasions when packs were visually located during winter, suggesting a relatively high frequency of predation.

Comparable rates of predation have been described for wolf packs in a number of other areas. Mech (1966) found that a pack of 15 wolves on Isle Royale killed an average of one moose every 3.0 days during late winter and, in subsequent studies, Peterson (1977) found that three packs averaging 12 members also killed an average of one moose per 3.0 days, with the largest pack (17 wolves) killing one moose per 2.6 days during late winter. Kolenosky (1972) found that a pack of eight wolves in Ontario killed one deer every 2.2 days during a 63 day period in an area having a high density of wintering deer. Burkholder (1959) found that during a 45 day observation period, a pack of 10 wolves inhabiting the Nelchina Basin in Alaska killed a caribou or moose every 1.2 days during late winter. Subsequent studies in the Nelchina Basin area under conditions of lower prey abundance indicated that wolf packs were taking moose and/or caribou at rates of one kill per four to 10 days per pack (Stephenson, 1978). The predation rates shown in Table 4.4 approach the relatively high rates observed in areas where prey were abundant. It should be noted, however, that the indicated rates are apparently not consistent throughout the year. For brief periods, predation rates exceeded one caribou per day, while at other times predation rates appeared to be somewhat lower than those shown in Table 4.4. Caribou taken by wolves during all seasons were almost completely consumed with only small amounts of edible portions remaining after a day or two.

Wolves were successful in killing caribou while hunting alone, as well as in packs with as many as 11 individuals. Our data suggest that yearling wolves were often successful in killing caribou while hunting alone. During summer 1978, for example, individual yearling wolves in the Iligluruk pack were observed at 13 recent or fresh caribou kills at which no other pack members were present and, in one instance, we watched the successful pursuit of an adult cow by an 11 month old wolf.

DISCUSSION

In many respects, the radio-marked packs studied in northwest Alaska conform to the generalized picture of wolf ecology that has emerged from recent work in North America and Europe. The seasonal patterns of den and rendezvous site use, parental behavior, food habits, pack cohesiveness and hunting behavior are not unusual but fall within the range of phenomena observed in other studies of wolf packs (Murie, 1944; Mech, 1966, 1970, 1977a; Peterson, 1977). In contrast, however, are the extensive winter movements undertaken by both the Iligluruk and Anisak packs in winter 1977, and again by the Iligluruk pack in 1978. Our observations show that on these occasions, the packs left their summer ranges in late October or November and moved initially 80-170 km to areas where caribou overwintered. In at least one case, subsequent movements were undertaken that extended at least 198 km from a denning area used in 1977. In each case, the wolves returned to their respective summer ranges by late April or early May and reared pups in the areas used the previous summer. The onset of these winter movements appeared to correspond closely to the low availability of caribou in the summer ranges following the autumn caribou migration. It is not clear whether these movements were prompted entirely by immediate necessity or were, in part, due to previous conditioning. The fact that in each case the packs left their summer range within a period of a few weeks in early winter and moved to the same general area suggests that the wolves may have been repeating a traditional pattern of movement rather than searching randomly for an area where prey were available. The lower Noatak valley has been used by wintering caribou quite consistently during recent decades (Skoog, 1968; Hemming, 1971), and appears to be the closest such area to the Iligluruk range and possibly to the Anisak area as well. It seems likely that the winter wolf movements we observed were a result of the long-term use of the area by wintering caribou. Of interest is the fact that the Anisak wolves left their summer range in 1977 despite the presence of a small resident population of moose in the area, while in 1978 they remained in the vicinity of their summer range when limited observations suggest caribou were available during the winter months.

Movements of a similar nature over distances of as much as 360 km by ear-tagged wolves have been reported in Canada by Kuyt (1972), and a general southward movement during winter has been reported for wolves on Alaska's North Slope by the Nunamiut (Stephenson and Johnson, 1973). In addition, the preliminary results of ongoing telemetry studies in the Northwest Territories, Canada, are showing a similarly strong association between wolves and caribou (George Calef, Government of the Northwest Territories, pers. comm.). It is apparent that wolves which are highly dependent on caribou herds regularly undertaking extensive seasonal migrations, have territorial habits that are quite different than those typical of wolves relying upon less migratory prey.

Acknowledgements

This work was funded by the U.S. Fish and Wildlife Service (National Petroleum Reserve-Alaska, 105c Studies) with support by the Alaska Department of Fish and Game. We thank H. Reynolds, P. Shepherd, J. Davis, P. Valkenburg and A. Magoun for their help in observing radio-marked wolves. J. Rood of Kotzebue deserves special thanks for his safe and effective flying during all phases of the study.

Winter Predation on Bison
and Activity Patterns of a Wolf Pack
in Wood Buffalo National Park

Sebastian M. Oosenbrug and Ludwig N. Carbyn

INTRODUCTION

Relationships of wolves to many prey species have received widespread attention in recent years. However, the interactions of wolves and bison have not been extensively investigated. One region still containing both wolves *(Canis lupus)* and bison *(Bison bison)* is Wood Buffalo National Park in Alberta and adjacent portions of the Northwest Territories, Canada. Early accounts of wolves and bison in Wood Buffalo National Park (Soper, 1941, 1945), along with political pressure, ultimately led to a number of wolf control programs during the 1950s (Fuller and Novakowski, 1955), although Fuller (1966) later concluded that wolf predation was not detrimental to general bison welfare.

Current wolf research by the Canadian Wildlife Service in Wood Buffalo National Park is conducted in conjunction with a series of studies on bison population dynamics and distribution, ongoing since 1972. These studies were initiated because of concerns for the long-term welfare of bison in the Park. Dramatic declines of bison have been documented in the Hook Lake area adjacent to Wood Buffalo National Park in the Slave River Lowlands, Northwest Territories. There, the bison population decreased from about 2,000 animals in 1972, to 750 in 1978 (VanCamp, pers. comm.).

Although the Slave River Lowlands are subject to native hunting, many local residents believed that wolves were the major reason for the decline of bison in the area. VanCamp considered that the decline had probably been brought on by disease (including brucellosis, anthrax and tuberculosis) and over-hunting, which were compounded by severe winter conditions during 1974–75, and then was continued by residual wolf predation and hunting pressure through 1978. During that latter period, the ratio of wolves to bison evidently permitted the wolves to effectively prevent bison recruitment.

43

The present study of wolves and bison was initiated in 1978 to determine the impact of wolf predation on bison in Wood Buffalo National Park. During winter 1978–79, a wolf pack containing several radio-collared individuals was studied to determine winter range size, movements, kill rates and prey selection. This pack, the Hornaday River pack, was one of several which occupied winter bison ranges in the Park.

STUDY AREA

Wood Buffalo National Park is situated within the Mackenzie Lowlands, a northern extension of the Great Central Plains (Figure 5.1).

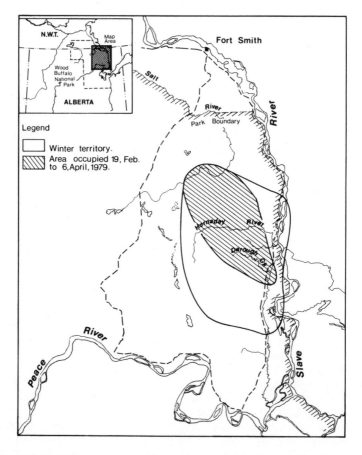

Figure 5.1: Winter territory (5 December 1978 to 29 April 1979) of the Hornaday River pack and the area covered during period of intensive tracking (19 February to 6 April 1979).

The Hornaday River pack's winter range is about 60 km south of Fort Smith, Northwest Territories, in the northeast corner of Wood Buffalo National Park. Most of the area is within the Salt Plains, one of four elevational units in the Park (Raup, 1935). The Salt Plains are characterized by flat, poorly drained terrain with marshes and slow-moving streams. It is a region of vast saline prairies separated by and interspersed with islands of white spruce *(Picea glauca)*, willow *(Salix sp.)* and aspen *(Populus tremuloides)*. The prairies, colonized by sedges and other salt-tolerant species, retain melt-water well into the summer in numerous ponds and marshes.

METHODS

Initial attempts to locate the Hornaday River pack were conducted by fixed-wing aircraft. Once the pack was located, selected wolves were darted from a helicopter using a technique similar to that described by Stephenson (1978). Telemetry data were subsequently obtained on one to four day intervals from 19 February to 6 April 1979. The pack was radio-located from the air and backtracked to the site of previous contact. Wolf travel routes were determined from aerial observations of the pack or its tracks. For some days, when consecutive daily coverage was not possible or wolf tracks were obliterated by snow or obscured by bison tracks, travel routes were partially determined by tracks and projected over topographic features most probably travelled. Where possible, ground personnel, on snowmobile or on foot, verified the travel routes of the pack. Field personnel tried to stay at least one day's travel behind the wolves to avoid disturbing them.

When kills were found, the following information was recorded: sex and estimated age, as determined by horn development and number of horn annuli, proportion of carcass consumed and, where possible, events preceding death were reconstructed. Kills were not examined until after the pack had voluntarily abandoned the kill site. Kills not completely consumed were periodically checked to determine the frequency of revisits by wolves.

Where possible, jaw and femur samples were collected from each carcass in order to accurately determine age and nutritive condition of bison killed. Second incisor teeth were cross-sectioned, decalcified, and cemental layers read (Armstrong, 1965). Bone marrow was removed from the femur and fat content determined using the methanol/chloroform extraction method (Verme and Holland, 1973). Skeletal remains were examined for anomalies, and suspect bone fragments were analyzed at the Smithsonian Institution, Washington D.C.

The size of the bison population in the winter range of the Hornaday River pack was determined by monthly aerial surveys from November 1978 to March 1979, conducted by the Park Warden

Service using a modified strip census technique (Tempany and Cooper, 1976).

RESULTS

The Study Pack

The Hornaday River pack was under almost daily surveillance from 19 February to 6 April 1979. During that 47 day period, aerial tracking was carried out on 35 days and radio contact was established on each of those days. Mean number of days between radio-fixes was 1.40±0.84. The pack, or part of it, was sighted on 34 of the days radio contact was made.

Five pack members (2 pups, 3 adults) were captured between 9 January and 1 April 1979. Three of these (1 adult male, 1 adult female, 1 female pup) were collared prior to the period of intensive tracking (19 February). The majority of observations are based on radio-fixes from these three wolves.

We were able to distinguish the alpha male and female on the basis of their dominant behavior within the pack. Neither of these animals was collared. We believe that there were four pups in the pack based on their playful disposition and subordinate behavior (Jordan et al., 1967). Two of these, along with three of the four remaining adults, were eventually collared. Thus, the 10 wolves whose activities were monitored included three adult males, two adult females, one male pup, one female pup, two pups of unknown sex, and one adult of unknown sex.

The Hornaday River pack numbered 14 members in early winter. Pack size counts between 19 February and 6 April averaged 10.04±0.66 wolves. The fate of the four wolves seen in early, but not in mid and late winter, is not known. Two wolves were reported trapped by natives in the area. During March, the pack split on two occasions into at least two groups. Although these groups rejoined about 10 days later, it is possible that a permanent splitting occurred with other members. Track and scat evidence suggested that at least three or four wolves travelled in the vicinity of Hay Camp when the collared wolves were 15–20 km farther west.

Movements and Winter Range

The composite winter range of the Hornaday River pack (5 December 1978 to 29 April 1979) was estimated as 1,250 km^2 (Figure 5.1), based on the minimum area method (Mohr, 1947). Between 19 February and 10 April (50 days), pack movements were restricted to an area of approximately 490 km^2, 39 percent of the total winter range (Figure 5.1). The reason for the decrease in area covered could not be determined, but may have been due to changes in bison distribution as well as snow conditions. Aerial surveys of bison during early winter have shown that bison do not settle into their winter ranges until early to mid-January (Couchie and Collingwood, 1978). Snow measurements taken by Atmospheric Environment Service personnel at Fort Smith indicated that snow depths increased throughout the winter and reached a maximum of 42 cm in March 1979. Snow depth possibly restricted bison movements in late winter and, as a consequence, the wolves travelled less. In a similar study in Riding Mountain National Park, a pack of wolves preying principally on elk *(Cervus elaphus)* ranged over 56 percent of its territory within a 19 day period during mid to late winter (Carbyn, 1981a).

The Hornaday River pack travelled 214 km (mean = 5.5 km/day) during 39 days of intensive tracking (Figures 5.2 and 5.3). Daily distances ranged from 0.0 to 18.7 km (Table 5.1). Daily distances travelled, excluding days spent at kills or carcasses, ranged from 2.3 to 18.7 km, and averaged 8.8 km.

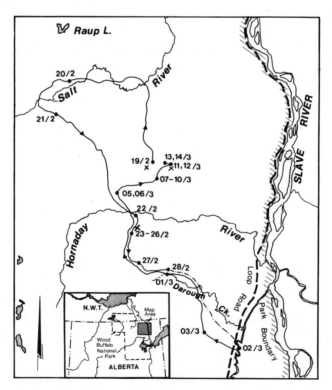

Figure 5.2: Travel routes and kill sites of the Hornaday River wolf pack as determined from radio telemetry tracking, aerial and ground tracking from 19 February to 14 March 1979. (Solid lines = distances tracked; broken lines = presumed travel routes based on topography and partial track evidence; x = kill sites.)

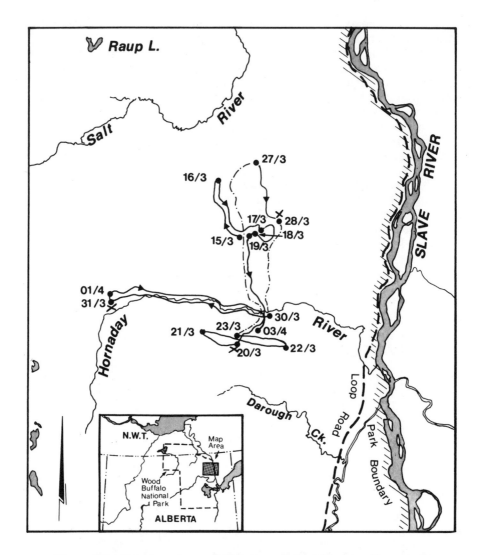

Figure 5.3: Travel routes and kill sites of the Hornaday River wolf pack as determined from radio telemetry, aerial and ground tracking from 15 March to 3 April 1979 (see Figure 5.2 for legend).

Predation Rates and Prey Selection

During the period from 12 February to 31 March (47 days), the Hornaday River pack killed six bison, an average of one every 7.8 days. The chronological sequence, sex, age and amount consumed of the bison killed by the pack during this time are shown in Table 5.2.

After killing a bison, the pack spent an average of 2.5 days at the kill site before moving on. For four of the kills, we were able to document revisits. Three of these kills were revisited once, the other twice. The wolves spent an average of 2.3 days on each revisit of kills they had made earlier in the winter.

Table 5.1: Mid to Late Winter Daily Wolf Activity Patterns of a Single Wolf Pack Studied in Wood Buffalo Park from 20 February to 3 April, 1979[1]

Date		Daily Distance Travelled (km)	Daily Distance Travelled Excluding Days Spent at Kills and Carcasses (km)
February	20	18.2	18.2
	21	8.8	8.8
	22	18.7	18.7
	23	2.3	2.3
	24	0.0	—
	25	0.0	—
	26	0.0	—
	27	5.6	5.6
	28	6.0	6.0
March	1	0.0[2]	—
	2	14.9	14.9
	3	5.4	5.4
	4	12.9[2]	12.9
	5	12.9	12.9
	6	0.0	—
	7	5.6	5.6
	8	0.0	—
	9	0.0	—
	10	0.0	—
	11	3.3	3.3
	12	1.2	—
	13	0.2	—
	14	0.0	—
	15	0.7	—
	16	5.8	5.8
	17	8.4	8.4
	18	7.5	7.5
	19	0.7	—
	20	10.1	10.1
	21	3.8	3.8
	22	8.6	8.6
	23	5.3	5.3
	28	6.5	6.5
	29	0.2	—
	30	9.8	9.8
	31	14.7	14.7
April	1	0.5	—
	2	7.5[2]	7.5
	3	7.5	7.5
Total		213.6	210.1

[1] Due to incomplete coverage during the periods 24–27 March and 3–6 April, distances travelled were not determined for these days.
[2] Estimated distances travelled.

Table 5.2: Chronological Sequence, Sex, Age and Utilization
of Prey Killed by the Hornaday River (HR) Pack
from 12 February to 21 March 1979

Date of Kill	Distance Travelled Since Last Kill (km)	% Fat in Marrow	Sex	Age[2]	Age by Tooth Cross-Section	Estimated Amount Consumed[1] First Exam	Second Exam	Third Exam	No. Days Pack was at Kill	No. Days Pack Scavenged Carcass
Feb. 18	–	–	M	5-6	6	3 (9)[3]	4 (35)	–	3	3
24	–	61	M	Mature[4]	7	2 (3)	3 (7)	4 (12)	2	1
Mar. 12	66.6	16	F	6-7	7	2 (4)	4 (23)	–	4	3
20	31.8	17	M	10+	4	4 (21)	–	–	1	2
28	–	28	F	10+	11	4 (8)	–	–	2	–
31	24.5	62	M	10+	9	3 (5)	3 (37)	–	3	–

[1] Based on Pimlott et al. (1969): 1 = 0-25%, 2 = 26-50%, 3 = 51-75%, 4 = 76-100%.
[2] Age estimated by examination of horn annuli.
[3] Figures in brackets are the number of days between the date of examination and the date of the kill.
[4] Indicates age could not be determined from horn annuli.

Percentage consumption was estimated when the kill was first discovered, and subsequently examined using criteria proposed by Pimlott et al. (1969). Although foxes *(Vulpes vulpes)* and ravens *(Corvus corax)* were present in the study area, observations indicated that wolves were the major consumers of bison carcasses. Kill utilization was very high; carcasses generally were totally consumed within a month. Although the extent to which carcasses were consumed prevented accurate field analyses, laboratory analyses of bone fragments, teeth and marrow samples provided an indication of the condition of bison killed. Three bison killed in March had very low fat reserves (Table 5.3); two adult males, one of which was a confirmed kill, suffered from osteomyelitis (G. Haynes, pers. comm.), a degenerative bone disease that can lead to paralysis and death.

Live weights for mature bison have been estimated as 570 kg for males and 420 kg for females (Banfield, 1974). Since four males and two females were killed, and assuming that 20 percent of any carcass is inedible (Pimlott, 1967), the total weight of bison consumed by the pack of 10 during the 47 day period from 12 February to 31 March 1979 was 2,500 kg, or 53 kg per day. The amount of food consumed per wolf per day thus equaled 5.3 kg.

In addition to actual predation, the wolves also visited four other carcasses. All were old (i.e. more than one month), and were already 75-100 percent consumed when the wolves visited them. Therefore, they were not considered to contribute to the amount of food available to the pack during the period of observation.

Monthly bison surveys indicated that within the range of the Hornaday River pack, an estimated maximum 191 bison were available to the wolves between mid-November and mid-March (based on the November count). Since to date we have no data to the contrary, we assumed that the observed rate of predation by the pack during February and March was representative of the overall winter predation rate. On this basis 19 bison, or 9.9 percent of the study area population, were consumed during winter 1978–79 (December to April).

Table 5.3: Laboratory Analysis of Age and Condition of Bison Carcasses Examined in the Hornaday River (HR) Pack Territory 19 January to 21 March 1979

Carcass No.	Sex	Estimated Age (yrs)[1]	Tooth Age (yrs)	% of Fat in Bone Marrow	Comments
1	M	7	6	NA	Killed by HR pack on 19 January.
2	M	Mature	NA	22.0	Old carcass, unconfirmed kill. Severe osteomyelitis on rib fragment - possible chronic infection could have led to paralysis and/or death.
3	F	Mature	9	19.0	Old carcass; unconfirmed kill.
4	M	5–6	6	NA	Killed by HR pack on 18 February. Osteomyelitis on rib fragment - possibly resulted from injury.
5	M	Mature	7	61.0	Killed by HR pack on 24 February.
6	M	7–8	8	46.0	Unconfirmed kill.
7	M	Old	12	NA	Old carcass; unconfirmed kill.
8	M	Mature	NA	NA	Old carcass; unconfirmed kill.
9	F	6–7	7	16.0	Killed by HR pack on 12 March.
10	M	10	4(?)	17.0	Killed by HR pack on 20 March.
11	M	10	11	28.0	Killed by HR pack on 28 March.
12	F	10	9	62.0	Killed by HR pack on 31 March.

[1] Minimum age.

Although the sex ratio of bison in the study area was not determined, aerial and ground observations indicated that, in most instances, adult males comprised the majority of bison in the immediate vicinity of the Hornaday River pack during the study period. Most of these occurred in widely dispersed groups of one to 10 animals. Thus, the availability of male bison may have been reflected in the sex ratio of bison killed. Of seven confirmed kills examined from 19 January to 31 March 1979, five were males and two were females (Table 5.3). All kills were of adult bison.

DISCUSSION

Our figures on food consumption are somewhat higher than those published for wolves preying on deer *(Odocoileus virginianus)*. In Ontario, Pimlott et al. (1967) and Kolenosky (1972) estimated a daily rate of consumption of 3.8 and 2.9 kg per wolf, and Mech and Frenzel (1971) calculated a rate of consumption in northeastern Minnesota of 2.7 kg per wolf per day (also see Stephenson and James, this volume). However, on Isle Royale, wolves preying on moose *(Alces alces)* were found to consume an average of 5.6 kg per wolf per day over three winters of observation (Mech, 1966). In Riding Mountain National Park, wolves consumed 6.7 kg of prey per day under unusual winter conditions and high elk densities (Carbyn, unpub.).

Our data indicate that each wolf in the Hornaday River pack consumed 5.3 kg per day, and are similar to those of Peterson (1977) who calculated winter food availability to vary from 6.2 to 10.0 kg per wolf per day during 1971-73 and 4.4 to 5.0 kg during 1974 on Isle Royale. Peterson (1977) argued that when degree of utilization was unknown, a measure of food availability rather than food consumption was obtained. Since our data described virtually complete utilization of kills by the Hornaday River pack, food availability in our case approached or equalled consumption.

Differences in food consumption among the various study areas mentioned above may be partially explained by size differences and energy requirements of the wolves under study. Based on the weights of our captured animals, wolves in the Hornaday River pack averaged 43 kg and were substantially larger than those found in eastern North America. However, differences in food consumption may be due more to annual variation in prey vulnerability and abundance (Peterson, 1977; Carbyn, 1981a), particularly as it relates to snow depths.

The mid to late winter predation rate of the Hornaday River pack was approximately one bison per 7.8 days, i.e. one wolf consuming one bison every 78 days. In the Slave River Lowlands, Van-Camp (in prep.) reported that over a six week period in early winter, a pack of 12 to 14 wolves consumed a minimum of six bison, or approximately one per week, although he felt that the winter predation rate for each pack (average size 11 wolves) more closely approached one adult every nine days and one calf every 15 days.

The mid-winter range of the Hornaday River pack during 1978-79 was approximately 490 km², larger than the winter ranges on Isle Royale which varied from 246 to 327 km² during 1971-74 (Peterson, 1977), in Manitoba which was 237 km² (Carbyn, 1981a), and in Ontario which was 224 km² (Kolenosky, 1972).

A comparison of our data on daily travel distances during winter with those reported in other studies suggests that movements of wolves in search of bison are generally less than those in search of deer (Stenlund, 1955; Kolenosky, 1972) and moose (Mech, 1966; Peterson, 1977). On Isle Royale, a pack of 16 wolves travelled

443 km over 31 days, or an average of 14.3 km per day (Mech, 1966). During days of actual travel (excluding stays at kills and carcasses when the wolves moved very little), the Hornaday River pack averaged 8.8 km per day. In Alaska, Burkholder (1959) followed a pack which averaged 24 km of travel per day for 15 days' travel, presumably including feeding periods. Kolenosky (1972) found that a pack of eight wolves in Ontario travelled 327 km over 46 days and averaged 7.1 km per day. More recently, Peterson (1977) calculated that the average daily distance travelled by the East and West packs on Isle Royale between 1971 and 1974 was 11.1 km (also see Stephenson and James, this volume).

In this study, pack movements between consecutive kills averaged 41 km (n = 3). Peterson (1977) found that on Isle Royale, the average distance travelled by packs between kills varied from year to year and ranged, during 1971–74, from 18.5 to 54.1 km over 1.8 to 4.6 days. He concluded that lowest travel per kill occurred in years when moose vulnerability was highest. In Ontario, Kolenosky (1972) reported an average distance of 14.7 km over 2.2 days between kills of deer during 1969. Although our data are inconclusive at this point, the relatively short travel distances between kills suggests that wolves preying on bison may use previous knowledge of their winter range (Peters, 1979) in order to locate and maximally exploit the clumped aggregations of prey.

The higher proportion of males among wolf-killed bison raises intriguing questions about differential predation mortality between the sexes. In our study, adult males comprised the majority of bison killed by the Hornaday River pack during the winter of 1978–79. Although the sample size is too small to be more than suggestive, it is worth noting that studies on deer by Stenlund (1955), Pimlott et al. (1969) and Mech and Frenzel (1971) reported a preponderance of males in the kill. The only other study on wolf-bison relationships provided data to the contrary (VanCamp, in prep.) and indicated that females and calves were the most vulnerable segment of the bison population. In the Hook Lake area where the sex ratio favored females, 86 percent of 52 bison killed by wolves were either mature females, subadult females or calves.

Adult male bison can, theoretically, be the greatest prey source for wolves if they form the majority of accessible prey within the winter range of a wolf pack, and if their solitary habits make them more vulnerable to wolf predation in winter. It seems likely that both of these conditions applied to predation on bison by the Hornaday River pack during the winter of 1978–79. Continued intensive winter studies should shed more light on prey selection by wolves in relation to bison abundance and distribution, and bison vulnerability to predation, particularly as it relates to winter severity.

Preliminary Investigations of the Vancouver Island Wolf
(*Canis lupus crassodon*)
Prey Relationships

Daryll M. Hebert, John Youds, Rick Davies,
Herb Langin, Doug Janz and Gordon W. Smith

INTRODUCTION

The Vancouver Island wolf (*Canis lupus crassodon*) is currently regarded as a separate subspecies (Jolicoeur, 1959; Lawrence and Bossert, 1967; Cowan and Guiguet, 1965), but apart from a limited examination of its taxonomic status, ecological investigations are nonexistent. The coastal ecosystems of Vancouver Island provide habitat, climate, topography, geological structure, timber harvest pattern and prey species which differ from published information for wolves in most other regions of North America. Historical information gathered from guides, hunters and Conservation Officers on Vancouver Island indicated that the Vancouver Island wolf was never numerous. Between 1900 and 1950, wolf abundance varied among watersheds on southern Vancouver Island. At that time, records from northern Vancouver Island were sparse due to a limited human population and restricted access and timber exploitation. However, reports indicate that wolves were relatively numerous on the islands east of Vancouver Island from Nanaimo to the northern end of Vancouver Island. Between 1950 and 1970, wolves were almost never seen on southern Vancouver Island, and although wolves were probably present in several watersheds on northern Vancouver Island, sightings were infrequent.

In 1970, the Vancouver Island wolf was included in the list "Endangered Wildlife in Canada" prepared by the Canadian Wildlife Federation. Canada signed the Convention on International Trade in Endangered Species in 1974 and included the Vancouver Island wolf in Appendix II with a reservation. By 1976 it was evident that the Vancouver Island wolf had increased its numbers substantially and requests from inside and outside the Fish and Wildlife Branch suggested that it be removed from rare and endangered status. In 1977, the Vancouver Island wolf was not included in the provincial "Threatened and Endangered Species Management Plan."

In this paper, we present recent data (1972–1978) on the distribution and abundance of the Vancouver Island wolf and examine relationships between these data and data collected on the wolf's two principal prey species, the Columbian black-tailed deer (*Odocoileus hemionus columbianus*) and the Roosevelt elk (*Cervus elaphus roosevelti*) (Scott and Shackleton, 1980).

STUDY AREA

Vancouver Island is a large island of approximately 32,137 km^2 on the west coast of British Columbia (Figure 6.1).

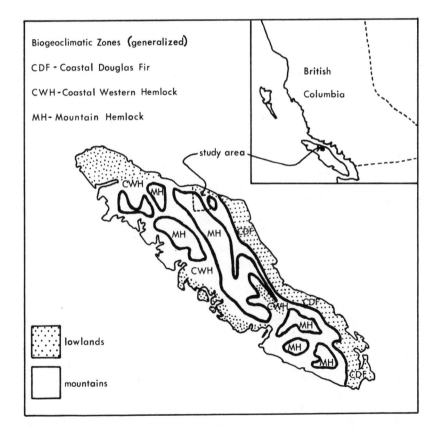

Figure 6.1: Location of study area on Vancouver Island, showing the major biogeoclimatic zones (Krajina, 1965) and topographic features.

There are three major biogeoclimatic zones (Krajina, 1965)—sub-alpine mountain hemlock, coastal western hemlock and coastal Douglas fir. The Island's mountainous interior is bordered by a lowland coastal plain which supports substantial populations of Columbian black-tailed deer and unevenly distributed populations of Roosevelt elk at the northernmost extreme of their natural distribution (Hebert, 1979).

Intensive predator and prey information was collected in the Adam and Nimpkish River valleys which lie on northern Vancouver Island. These rivers drain into the ocean on the east side of Vancouver Island and flow between mountain ridges which rise to approximately 2,200 m.

METHODS

During the period 1972–1976, the British Columbia Fish and Wildlife Branch collected random wolf sightings from the general public. Postpaid sighting forms were developed in 1972 and distributed mainly to interested members of the public. These sighting cards requested information on number of wolves, wolf age (mature or immature), location, vegetation type, method of observation, observation type and activity. No measures of search effort or the frequency of reported sightings were available; therefore, only the presence of wolves in a particular area can be determined from the data. Between 1976 and 1978 a wolf sighting questionnaire was enclosed with the annual deer harvest questionnaire, with a second send-out to nonrespondents. Sightings were tabulated by management unit (MU) and subunit (watershed). An estimate of search effort is available from the hunter sample data enabling relative density estimates to be calculated for each subunit. Wolves were radio-collared in the Adam River (Scott, 1979; Scott and Shackleton, this volume) to obtain information on pack size and distribution to determine the reliability of using hunter sightings to estimate wolf density. Home range boundaries were determined for 50, 90 and 100 percent of the radio locations. Pack density estimates were adjusted to management subunit density estimates based on the degree of overlap between pack home range boundaries and management subunit boundaries.

The relationship between wolf density and prey numbers and/or density was determined from prey indices such as night counts, percent juvenile carry-over and pellet group indices. In addition, deer harvest statistics were collected at hunter checks held on the opening weekend and weekends within the antlerless season and by deer harvest questionnaires. Elk population parameters (cow/calf ratios) were collected mainly by aerial survey methods.

RESULTS

Wolf bounty records for Vancouver Island (Figure 6.2) generally coincide with historical information collected from long-time Island residents. The bounty on wolves was introduced by 1906, although the exact date is unknown. Bountied wolves declined from 140 in 1909 (first records) to less than 30 by 1911 and, with one exception, remained below 20/year until 1955 when the bounty was terminated.

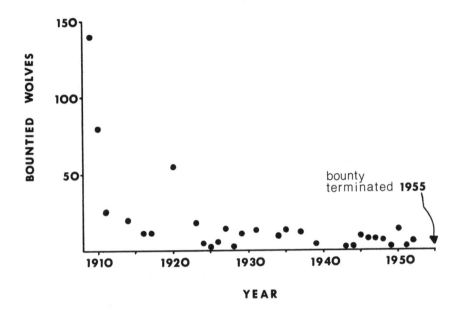

Figure 6.2: Wolf bounty records for Vancouver Island (1909–1952).

Wolf Sightings

Public Wolf Sightings: Wolf sightings contributed by the public increased in number and distribution between 1972 and 1976. In 1972, wolves were sighted in MUs 5, 6 and 10, while in 1976 they were distributed through most of the 13 MUs on Vancouver Island (Figure 6.3). The total number of sightings increased from 14 (37 wolves) in 1973 (first full year of the sighting program) to 51 (88 wolves) in 1976.

Hunter Wolf Sightings: Wolf sightings indexed by hunter efforts were partitioned by subunit but summarized and analyzed by MU (Table 6.1).

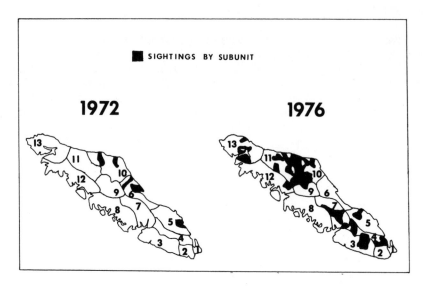

Figure 6.3: Distribution of public wolf sightings, 1972 and 1976.

Table 6.1: Wolf Sighting Index (Wolves/Hunter Day x 100)

MU	1976	1977	% Change	1978	% Change
1	0.12 (6)*	0.05 (2)	− 58	—	—
2	0.30 (42)	0.66 (81)	+120	0.56 (87)	−15
3	0.41 (32)	0.26 (19)	− 37	0.28 (31)	+ 8
4	0.43 (57)	0.39 (41)	− 9	0.26 (39)	−33
5	0.23 (58)	0.20 (43)	− 13	0.12 (35)	−40
6	0.57 (91)	0.77 (114)	+ 35	0.50 (121)	−35
7	1.11 (84)	1.15 (65)	+ 4	1.10 (92)	− 4
8	1.21 (13)	3.40 (19)	+181	0.29 (4)	−91
9	1.86 (70)	1.54 (36)	− 17	1.92 (54)	+25
10	0.99 (210)	1.28 (217)	+ 29	1.14 (256)	−11
11	2.03 (245)	3.78 (371)	+ 86	3.69 (439)	− 2
12	2.10 (132)	2.70 (102)	+ 29	3.31 (170)	+23
13	0.84 (100)	1.85 (194)	+120	2.04 (280)	+10
Total	0.78 (1140)	1.00 (1304)		0.97 (1608)	
Hunters:	25,500	23,178		25,995	
Hunters/ days:	145,768	130,482		165,802	

*() = total wolves sighted.

In 1976 the wolf index varied among MUs, but was highest on northern Vancouver Island (MUs 9-13). Between 1976 and 1978 the wolf index increased on northern Vancouver Island (only two MUs declined in 1978), but it stabilized or declined on southern Vancouver

Island during the same period (four MUs were positive and four negative in 1977, but only one was positive in 1978). Stabilization of the wolf index on southern Vancouver Island may be partially due to the more limited sample size.

The total number of wolves sighted increased during the period 1976-1978 (Table 6.1). Wolves/hunter day (x 100) averaged approximately three times higher for northern Vancouver Island (MUs 9-13) than for southern Vancouver Island (MUs 1-8) in 1976 and 1977, and about six times higher in 1978 (Table 6.2).

Table 6.2: A Comparison of the Wolf Sighting Index (Wolves/Hunter Day x 100) for Northern (MUs 9-13) and Southern (MUs 1-8) Vancouver Island, 1976-1978

MUs	1976	1977	1978	1976-1977 % Change	1977-1978 % Change
1-8 (southern Vancouver Island)	0.55	0.86	0.44	+56	−49
9-13 (northern Vancouver Island)	1.56	2.28	2.42	+46	+ 9
% difference (north Island – south Island)	+184	+165	+450		

Wolf Density

Adam River Intensive Study Area: Radio-collaring and observational data suggested that three separate wolf packs utilized the Adam River drainage during 1977 and 1978. Pack sizes (Table 6.3) were approximated by direct count where possible, and by direct count and sighting form data for the Schoen Lake pack. The pup component of each pack was determined by direct count and by estimates of litter size from the literature (Mech, 1970).

Table 6.3: A Summary of Pack Size and Wolf Density for the Adam River Watershed

Wolf Pack	Estimated Number . . .of Wolves . . . 1978	1979	. . . Km^2 per Wolf . . . Total Area	<760 m*
Upper Adam and Schoen Lake	16	22	6.3-8.6	2.8-3.8
Lower Adam	5	8	26.3-41.7	13.5-21.7
Combined packs	21	30	11.7-16.7	5.6-8.0

*Land area under 760 m elevation.

Wolf density was calculated for both the entire watershed and the lower elevations most likely frequented by wolves. (Table 6.3: Although 760 m is an arbitrary value, it was considered a relative approximation of the upper elevational boundary of wolf pack use.) The density for the entire watershed was approximately one wolf per 12-17 km^2 using both 1978 and 1979 pack size estimates. The density for that portion of the watershed below 760 m elevation was about double, with a maximum density of one wolf per 3 km². Scott and Shackleton (this volume) calculated densities of one wolf per 15 km² and 6.4 km² for the lower and upper Adam packs (excluding Schoen Lake) respectively, using data obtained from radio-collared wolves. However, these radio-locations were obtained only during summer and early fall, and were almost exclusively diurnal. Thus, underestimates of pack home range sizes are likely.

Estimation of Wolf Density: Based on the juxtaposition of pack home ranges and management subunits, wolf pack density estimates for the Adam River study area were converted to wolf density estimates which would correspond to management subunits in the area (Table 6.4). This procedure was used so that comparisons between the hunter sighting index (relative density estimate) and the Adam River wolf density estimates could be made on a management subunit basis. Presently, only two data points are available (Figure 6.4); however, future years will provide more points and, ultimately, allow derivation of an absolute density-relative density relationship which may provide a means of estimating absolute wolf density in other watersheds in Vancouver Island given only the hunter sighting index.

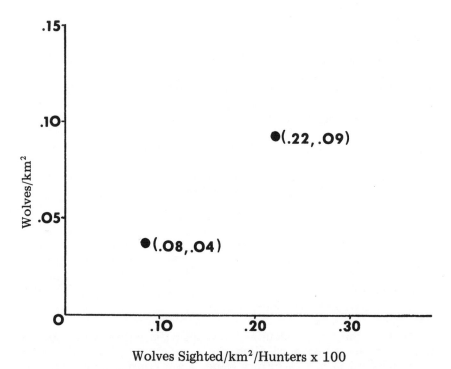

Figure 6.4: Initial data points for establishing the relationship between relative wolf density (X) and absolute wolf density (Y) for Vancouver Island.

Table 6.4: Wolf Pack Density Estimates Derived for Adam River
Management Subunits for 1978
(The Absolute Density Estimates, Wolves/km^2, Are Compared
with Relative Density Estimates in Figure 6.4.)

Subunit	Number of Wolves	Area (km^2)	Density (Wolves/km^2)	Wolf/km^2
10-12 (upper Adam)	13	138	0.0942	1 wolf/10.6
10-13 (lower Adam)	8	213	0.0376	1 wolf/26.6

Wolf-Deer Relationships

Night Count Index: Black-tailed deer population indices were determined in MU 10 and 11 (high wolf density) and MU 5 (low wolf
density). In general, areas with low wolf density showed increasing
deer populations during the mild winters of 1975 to 1978 when only
antlered animals (Nanaimo River) or both antlered and antlerless animals (Nanaimo Lakes) were harvested by hunters (Figure 6.5). In
contrast, high wolf density areas (Adam River, White River, Eve
River) indicated declining populations regardless of whether antlered
(White River) or antlered and antlerless animals (Adam and Eve
Rivers) were harvested by hunters.

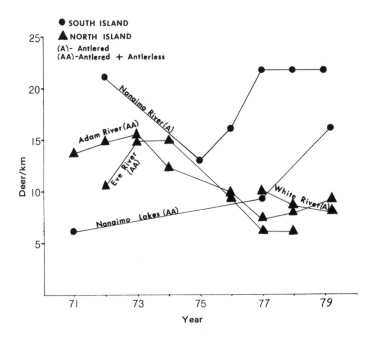

Figure 6.5: A comparison of the deer population index between north and south
Island watersheds and between those with antlered and antlerless harvests.

Within MU 11, three subunits were selected which showed an increasing deer index (night counts showing deer/km) (11-6), a stable deer population (11-4) and a declining population (11-9) (Figure 6.6). These data indicate that the declining deer index area had a high wolf index during the period of measurement (1976 to 1978), while the increasing deer index area had a relatively low wolf index. The deer index established in subunit 11-4 appeared relatively stable between 1970 and 1976, but declined rapidly between 1976 and 1979 as the wolf index increased markedly. The negative correlation between deer and wolf indices was significant (r = –0.90, p < 0.01).

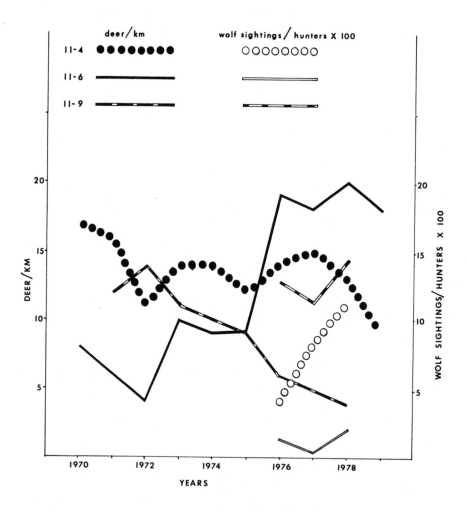

Figure 6.6: Deer population indices from night counts (deer/km^2) in relation to the wolf sighting index for three subunits in the Nimpkish watershed (MU 11).

It was not possible to establish this relationship for an entire MU due to the high cost of yearly deer counts. The large areas involved and a combination of factors (logging, second growth, slash, silviculture treatment) affecting local deer populations do not permit extrapolation of the relationship on an MU basis.

Wolf pack home range boundaries (Scott and Shackleton, this volume) were fitted graphically with deer night-count transects in the Adam River (Figure 6.7). The comparison indicates that deer night-count indices from transects within the 90% home range boundary of the lower and upper Adam wolf packs declined an average of 8 and 33%, respectively, from 1978 to 1979. Transects outside the home range boundaries or near the perimeter of the 100% boundary increased an average of 18% during the same period. The greatest increase in a deer night-count index (25%) occurred outside and between pack home ranges.

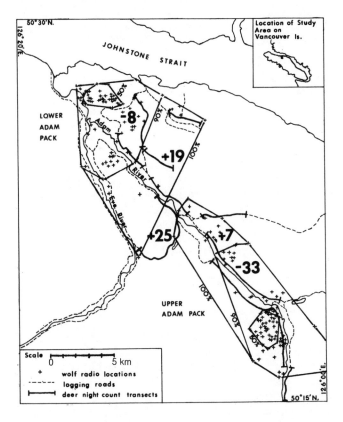

Figure 6.7: The percent change in deer night count population indices from 1978-1979 in relation to wolf pack home range boundaries (50%, 90% and 100%) for the Adam River watershed (home range boundaries from Scott and Shackleton, this volume).

The above indicate a significant negative relationship between wolf and deer population indices. The potential impact of wolf predation on black-tailed deer was calculated for the Adam River watershed using an intake rate of 2.6 kg/wolf/day (Mech and Karns, 1977), and a relative weight in diet of 63.2% for adult deer and 8.1% for fawns (Scott and Shackleton, 1980). This indicated that the three wolf packs could utilize 400 to 600 deer annually. However, Scott and Shackleton's figures may overestimate adult utilization while underestimating fawn utilization because fawn hair could not be accurately separated from adult hair after the June-August period. To compensate for this potential bias on an annual basis, the estimate was increased 20 to 30% (Table 6.5), producing an increase in the predation rates. These estimates compare to the hunter harvest of 200 to 400 deer/year.

Table 6.5: The Potential Number of Deer Consumed by Three Wolf Packs in the Adam River Watershed, Showing the Dependency of Total Kill on the Percentage of Fawns in the Diet.
Total Wolves: 21 (1978), 30 (1979).

Utilization Based on Relative Weight		Potential Deer Kill		
% Adult Deer	% Fawns	Fawns	Adults	Total
63.2	8.1	108–154	315–450	423–604
51.3	20.0	266–380	256–365	522–745
41.3	30.0	399–569	206–294	605–863

Deer Harvest Index: Deer harvest data from north and south Island game checks were analyzed to assess the potential impact of wolf predation on deer populations and hunter success. It must be kept in mind, however, that any impact of wolves on deer harvest can be obscured by hunting conditions, winter weather, logging patterns and changes in hunter effort.

In general, a sharp decline in hunter day success occurred between 1973 and 1974 as winter range declined, the length of the antlerless season was reduced, and the antlerless bag limit was reduced from two to one. South Island watersheds (MU 5), with extensive logging but lower wolf sighting indices, had lower success rates than north Island watershed (Table 6.6). Similarly, success levels were higher during 1966-1973 than during 1974-1978 in MU 10 and 11. No apparent declines in hunter day success occurred between 1974 and 1978 when wolf sighting indices were increasing or stabilized at high levels.

The ratio of yearling females to total females may be an indication of annual recruitment. No significant change occurred in the population of yearling females checked in MU 10 following the increase in the wolf sighting index (Table 6.7). Similarly, the fawn/100 doe ratio in MU 11 declined during 1972-1974 and was paralleled by a low yearling proportion between 1973-1978. These results may indicate low productivity and/or survival from a declining population between 1973 and 1975 with recovery in productivity between 1975 and 1978 from a reduced population base. In addition, age structure of hunter harvested deer collected from large land units (MUs) could be masked by inter-watershed differences. For example, subunits 11-4, 11-6 and 11-9 differed markedly in their deer and wolf sighting index. As might be expected, no simple relationship exists between hunter check information and changes in wolf density or wolf sighting index.

Table 6.6: Average Hunter Day Success in Low and High Wolf Density Areas

| | . . . . Deer/Hunter Day . . . . | | | |
| | . High Density . . | | . Low Density . . | |
Period	10*	11	5-4	5-3
1966–73	0.22	0.35	0.08	0.07
1974–78	0.14	0.16	0.09	0.05

*Management unit or subunit.

Table 6.7: Yearling and Fawn Ratios in the Checked Deer Kill, 1966–1978

| | . . . . MU 10 . . . . | | | . . . . MU 11 . . . . | | |
Year	Total Class.	% Female Yearling of Class. Does	Fawns/ 100 Does	Total Class.	% Female Yearling of Class. Does	Fawns/ 100 Does
1966	82	40.2	36.0	1	—	39.0
1967	31	35.5	59.5	30	33.3	61.0
1968	167	33.5	56.8	143	31.5	51.4
1969	116	29.3	57.5	55	10.9	64.5
1970	65	32.3	58.2	29	41.4	106.8
1971	93	24.7	31.3	51	39.2	60.4
1972	73	21.9	53.5	44	40.9	36.8
1973	93	26.9	68.3	146	19.2	40.8
1974	69	27.5	29.5	58	24.1	39.8
1975	107	31.8	43.0	38	10.5	69.1
1976	100	37.0	7.5	64	32.8	67.2
1977	108	30.6	48.9	118	28.0	43.7
1978	150	33.3	44.0	71	23.9	50.0

Pellet Group Index: Black-tailed deer pellet group surveys were conducted on mature timbered ranges in MUs 5, 10 and 11 in 1972–1974, and again in 1979. Winter conditions were moderate to severe in 1972–1974 and mild in 1979. Deer concentrate in mature timber during severe winters, increasing pellet plot averages. During mild winters, deer are distributed widely and rely less on mature timber for cover. Plot averages declined in the MUs examined (Table 6.8). Pellet-plot data for slash/second growth timber areas in MU 11 declined 75% between 1972–1979 (Jones, pers. comm.). Furthermore, it appeared that declines in pellet-plot density occurred mainly in the developed or accessible portion of each MU (Table 6.8). Average plot density from underdeveloped and relatively inaccessible portions of each MU did not change (MUs 5 and 10), or increased (MU 11). However, these areas made up a small proportion of each MU. Pellet-plot information from two adjacent watersheds (Koprino and Claud Elliott) which were currently undeveloped, unhunted and relatively inaccessible indicated declines of 60 to 80% from 1974 to 1979.

Table 6.8: The Percent Change in Number of Deer Pellet Groups
per Plot Between 1972–1974 and 1979
in Three Management Units on Vancouver Island*

| | . . . . .Total Unit . . . . . | | Accessible Portion of Unit | |
MU	Change	Range	Change	Range
5	–19%	0 to –33%	–28%	–25 to –33%
10	–41%	0 to –70%	–57%	–45 to –70%
11	–48%	58 to –85%	–65%	–35 to –85%

*No. of major winter range groups surveyed: 3 in MU 5, 5 in MU 10,
5 in MU 11.

Wolf-Elk Relationships

Systematic late winter surveys initiated in 1974–1975 obtained cow-calf ratios and carry-over (percent juvenile of total count) for Vancouver Island Roosevelt elk. Prior to 1975, except for the period 1967–1971, data were inadequate to determine annual trends (Figure 6.8). Information to date suggests that northern Vancouver Island (MUs 9–13) is currently more productive than the south Island (MUs 1–8). During the period of low wolf populations (1950–1970), cow/calf ratios and carry-over appear to reflect the influence of extensive harvesting and several moderate to severe winters (1953–54, 1956–57, 1961–62, 1964–66, 1968–69, 1971–72 and 1974–75).

During the period of increasing wolf populations (1975–1979), logging of mature timbered winter range had proceeded to the final stages in most south Island watersheds and was 30 to 70% complete in most north Island watersheds. In response to mild winter conditions during 1976–1979, north Island calf/cow ratios and carry-over increased to almost 40/100 cows and 24%, respectively. South Island ratios increased to almost 35/100 cows and 22% in 1977, but they have since declined. North Island productivity and survival figures collected in MUs 9 and 10 have increased and remained stable in spite of increasing wolf populations.

Systematic elk surveys have not been conducted in the Nimpkish watershed (MU 11), but elk sightings and preliminary classified counts (G. Jones and D. Lindsay, pers. comm.) were tabulated for 1975–1977. The number of calves per 100 cows declined from 45 and 50 in 1975 (n = 57) and 1976 (n = 96), respectively, to 18 in 1977 (n = 93) in spite of mild winters. The proportion of yearling males also declined from 9 to 3% during the same 3-year period. Systematic surveys in the Nimpkish were increased during 1979 to 1980 in an attempt to verify possible declines in productivity and survival which may coincide with increased wolf populations and decreased deer numbers (Figure 6.6). These data were not analyzed prior to this writing.

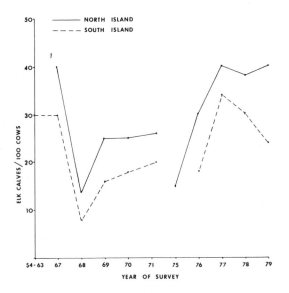

Figure 6.8: Late winter age ratios for Roosevelt elk from Vancouver Island.

Elk are a component of the wolves' diet in the Adam River watershed (Scott and Shackleton, 1980). On an individual prey basis, adult elk comprise 4.4% and elk calves 14.0%, while adult deer and deer fawns comprise 59.1 and 20.4%, respectively. Similarly, on a relative weight basis, adult elk and elk calves comprise 14.1 and 13.9%, while adult deer comprise 63.2% and deer fawns only 8.1%.

Although these data suggest significant utilization of elk in the diets of the Adam River packs, elk herds in the drainage do not presently appear to be affected detrimentally. However, the wolf food habits data are limited in many respects. First, the hair of ungulate young is not discernible from that of adults after August and therefore is included in the adult portion for the remainder of the year. In addition, the proportion of adult and calk elk in the diet of the lower and upper Adam wolf packs is inconsistent with elk distribution data collected between 1975 and 1979. The data indicated that elk calves comprised 25, 57 and 57% of the lower Adam pack diet (relative weight) during June, July and August, respectively (Scott and Shackleton, 1980). However, the majority of elk and elk movements within the Adam River watershed do not overlap with the lower Adam wolf pack home range and, generally, radio-collared cows and calves have left the home ranges of both the lower and upper Adam wolf packs by July for summer ranges.

DISCUSSION

A dramatic increase in wolf numbers occurred over much of Vancouver Island between 1972 and 1979. Prior to this period, a concerted effort by government agencies and private groups had been made to place this species on the rare and endangered list. Low wolf numbers along the east coast of Vancouver Island during the 1955 to 1970 period generated this concern. Recovery occurred to the extent that the density estimates for portions of the north end of the Island are among the highest reported in North America (see Mech, 1970).

Assessment of wolf population changes concomitant with potential impact on prey was complicated by restrictive budgets, existing data collection programs not designed to examine this problem, and lack of wolf inventory methodology for dense coastal vegetation. Analysis of public and hunter sightings was the only method available to assess wolf distribution and density over a wide area. Estimates of hunter effort within a land unit were used to standardize wolf sighting information from hunters and to produce comparable density indexes among watersheds. An intensive study in one watershed (Adam River) produced wolf density estimates which could be related to the wolf sighting index. The sighting program may be useful in establishing wolf density estimates over large land areas; however, further data are required in order to establish a relationship.

Wolf sighting information was consistent with general knowledge of wolves on Vancouver Island, and indicated higher wolf numbers on northern Vancouver Island where deer densities are generally higher and habitat deterioration by logging is less (Hebert, 1979). Preliminary estimates of wolf density in the Adam River watershed suggest one of the highest densities in North America (one wolf per 12 km^2 in entire watershed, one wolf per 6 km^2 in upper Adam). Comparable estimates are one wolf per 12 km^2 on Isle Royale (Peterson, 1977), and one wolf per 26 km^2 in northeastern Minnesota (Mech, 1977c). Currently, wolf distribution and density estimates coincide with information on annual harvest and general population levels for black-tailed deer. Annual deer harvests on southern Vancouver Island have declined by 50% since 1964 due to extensive removal of winter ranges (Hebert, 1979). Deer populations in second-growth forests in this area are substantially lower than those in the remaining mature forests of northern Vancouver Island. It appeared that many north Island populations began to decline in 1974 as mature timbered ranges were removed and wolf numbers increased. However, between 1975 and 1979 the relative deer population levels from the north and south Island were maintained by four mild winters.

Inadequate data systems do not allow stratification of all data sources to the same land base. Thus, deer harvest estimates from questionnaires and the wolf sighting index are summarized by MU while wolf sightings and deer count indexes may be summarized by subunit (watershed), and pellet group counts and some deer count indexes may be from portions of a watershed. The larger the land unit from which information is collected (at the same level of sampling effort), the greater the masking effects from the variety of conditions available. Generally, masking effects occur at the MU level from variations in logging, hunting pressure, predation and climate.

Information to date suggests significant relationships between wolf sighting distribution, wolf sighting index and deer population estimates from a night-count methodology. Similarly, preliminary examination of pellet group information as an indicator of black-tailed deer populations suggests significant relationship with wolf population indexes. There are indications that areas that are least accessible to logging, hunting and possibly to wolf movement have maintained higher deer populations. Often these steep areas provide optimum deer range, especially during severe winters (Smith and Davies, 1975). However, information on hunter success and fawn and yearling deer age ratios collected at hunter game checks from large and often unknown land units shows little relationship to wolf population indexes. Variability in deer and wolf densities within a land unit combined with hunter mobility could effectively mask the results from game checks.

The high prey density of black-tailed deer suggests the potential for a relatively high predator density. Subsequent predation has the potential to severely affect prey populations, especially where logging reduces or totally removes mature timbered winter ranges causing deer to concentrate to 150 to 200/km^2 in small timbered areas during moderate to severe winters.

The potential impact of wolf predation on elk populations is less certain. To date, it appears that wolves are utilizing elk as a food source but at a rate, in most areas, considerably below recruitment. However, preliminary examination of the MU 11 elk data, particularly from subunit 11-4, indicates that in areas where deer populations are depressed or declining (Figure 6.6), the impact of wolf predation on elk may increase. Also of concern is that elk vulnerability may increase considerably during moderate to severe winters as the elk concentrate in low elevation timber stands.

Pack dynamics of the Vancouver Island wolf require more intensive study. Extensive movements (>70 linear km) have been recorded for two wolves which were originally trapped and radio-collared in the Adam River watershed (Scott and Shackleton, this volume). Also, incidental signs and recent radio-collar information indicate shifts in pack home ranges (unpub. data). It is possible to speculate that the relatively continuous distribution of black-tailed deer will produce less stable home range distribution and pack cohesiveness than is seen in most North American wolf packs.

The role of the wolf in coastal ecosystems is largely unknown. During extensive and rapid habitat changes due to logging, wolf population status and impact on prey populations are difficult to assess. The potential for significant impact of wolves on deer on Vancouver Island appears high. A severe winter combined with the present high wolf densities and reduced winter range could produce a substantial decrease in deer numbers. The likelihood of this combined effect is supported by the work of Mech and Karns (1977), who showed that a decline in a white-tailed deer (*Odocolius virginianus*) population could be attributed to high wolf densities coupled with deteriorating winter range quality.

Acknowledgements

Since this paper is composed of information on predators, prey and habitat from a large portion of Vancouver Island, many people contributed to its preparation. In particular, we would like to thank B. Mason and G. Jones for the use of their information from MU 11. In addition, we used specific pieces of information from Ms. B. Scott's Master's thesis which was completed for the Fish and Wildlife Branch in 1979. Ms. B. Schenker typed drafts and final copy of the manuscript. W. Kale designed and initiated the wolf sighting questionnaire.

Gray Wolf-Brown Bear Relationships in the Nelchina Basin of South-Central Alaska

Warren B. Ballard

INTRODUCTION

The wolves *(Canis lupus)* of southcentral Alaska have been a focus of interest and study for over 30 years. From 1948 to 1953 poisoning and aerial shooting by the federal government reduced populations of wolves to low levels. In 1953 only 12 wolves were estimated to survive in the Nelchina study area described by Rausch (1969). Bears *(Ursus arctos)*, wolverines *(Gulo gulo)*, other carnivores and some omnivores were also probably reduced by poison. The wolf population gradually increased and reached a peak of 400–450 animals in 1965 (Rausch, 1969).

Rausch (1969), Bishop and Rausch (1974) and McIlroy (1974) described the history of the Nelchina Basin moose *(Alces alces)* population. The moose population began declining after the severe winter of 1961–62. The decline continued with severe winters occurring in 1965–66 and 1971–72. Although wolf predation was not suggested as the main reason for the population decline, it was thought to have at least accentuated the decline and, perhaps more importantly, prevented recovery during mild winters (Bishop and Rausch, 1974). Stephenson and Johnson (1972, 1973) found high percentages of calf moose in wolf scats which suggested that wolf predation on moose calves was preventing the moose population from recovering. Consequently, in 1975 a series of studies on wolf-moose relationships were initiated. These studies were later expanded to include brown bear-moose relationships. Information pertaining to these studies has been reported by Stephenson (1978), Ballard and Taylor (1978a,b), Ballard and Spraker (1979), Spraker and Ballard (1979) and Ballard et al. (1980, 1981). Considerable attention was focused on gathering information on wolf food habits during the late spring and summer when most moose calf mortality occurred. The purpose

of this paper is to report on encounters between bears and wolves observed during these studies, and to discuss the significance of these
interactions.

MATERIALS AND METHODS

Radio-collared wolves were tracked and visually observed, when
possible, from fixed-wing aircraft, as described by Mech (1974b).
Monitoring intensity varied among seasons and packs but consisted
of at least bi-monthly efforts during winter months. Two wolf packs
were intensively studied during the summers of 1977 and 1978,
and were located either once or twice daily from late May to mid-
July 1977 and to late June 1978.

Age of captured wolves was determined on the basis of tooth
eruption and wear. During radio-tracking, ages of unmarked wolves
were occasionally estimated on the basis of relative size and also by
criteria described by Jordan et al. (1967). The age-sex structure of
certain packs was not ascertained until the animals had been killed
by hunters and trappers. Hunters and trappers were encouraged to
provide us with wolf carcasses taken in the study area by offering
$10 per carcass. Ages of harvested wolves were determined by
examining epiphyseal cartilage of the longbone according to methods
described by Rausch (1967).

Moose were classified as calf, yearling or adult from fixed-wing
aircraft based on a combination of size, pelage and antler growth.
When practical, wolf kills were examined on the ground. Cause of
death was determined according to methods described by Stephenson
and Johnson (1973) and Ballard et al. (1979). Observations were
summarized in flight on a portable cassette recorder and later transcribed.

Wolf summer home ranges were determined by plotting all radio
locations for individual packs and then connecting the outermost
observations. Size of home ranges were determined with a compensating polar planimeter.

Study Areas and Wolf Pack Histories

The study was conducted in the Nelchina Basin, Game Management Unit 13 of southcentral Alaska, an area of approximately
61,595 km^2 of which 18,798 km^2 is over 1,200 m elevation. Year-
round studies involved up to 16 wolf packs inhabiting portions of the
Unit lying north of the Chugach mountain range and east of the
Talkeetna mountains. Only data pertaining to wolf-bear relationships
and their implication to wolf summer food habits will be presented
here.

Mendeltna Pack

The Mendeltna pack area (Figure 7.1) is basically a level plateau of wet muskegs interspersed with numerous ponds and lakes. Drier rolling hills on the western portion range from about 600 to 1,170 m. Lowland areas are vegetated with sparse to dense stands of spruce *(Picea mariana, P. glauca)* interspersed with wet muskegs. The higher, western portion is a transition between spruce-muskeg and subalpine tundra. Willows *(Salix sp.)* and birches *(Betula sp.)* occur along stream courses and, with aspen *(Populus tremuloides)*, on drier, better drained soils. Understory is comprised of cranberry *(Vaccinium vitis-idaea, Viburnum edule)* and blueberry *(Vaccinium ovalifolium, V. uliginosum)*. Further details can be found in Skoog (1968).

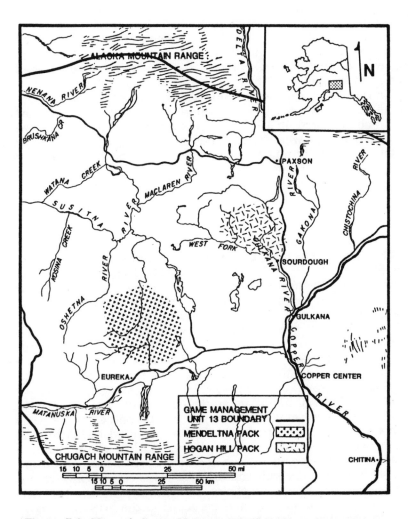

Figure 7.1: Map of Game Management Unit 13 and year-round territories of two wolf packs intensively studied during summers 1977 and 1978 in the Nelchina River Basin of southcentral Alaska.

During the 1977 season, the Mendeltna pack numbered seven adults, of which two to four were yearlings. Two adults and one yearling were radio-collared. This pack occupied a summer home range of approximately 829 km^2. During the 1977 season, the pack had two den sites at which two litters were raised, totaling at least eight pups.

Hogan Hill Pack

The northern two-thirds of the Hogan Hill pack's range (Figure 7.1) consists of low, rounded hills reaching about 1,600 m. The higher elevations are characteristically subalpine tundra; lower elevations are thickly wooded with white and black spruce. Several creeks bisect the area. The southern slopes, which were predominantly used by the pack during summer 1978, are thickly vegetated with spruce and willow along stream bottoms and adjacent to ponds. The lowland areas are similar to those of the Mendeltna pack's range.

During 1978, the Hogan Hill pack was comprised of eight adults, of which at least two were yearlings. One adult and two yearlings were radio-collared. In 1978, the pack maintained one den site where at least five pups were raised. They ranged over an area of approximately 570 km^2 during early summer 1978. The boundaries of their year-round territory are shown in Figure 7.1.

RESULTS AND DISCUSSION

From June 1976 through June 1978, 16 wolf packs were observed on 130 kills, of which approximately 75 percent were moose (Ballard and Spraker, 1979). Of that total, 17 (13.1%) were contested by brown bears. In most instances, I was unable to determine which predator species had made the kill. During the summers of 1977 and 1978, I intensively studied one different wolf pack each year and was able to document some of the circumstances surrounding bear-wolf encounters at, and away from, kill sites. Because such observations are rarely witnessed, my notes and interpretations of bear-wolf encounters for the Mendeltna pack during the summer of 1977 are summarized chronologically below.

On 11 June at 21:00, an adult female wolf was observed being chased by a cow moose which was exhibiting aggressive behavior (mane ruffed-up and ears down). The wolf would veer off a straight line in what appeared to be an attempt to lose the pursuing cow. When the wolf seemed to lose the cow by crouching in brush, the cow searched for the wolf and, in three instances, was able to find it. The cow gained ground on the wolf, which appeared to tire. On one occasion, the wolf stopped and crouched in the brush. The cow ran over to the area and appeared to trample the wolf. The wolf then continued running, but at a much slower pace and with a limp. The chase lasted about 15 minutes, at which time the wolf headed for a den site while the cow began travelling back in the direction from which it had come. The cow continued to exhibit aggressive behavior. It began swimming across a pond toward a brown bear sow with three yearling cubs. The cubs were huddling over and dragging around a calf moose carcass. The cow ran around the bears within a 40-50 m radius. A yearling male wolf was present in dense spruce some 150-175 m from the bears, but was not observable. I returned to the calf kill at 22:30 via helicopter and frightened the bears away from the site. The kill had puncture marks on the neck and either puncture or claw marks on the anus. Only the head had been fed on. The skull was cleaned out, leaving only the skin casing. Tongue, eyes and ears had been eaten, which was characteristic of bear-killed calves (Ballard et al., 1979). Imprints were noted in the area and bear hair was noticeably evident on surrounding brush. Interpretation: Bears made the kill and wolves were attempting to scavenge, but were chased away by cow or bears, or both.

On 12 June at 09:10, an adult male and a smaller adult wolf of unknown sex were observed chasing and harassing the same bears observed the previous evening. The wolves stayed fairly close to each other while chasing the fleeing bears. When the bears stopped running, one wolf typically crouched and approached the sow. The sow would charge the approaching wolf at which time the other wolf would charge and chase the yearling cubs, causing the sow to charge the second intruding wolf. On one occasion, the wolves treed all three cubs. The wolves appeared to press their charge when the bears' direction of movement was toward the wolves' den, less than 2 km away. On one occasion, the radio-collared wolf was observed sneaking around and crouching down in front of the bears' direction of movement. Apparently the sow detected this action because when she was approximately 10 m away, she charged the crouched wolf and almost caught it by the hind quarters. It appeared that when the bears finally established a trend of movement away from the den, the wolves no longer pursued and began heading back toward the den. These activities lasted 15 minutes and covered 0.6 km from where we first observed the bears. Interpretation: Wolves discouraged bear movement toward wolf den.

On 14 June at 17:20, a yearling male wolf was observed alone, resting on a sand bar. Approximately 60 m away, a single adult brown bear was feeding on an adult moose kill estimated to be 80 percent consumed. The wolf appeared to have a swollen abdomen, indicating it also had fed on the kill. Interpretation: Kill was made by wolves, and wolves were either displaced by bear or abandoned the kill before bear arrived.

On 15 June at 08:50, an adult male and two yearling wolves were observed approaching a moose calf kill which had one sow and one yearling brown bear feeding on it. The kill was estimated to be 80 percent consumed with guts and hide remaining. The approach of the airplane and perhaps the wolves frightened the bears, causing them to run from the kill. The wolves went directly to the kill and began feeding. Interpretation: Kill made by bears, observer approach and/or wolves caused bears to leave kill, which was taken over by wolves.

On 16 June at 19:45, an adult male, an adult of unknown sex and two yearling wolves were observed attacking an adult brown bear which possessed an adult moose kill. Initially, three wolves were observed equally spaced around the bear. One of the wolves attempted to nip the bear in the rump. The bear made several charges at the wolves which were approaching to within 3-5 m. The wolves easily outmaneuvered the bear and three of the wolves appeared to keep the bear away from the kill as a fourth wolf fed on it. The bear's direction of movement was toward the kill and, after 15 minutes of encountering the wolves, the kill was reached. When the bear reached the kill, the wolves stopped harassing the bear and began travelling in the direction of the main den. The kill was estimated to be 50 percent consumed. Interpretation: Either the bear or the wolves made the kill and the wolves were attempting to displace the bear.

On 22 June at 08:29, an adult male and another adult wolf were observed feeding on what I tentatively identified as a moose calf. Ground inspection of the kill site at 12:00 revealed the kill was actually a yearling brown bear. A portion of the carcass had been buried, but most had been consumed. The kill site contained tracks of a small bear and wolf. Interpretation: Yearling bear was killed by wolves.

On 24 June at 16:55, an adult male wolf was observed resting alone approximately 10 m from an adult moose kill with one adult brown bear feeding on it. Only the head, rear quarters, guts and skin remained. Interpretation: Kill was made by either wolves or bear. The wolf may have been attempting to scavenge and/or displace bear. On 29 June, three wolves were observed feeding at the kill site.

On 27 June at 22:00, an adult male, a second adult of unknown sex and one yearling wolf were observed resting close to an adult brown bear which was feeding on a calf moose kill. I estimated the kill to be 50 percent consumed with head and front quarters missing. The bear seemed unconcerned by the wolves' presence. The bear was still present on 28 June at 10:00, and had eaten most of the carcass. Interpretation: Kill was made by either bears or wolves. Wolves were attempting to scavenge and displace the bear.

On 8 August at 07:30, an adult male, another adult of unknown sex and two yearling wolves were observed scattered around an adult moose kill with one brown bear feeding on it. Three of the wolves huddled together touching noses and wagging tails, then separated and charged the bear, displacing it from the kill. Another wolf, hidden by a large spruce, ran to the kill and tore off a large chunk of flesh as the returning bear charged. Another wolf followed, carrying the meat into the dense spruce. Several other bear charges were observed. The bear remained in possession of the kill. Interpretation:

Either the bear or wolves made the kill. The wolves attempted to scavenge and/or displace the bear.

Aggressive behavior between the two predator species occasionally results in mortality to the participants. Joslin (1966) reported an adult female wolf killed close to her den by a black bear *(Ursus americanus)*. In September 1976, a member of the Mendeltna pack was killed by a brown bear, probably as a result of competition over an adult moose kill. Details of this particular observation were presented by Ballard (1980). Mech (1970) thought that occasionally wolves killed bears, but that the victims were probably cubs, young bears or older weakened bears. Murie (1944) suggested that wolves were more aggressive toward bears near wolf dens. The Mendeltna wolves exhibited agonistic behavior toward bears both at kills and in areas close to den sites. My observations indicated that wolves do occasionally kill bears. The result of brown bear-gray wolf encounters, therefore, may be an additional source of natural mortality not previously documented for either predator species. Whether it is a significant source of mortality for either species remains unknown.

Reason for Contested Kills

During the summer of 1977, the Mendeltna pack had six of 11 kills contested in addition to several bear encounters away from kill sites. In contrast, during a similar period in 1978, the Hogan Hill pack had none of its six kills contested, and no bear encounters away from kills were observed. The larger number of contested kills for the Mendeltna pack may be related to several factors including: (1) observability, (2) predator density, and (3) prey density.

If there was a difference in observability between the areas, it was not detectable. During the summer of 1977, the three radio-collared members of the Mendeltna pack were observed on 188 of 224 (83.9%) occasions they were located. In comparison, the three radio-collared members of the Hogan Hill pack were observed on 97 of 114 (85.1%) occasions.

Differences in bear density are unlikely to have caused the disproportionate number of contested kills in the Mendeltna area. Although no accurate estimates of bear density exist, tagging data and sightings of bears (Ballard and Taylor, 1978a; Spraker and Ballard, 1979) suggest the study areas had similar densities, approaching one bear per 39 km^2. There were, however, differences in wolf densities (Table 7.1). Based upon areas occupied during summer, wolf densities ranged from one wolf/73 km^2 for the Hogan Hill pack, to one wolf/119 km^2 for the Mendeltna pack. Thus, the area with the lowest wolf density had the largest number of kills contested by bears. Differences in wolf density may have been partially related to the maintenance of two den sites 8 km apart by the Mendeltna pack, but was more likely related to differences in prey density.

Number of moose counted in fall sex and age composition surveys from 1976 through 1978 were utilized to calculate a crude approximation of moose density (Table 7.1). In the one containing the Hogan Hill pack, 0.42 moose/km^2 were counted, while 0.34 moose/km^2 were counted in the two count units containing the Mendeltna pack territory. Therefore, the area with the highest wolf density also had the highest moose density.

Table 7.1: Summary of Predator-Prey Statistics for Two Wolf Pack Areas Intensively Studied During Early Summer 1977 and 1978 in the Nelchina Basin in Southcentral Alaska

Pack	Study Period	Wolf Density within Summer Range (wolf/km^2)	Moose Density (moose/km^2)	Number of Bear-Wolf Contested Kills	Known Kill Rate (days/kill)	Feeding Rate (days/feed)	Available[1] Prey Biomass (kg)	Kg Prey Biomass per Adult Wolf/Day
Hogan Hill	28 May-21 June 1978	1/73	0.42	0 (6)[2]	4.0	4.0	841	4.4
Mendeltna	27 May-15 July 1977	1/119	0.34	6 (11)	10.0	4.6	2,173	4.7-6.2[3]

[1] Biomass of available food based upon the following assumptions: Weight of adult moose = 427.5 kg (from Franzmann and Bailey, 1977) and yearling moose = 197.5 kg (from Franzmann and Arneson, 1973, 1975), of which approximately 75% (from Peterson, 1977) available as food yielding 321 kg and 148 kg, respectively. Newborn calf moose weighs 13.3 kg (from Ballard and Taylor, 1978a,b), and gains weight at a rate of 1.3 kg/day (from Franzmann and Arneson, 1973). Therefore, 15 day old calf weighs 32.7 kg, of which 90% is consumable yielding 29.5 kg. Yearling brown bear weighs 45 kg (from Spraker and Ballard, 1979), of which 75% is consumable yielding 34 kg. Snowshoe hare weighs 1.4 kg (from Burt and Grossenheider, 1964), of which all is consumable yielding 1.4 kg.

[2] Total number of kills with wolves present.

[3] First value assumes 50% of contested kill was available to wolves, while the second assumes all was available.

I speculate that the bear-wolf encounters observed while studying the Mendeltna wolf pack were due primarily to lower moose densities in that area. This speculation was supported by predation data collected by monitoring radio-collared bears. These data indicate that bears took substantial numbers of moose in all the areas studied (Ballard et al., 1981). Thus, in the Hogan Hill area where moose were more abundant, no kills were contested because sufficient moose were probably available for each predator during the study period.

Speculation that low prey density was responsible for the disproportionate number of contested kills for the Mendeltna wolf pack is also supported by data on the chronology of moose calf mortality. During 1977 and 1978, 79 percent of radio-collared calf mortality was attributed to predation by brown bears (Ballard et al., 1981). During 1977 when the Mendeltna pack was being intensively monitored, 53 percent of all calf mortalities had occurred by 11 June. This corresponds with the date of the first observed contested kill, suggesting that ample moose were available for both predator species until mid-June, but not afterwards. If correct, then bear-wolf encounters at kills for the Hogan Hill wolf pack would be expected to occur at a later date had a declining prey base influenced its occurrence. Although daily contact with the Hogan Hill pack terminated on 21 June, two and possibly three of four kills observed between 1 July and mid-November were contested by bears after 4 August.

Significance of Contested Kills to Predator Ecology

During this study I was unable to quantify the volume consumed by each predator species at a particular kill site because the observation periods were too short and, in some cases, my presence may have interfered. In many cases, however, it was apparent that both species were able to feed at many of the kills for varying lengths of time. The amount consumed by wolves at a particular kill site could thus alter kill rates and influence how kill data are interpreted.

From 27 May through 15 July 1977, during intensive monitoring of the Mendeltna wolf pack's activities, all or some Mendeltna wolves were observed on 11 kills. The kills included six adult moose, three calf moose, one yearling moose and one yearling brown bear. In comparison, from 28 May to 21 June 1978, members of the Hogan Hill pack were observed on six kills comprised of two adult moose, two calf moose, one yearling moose and one of unknown species. Based upon these data, wolf kill rates were calculated for kills which were known to have been made by wolves and for kills when bears were involved (Table 7.1). The latter rates are referred to as feeding rates. For known wolf kills, there was a large difference in kill rates: One kill every 4.0 days for the Hogan Hill pack (with 8 adults) versus 10.0 days for the Mendeltna pack (with 7 adults). However, when bear-contested kills were added, the Mendeltna pack kill rate increased to one every 4.6 days, while the Hogan Hill pack rate remained unchanged.

The amount of prey biomass available per adult wolf was calculated for both study packs (Table 7.1). Two values were calculated for the Mendeltna pack: The first value assumed that only 50 percent of the biomass on bear-contested kills was available to wolves, whereas the second value assumed that all of the prey biomass was available to wolves, even though bears were present on some kills. Although I could not determine how much was eaten by either predator at a kill site, I did observe that both usually fed on some quantity.

Both the kill and consumption rates during summer for the Mendeltna and Hogan Hill wolf packs fall within the range of values reported in the literature for the winter season. Mech (1970) reported that a pack of 15 to 16 wolves had a kill rate of one moose per 3.0 to 3.7 days on Isle Royale. Fuller and Keith (1980) reported a kill rate of one moose per 4.7 days for a pack of nine wolves in northern Alberta. Mech (1966) and Peterson (1977) reported food availability of 3.8 to 10.0 kg/wolf/day for Isle Royale wolves. Mech (1977a) determined that one Minnesota wolf pack declined after a winter when only 3.0 to 3.4 kg/wolf/day of food was available, increased at 5.8 kg/wolf/day, and remained stable at 3.6 kg/wolf/day. Mech (1970) reported that a higher kill rate occurred when calf moose comprised a larger percentage of the prey taken in winter. The same appeared to be true during this study in summer.

Stephenson (1978) speculated that competition from bears at wolf kills could result in an increased wolf predation rate. Data presented from this study indicate that, if true, the increase may not be detectable with the study methods used. Competition at kill sites could also increase bear predation rates.

Within recent years, scat analyses have been used to determine wolf food habits. Although most such studies have acknowledged that the derived data represent what was eaten rather than what was actually killed, the observations of wolf-bear encounters further emphasize the need for caution when analyzing both wolf and bear scat data and interpreting their significance to predator-prey relationships. If both species were feeding on the same kill, the resulting food data could only be viewed as that obtained by both scavenging and direct killing.

Black and brown bears have only recently been identified as significant predators of cervids (Schlegel, 1976; Franzmann and Schwartz, 1978; Ballard and Taylor, 1978a; Ballard et al., 1980, 1981). The fact that both predator species have potential to not only prey upon ungulate species, but also to scavenge and interact with one another, could greatly complicate our attempts to understand predator-prey relationships.

SUMMARY

From June 1976 through June 1978, 16 wolf packs were observed at 130 kills in the Nelchina Basin. Seventeen of these kills were contested by brown bears. Nine of the 17 kills contested by both bears and wolves were observed in conjunction with studies of the Mendeltna pack. Comparisons of predation rates, predator densities and prey densities were made between two packs which were intensively studied during late May and June 1977 and 1978. It appears that the disproportionate number of bear-wolf encounters at kill sites for the Mendeltna pack was primarily the result of a lower moose density in the Mendeltna area. The possible significance of these observations to predator-prey relations was discussed.

Acknowledgements

The study was funded, in part, by Alaska Federal Aid in Wildlife Restoration Project W-17-R.

Sterling Miller, Karl Schneider, Donald McKnight and Karen Wiley, all of the Alaska Department of Fish and Game, reviewed earlier drafts of the manuscript and made many helpful suggestions. Appreciation is also expressed to Rolf Peterson, Michigan Technological University, for reviewing the manuscript.

Patterns of Homesite Attendance in Two Minnesota Wolf Packs

Fred H. Harrington and L. David Mech

INTRODUCTION

Summer is a crucial time for a wolf *(Canis lupus)* pack. During the three to four months after parturition, the pups must be adequately fed to ensure that their physical development is sufficient to survive the rigors of late fall and winter. The pups spend most of the summer and early fall at a number of dens and rendezvous sites (Murie, 1944; Joslin, 1967), collectively called "homesites" (HS). They are fed at HSs by adults and/or yearlings, which forage away from the HSs and return periodically to regurgitate food to the pups (and occasionally to older animals) (Murie, 1944; Rutter and Pimlott, 1968; Haber, 1977). Since most pup mortality occurs by six months of age, at least in Minnesota (Van Ballenberghe and Mech, 1975), events at the HSs largely determine pup survival and, hence, the reproductive success of the pack.

Previous studies of wolf behavior in forested areas of Minnesota, Isle Royale, Ontario and elsewhere have largely focused on the pack as a unit because of the logistical problems of observing individual wolves during summer. Therefore, much is known about HSs and the general characteristics of wolf movements associated with them (Jordan et al., 1967; Joslin, 1967; Kolenosky and Johnson, 1967; Rutter and Pimlott, 1968; Pimlott et al., 1969; Theberge and Pimlott, 1969; Carbyn, 1974; Van Ballenberghe et al., 1975; Peterson, 1977; Harrington and Mech, 1978). However, little is known about adult and yearling activity around HSs, although some qualitative data exist for wolves inhabiting tundra (Murie, 1944; Clark, 1971; Haber, 1977). Knowledge of how individuals behave is important because the individual is the *principal* unit of selection (Williams, 1966; Lewontin, 1970).

An individual's evolutionary success, or *inclusive fitness*, is the combination of its own fitness, measured in terms of its offspring, plus its effects on the fitness of its relatives (Hamilton, 1963, 1964; Maynard Smith, 1964). *Individual selection* operates at the level of an individual's own reproductive success, whereas *kin selection* operates through an individual's relatives (Maynard Smith, 1964). Although the concept of *kin selection* is merely the logical extension of individual selection, both being based on the concept of genetic relatedness (r) (Dawkins, 1976), a practical distinction can be made between the two for behaviors that might increase fitness at one level while decreasing it at another (Hamilton 1964). Thus, altruistic behaviors, which decrease individual fitness while increasing kin fitness, can be distinguished from selfish behaviors, which increase individual fitness, even though the ultimate effect of both is to increase the individual's inclusive fitness. In the wolf, the importance of kin selection in the evolution of social behavior, with reference to pup care, has recently been stressed (Bulger, 1975, Haber, 1977), although solid empirical evidence distinguishing between the individual and kin benefits of sociality are essentially lacking.

More detailed quantitative data are needed concerning differences among the behavior of individuals. In the present context, a knowledge of how individual pack members differentially contribute to pup survival should reveal what factors influence their behaviors around HSs, and indicate at what level each animal may be maximizing its inclusive fitness. This information will permit us to better understand the dynamics and evolution of wolf sociobiology.

The present study quantifies the presence or absence of individual wolves at two wolf pack HSs in Minnesota. Because visual observation was not possible, we monitored the wolves via radio-telemetry. From the patterns of HS attendance that emerge, we will draw inferences on probable individual and kin benefits.

STUDY AREA

This study was conducted in the Superior National Forest of northeastern Minnesota (92°W. longitude, 48°N. latitude) during summer 1973. Additional data drawn from a study of elicited howling during 1972 and 1973 (Harrington, 1975; Harrington and Mech, 1979) will be used where noted. The forest supported a population of about one wolf per 26 km² (Mech, 1973). The wolf's primary prey, white-tailed deer *(Odocoileus virginianus)*, was declining drastically (Mech and Karns, 1977); beaver *(Castor canadensis)* and moose *(Alces alces)* were alternate prey during summer (Mech and Frenzel, 1971). The wolves apparently were food-stressed during this study, as indicated by a relatively high proportion of deviant blood values, especially in pups (Seal et al., 1975; Van Ballenberghe and Mech, 1975). Further details about the study area and wolf population may be found elsewhere (Ohmann and Ream, 1971; Mech and Frenzel, 1971; Mech, 1972, 1973, 1974b, 1975, 1977a,

1977c, 1977d, Peters and Mech, 1975; Harrington and Mech, 1979; Rothman and Mech, 1979).

METHODS

Several members in the Jackpine (JP) pack and the Harris Lake (HL) pack were outfitted with radio-collars (Mech, 1974b) (Table 8.1). Using only the radioed members, data on HS attendance were collected by two methods, as follows.

Table 8.1: Pack Composition and Radio-Collared Animals in the Study Packs. For Radio-Collared Animals, Their Number and the Portion of the Study Period (April through August 1973) They Were Followed Are Given

Harris Lake Pack	Jackpine Pack
Alpha male 2499 (7 May-20 August)	Alpha male***
Alpha female 2407 (entire period)*	Alpha female***
Yearling male 2247 (up to 10 May)**	Adult male 2449 (up to 24 August)†
Yearling male 2489 (7 May on)	Yearling male 2443 (entire period)
Two pups***	Yearling female 2445 (entire period)
	Yearling††
	Six pups***

*Due to partial transmitter failure, her radio-signal was too weak to be monitored, but it still could be reliably spot-sampled.

**Radio expired and subsequent presence with pack unknown.

***Unradioed, but known to be present.

†Radio expired, but wolf known to remain with pack.

††Unradioed, but presence suspected.

Spot Sampling

A monitoring station was established about 400 m from each pack's HS and visited daily, if possible, for checking. At each visit, each radioed wolf's presence or absence was checked with an LA-12 (AVM Instrument Co.) receiver. The majority of such checks occurred from 0600-1100 and 2000-0200 hours, although mid-day checks were also made.

Monitoring

Several weeks after spot sampling was initiated, continuous recording of radio-signals was begun (Gilmer et al., 1971). For the

JP pack, an LA-12 receiver monitored a single animal's radio, and the signal was recorded on a Rustrak Model 288 strip-chart recorder operated at 5 cm/hr. Typically, the receiver was tuned to whichever radioed wolf was present during a spot check, and left there until the next daily check. If no other radioed animal, or no radioed animal at all was present the next day, the receiver was left untouched because retuning was impossible without the animal being present. However, if another radioed animal was present, the receiver was retuned to that individual. Because an animal had to be present to be monitored, the data are biased toward animals that are often present, and they overestimate time spent at the HS, underestimate the duration of stays, and do not indicate arrival times at the HS. However, the data can be used to determine departure times to assess data gathered by spot samples.

Before 30 May, a similar system recorded the presence or absence of the HL pack alpha female around the den. After that date, a scanning receiver (Realistic Pro-8) was substituted. The scanner was modified to monitor each of three radio-collared animals for six minute periods. In addition, a reference channel monitored a transmitter placed nearby, while a control channel monitored atmospheric electrical activity so that it could be distinguished from radio signals. Thus, each radioed wolf was sampled twice per hour for a total of 12 minutes. If an animal was present or absent for two consecutive periods, we assumed it was present or absent during the intervening period as well. If the individual left or arrived between intervals, as usually happened, the departure (arrival) time was considered to be at mid-interval. When atmospheric radio activity obscured radio signals, such periods were classified as "unknown."

The HL pack occupied at least one den and a nearby rendezvous site during 1973 (Table 8.2). On 2 August, this HS area was aban-

Table 8.2: Homesites Used by the Harris Lake and Jackpine Packs in 1973

Homesite	Period Used	Monitored?
Harris Lake Pack:		
Harris Lake East*,**	3 April-2 August	Yes***
Jackpine Pack:		
Den*	Mid-April-12 June	Yes†
Snake Creek**	12 June-31 July	Yes†
Sphagnum Road**	27 July-2 August	No
Gesend Pond**	4 August-6 August	No
Snake Creek**	8 August-28 August	No
Dragon Lake**	16 August-22 August	No
Kitigan Creek**	5 September-25 September	No

  *Den
  **Rendezvous site
  ***Monitored with LA-12 receiver between 28 April and 28 May, and
      with scanner between 29 May and 2 August
  †Monitored with LA-12 receiver between 30 May and 16 July

doned, after which the pack rarely attended HSs. In the JP pack, the den area was occupied until 12 June, when the pack moved to a rendezvous site that had also been used a year earlier (Table 8.2). Four other rendezvous sites were used, although the first site was returned to periodically throughout the summer. Spot sampling was done throughout the period that each HS was used; the periods monitored are indicated in Table 8.2.

RESULTS

Seasonal Changes in Attendance

In the HL pack during April (prenatal period and parturition) and May (neonatal and preweaning period) (Mech, 1970), alpha female 2407 spent most of her time at the HS (Figure 8.1). However, she was absent between 12 and 20% of the time for periods as long as 17 hours (median 4 hours) (Figure 8.2). During the last week of May, however, at about the usual age of weaning (Mech, 1970), her den attendance dropped sharply, and during June and July she only spent about one-third of her time at the HS (Table 8.3). Alpha male 2499's pattern of HS attendance paralleled 2407's, although his attendance dropped a week or two later and, overall, he was present only half as often (Table 8.3, Figure 8.1).

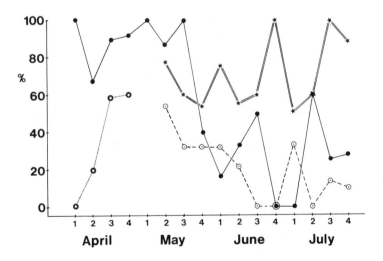

Figure 8.1: The percentage of spot samples in which radio-collared wolves were located at the Harris Lake pack homesite, 1973. Individuals (and mean weekly samples sizes): ●———● = alpha female 2407 (6.3); ʘ – – – – – –ʘ = alpha male 2499 (7.7); ❍·············❍ = yearling male 2247 (7.5); ★═══★ = yearling male 2489 (6.6).

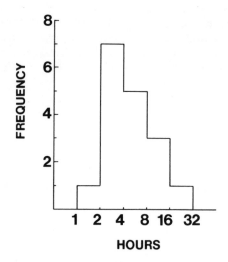

Figure 8.2: The duration of alpha female 2407's absences from the den area during April and May 1973.

Table 8.3: Changes in Homesite Attendance During Major Periods of Pup Development and Their Statistical Significance[a]

| Wolf (Pack) | | ....Change from:......... | |
		Pre-Parturition to Nursing Period	Nursing to Post-Weaning Period
Alpha male 2499	(HL)	–	-29%[b]
Alpha female 2407	(HL)	+12% NS	-63%[d]
Yearling male 2247	(HL)	+25% NS	–
Yearling male 2489	(HL)	–	0% NS
Adult male 2449	(JP)	+29%[c]	+7% NS
Yearling male 2443	(JP)	+52%[d]	-25% NS
Yearling female 2445	(JP)	+42%[d]	+13% NS

[a] The pre-parturition period includes the first three weeks of April, the nursing period covers the next month, and the post-weaning period begins during the last week of May. G-tests were used to test the significance of attendance changes between periods.

[b] p < 0.05

[c] p < 0.01

[d] p < 0.001

NS = not significant at p = 0.05

The HS attendance of HL pack yearling males differed greatly from that of their parents. Wolf 2489 spent over half his time at the HS, and this proportion increased as summer progressed (Figure 8.1). His littermate, 2247, spent about half his time at the HS during April, before his transmitter failed.

In the JP pack, monthly HS attendance patterns for subordinate male 2449 and yearling male 2443 were similar (Figure 8.3). Neither male was found at the HS until the pups were born in late April, at which time their HS attendance increased significantly and remained relatively high until mid-September when HSs were abandoned. Both males showed a drop in attendance in mid-June, but this decrease was not significant (Table 8.3). By late July, both males were spending about half their time at the HS.

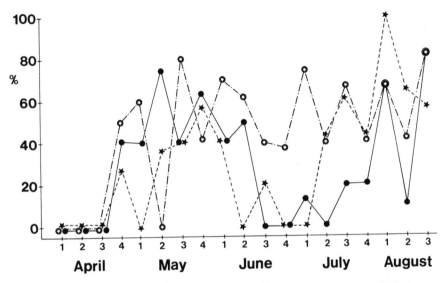

Figure 8.3: The percentage of spot samples in which radio-collared wolves were located at the Jackpine pack homesite, 1973. Individuals (and mean weekly sample sizes): ★----------★ = subordinate adult male 2449 (6.2); ●——————● = yearling male 2443 (6.5); o——·——o = yearling female 2445 (6.4).

Yearling female 2445's pattern of HS use differed from that of her brothers, although not significantly. Like her brothers, she spent little time at the HS prior to the pups' birth, and her attendance increased significantly in May. But then her HS attendance continued to increase to over 50% throughout the remainder of summer.

The major difference in HS attendance between packs was that JP pack subordinates spent little time at the HS until pups were born, whereas HL pack subordinates spent nearly half their time at the HS during the month prior to parturition. This difference could be related to the presence of two large prey carcasses near the JP pack HS that were utilized for extended periods. On 12 April, the JP pack located a moose carcass 1.3 km from the HS. Five days later, the pack killed a deer 4.5 km from the HS; that carcass floated in a lake for at least a week and was only occasionally available. Both carcasses were periodically visited; 61% of the time between the discovery of the carcasses and parturition (n = 38), subordinate wolves were located at one of these carcasses. In the HL pack, no similar localized resources were available during the same period. Thus, the carcasses and whatever food remained served as a more powerful attraction to the subordinate wolves than did pre-parturition activity at the HS.

Diurnal Patterns

Two diurnal HS attendance patterns were evident (Figure 8.4A-E).

Figure 8.4A-E: The percent of time various radio-collared wolves were located at homesites as a function of time of day, based on monitoring data. The stippled band between presence and absence represents the uncertainty in interpreting the records; thus, the values for absence and presence represent minima. Individuals (and total hours monitored): (A) alpha female 2407 during April (548 hours); (B) alpha male 2499 during June and July (1,509 hours); (C) yearling male 2489 during June and July (1,439 hours); (D) yearling female 2445 during June and July (579 hours); (E) yearling male 2443 during June and July (164 hours).

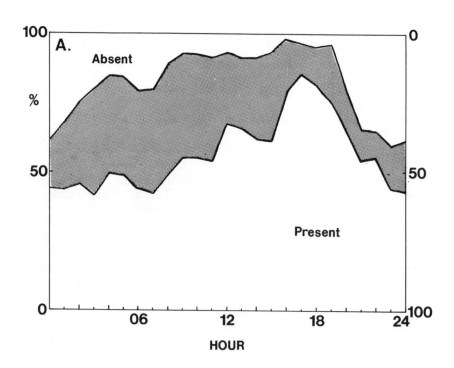

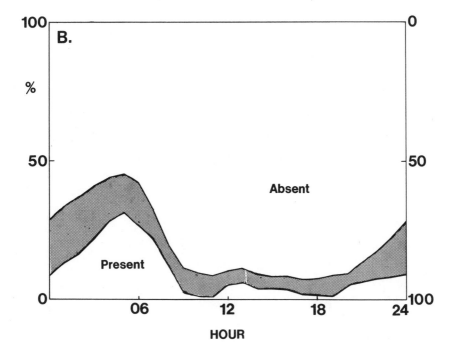

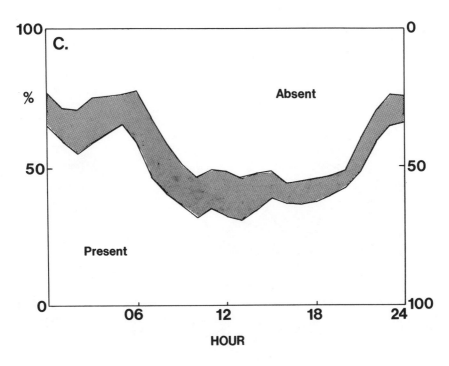

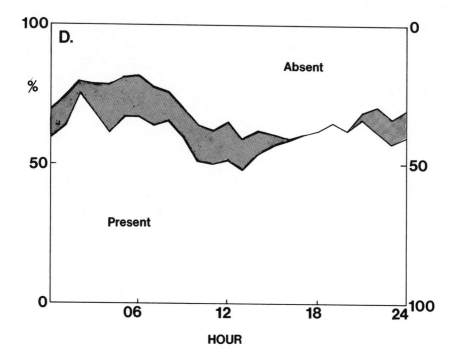

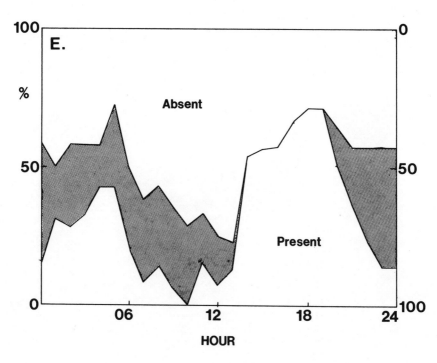

In May, alpha female 2407 of the HL pack was either present uniformly throughout the day, or was briefly absent around dusk or early evening. The other wolves were studied during June and July and showed a pattern of night attendance and day absence. Alpha male 2499 was most often present around dawn, left before 0900, and returned again between dusk and dawn. Yearling male 2489 behaved similarly, although he usually returned just after dusk. JP pack yearling female 2445 followed a pattern similar to that of 2489, although she was much more likely to be present during the day. JP pack yearling male 2443 apparently followed a pattern similar to the above three individuals, although data from him were too few to draw any firm conclusions. The two patterns noted may represent either seasonal differences (May versus June/July) or role differences (nursing versus non-nursing animals).

The basic pattern of nighttime attendance and daytime absence of adult and yearling wolves at the HS was supported by independent data collected through howling responses from the same packs (Figure 8.5).

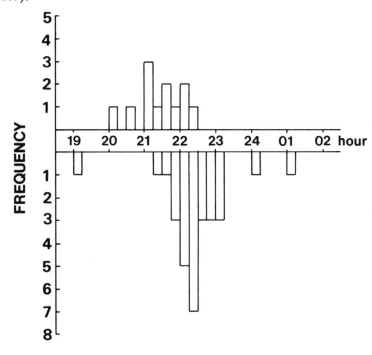

Figure 8.5: The number of howling sessions when pups were either located alone (upper graph) or with at least one adult or yearling (lower graph). Only one howling session was conducted per night. The determination of pack composition was based on tape-recorded howling replies using criteria from Harrington (1975).

The above data can be used to calculate the probability of finding various sized groups of radioed animals at the HS as a function of time of day. The assumption necessary is that the patterns observed reflect those generally found in the local population, allowing us to combine data from the two packs. By combining, in the following order, data from the HL pack alpha male and yearling male, and the JP pack yearling female and yearling male, we have calculated the probability of finding groups of two, three and four animals at the HS at the same time (Figure 8.6). The resulting probability distribution suggests that groups of animals would most likely be found together around dawn and, less frequently, again at dusk. One would expect to find few large groups present during the day.

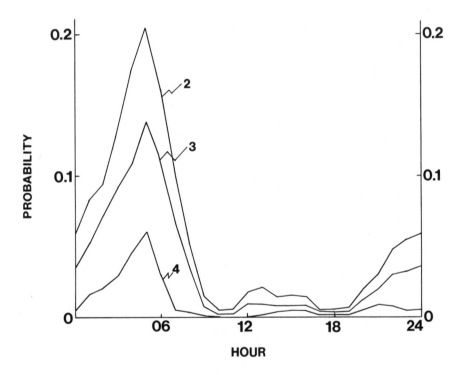

Figure 8.6:   Probability distributions of finding groups of two, three and four adult/yearling wolves together at the homesite as a function of time of day. These distributions are calculated from data in Figure 8.4 as follows: Groups of two from wolves 2489 and 2499; groups of three from wolves 2489, 2499 and 2445; groups of four from wolves 2489, 2499, 2445 and 2443.

Durations of Attendance

Data on durations of attendances and absences from HS were available from two wolves (Figure 8.7). HL pack alpha male 2499 rarely remained at the HS for a long period, the longest being 14.7 hours. His attendances were less than half as long as those of yearling male 2489 of the same pack (Mann-Whitney U Test: two-tail, $p < 0.005$); a third of 2489's attendances were longer than 14.7 hours, the longest being 86.5 hours.

Similar differences were noted for absences: alpha male 2499 remained away for significantly longer periods (Mann-Whitney U Test: two tail, $p < 0.001$). The median absence for 2499 was nearly 20 hours; 2489's median absence was less than half that. Forty-three percent of 2499's absences were longer than 24 hours, but only 25% were greater than 48 hours, the longest being 108.6 hours. It is possible, although unlikely, that some absences may have been interrupted by undetected 15 to 30 minute visits. Although we cannot unequivocally state the precision of these longer absences, the basic pattern, that of the alpha male's longer absences, remains unquestioned.

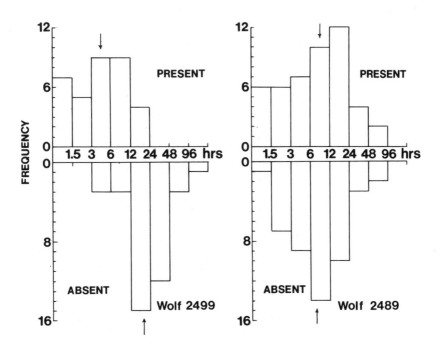

Figure 8.7:   The durations of absences from, and presence at homesites for two Harris Lake pack wolves. Medians are designated by arrows above the graphs.

Arrival and Departure Times

Both 2489 and 2499 departed from the HS predominantly up to four or five hours after dawn (G test: p [two-tail] <0.001, using 4 hour time classes) (Figure 8.8). However, at times, they did depart throughout the day, especially 2489.

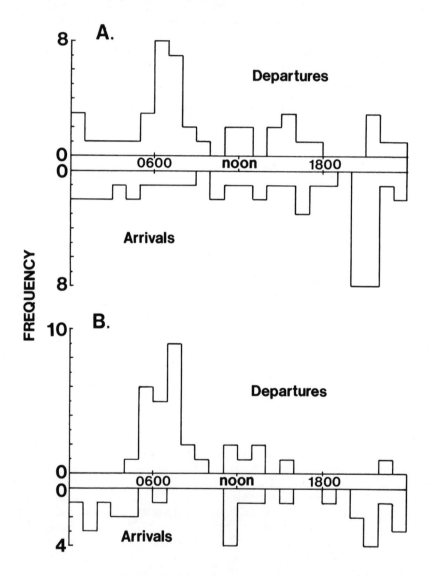

Figure 8.8:   Times of departures and arrivals at the Harris Lake pack homesite: (A) yearling male 2489; (B) alpha male 2499.

Considering arrivals, however, only yearling male 2489 showed a nonrandom diurnal preference (G test: p [two-tail] <0.001) (Figure 8.8). He usually arrived within two hours after sunset. Alpha male 2499, on the other hand, seemed as likely to arrive at almost any time of day or night.

There are too few data from the JP pack to obtain a clear picture of arrival times since animals were usually monitored *after* their arrival at the HS (see Methods). However, the departure data seem to parallel those of the HL pack with a greater number of early morning departures (Figure 8.9).

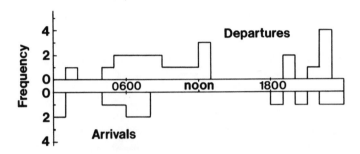

Figure 8.9:   Times of departures and arrivals at the Jackpine pack homesite. Data are pooled from all three radio-collared animals.

Coordination of HS Attendance

In this section, we test the assumption that an individual's pattern of HS attendance was independent of that of other adults or yearlings, using both spot sample and continuous data.

Using spot-sample data from both packs, individual probabilities of being located at the HS were calculated. Thus, if an individual had been located at the HS during three of 20 spot samples, its probability of attendance was 0.15. Using the binomial expansion, the expected probabilities of finding none, one, two and all three radio-collared animals at the HS during any one particular spot sample were calculated and compared with the observed frequencies.

The JP pack pattern differed significantly from random and indicated clumping; individuals were found together more often and alone much less than expected by chance (Table 8.4). Similar trends were evident in the HL pack data, but they were not significant.

Using continuous sample data, coordination between the HL pack alpha and yearling males was studied in a similar way. Because these animals were actually sampled twice an hour, the data are expressed both in hours and in sample period (Table 8.5). Again, there was a significant deviation from expected patterns, with the deviation toward clumping. These data suggest that pups may be left alone more often than one would predict by chance.

Table 8.4: The Frequency with Which Various-Sized Groups
of Radio-Collared Wolves Were Located at Homesites During Spot
Samples Compared with the Expected Frequency of Finding Such
Groups Under the Assumption That Individual Homesite
Attendance Probabilities Are Independent of One Another.

Number of Radioed Wolves	Harris Lake Pack			Jackpine Pack		
	......Frequency......		X^2	........Frequency.........		X^2
	Observed	Expected		Observed	Expected	
0	11	7.8		13	12.1	
1	20	25.0	3.45 NS	33	30.0	9.09 (p < 0.05)
2	20	20.5		12	20.6	
3	7	4.7		9	4.3	

Table 8.5: The Amount of Time That Alpha Male 2499
and Yearling Male 2489 Were Located Together, Singly
or Not at All at the Homesite Compared with the Amount
of Time One Would Expect, Assuming That Their Homesite
Attendance Probabilities Were Independent of One Another.
The Number of Sampling Periods, Presented in Parentheses,
Comprise the Data Tested Statistically.

Wolves Present	Hours Observed	Hours Expected	X^2
Neither	554 (1108)	513 (1026)	
2499 or 2489	454 (908)	536 (1072)	82.8 (p < 0.001)
2499 and 2489	111 (222)	69 (138)	

However, a non-radioed yearling may have been present in the
HL pack, and at least two, and possibly three non-radioed animals
were present in the JP pack (Table 8.1). These individuals could have
been present when the three radioed animals were not. However, if
that were so, then the non-radioed individuals in each pack would
have had to have differed fundamentally from those studied; since
the patterns of observed pack members were either independent or
clumped, the patterns of *unobserved* members would have had to
have been overdispersed *if* the amount of time pups were left alone
was to be minimized. That situation would lead to the unlikely
possibility that the crucial pup-caring animal for the HL pack was a
yearling male, whereas, for the JP pack, it was one or both of the
alpha animals. However, the radioed alpha pair of the HL pack spent
little time at the HS, suggesting that in the JP pack, nonradioed
animals were not disproportionately covering the periods when all
three radioed members were absent from the HS.

One more set of data was used to estimate how often the pups
actually were left alone. During the HS season, virtually all pack
members reply frequently to simulated howling when present at a HS
(Harrington, 1975; Harrington and Mech, 1979). By eliciting howling
when no radioed animals were present, and later analyzing recordings

of the replies, we determined how often the pups were accompanied by non-radioed adults. For both packs, we found pups unattended on a significant number of nights (Table 8.6). The proportion of occasions when pups were left alone decreased from 1972 to 1973 for both packs, coincident with increases in the number of adults and/or yearlings in both packs. With more adults and yearlings in the pack, the pups by chance were left alone less.

Table 8.6: The Percent of Howling Sessions When Pups Were Not Accompanied by Non-Radio-Collared Adults or Yearlings. These Data Are from Sessions When No Radio-Collared Adults or Yearlings Were at the Homesite.

Pack	1972 (n)	1973 (n)
Harris Lake	50.0 (4)	25.0 (4)
Jackpine	42.1 (19)	28.6 (35)

It appears, from the above analyses, that adult and yearling wolves may sometimes coordinate their travels to be with other pack members and, as a result, often leave the pups alone.

Effect of Large Prey Carcasses on HS Attendance

As mentioned earlier, the presence of large prey carcasses appeared to draw subordinate wolves away from the JP pack HS for several weeks prior to parturition. Later in summer, another large carcass was exploited by the JP pack. On 3 August, the entire pack travelled approximately 16 km to a moose carcass located in an extreme corner of the territory. (Actually, the site might be more properly considered part of the neighboring HL pack territory, where it was only 4.5 km from the pack's current HS.) The carcass was consumed during the next two days, after which the JP pack returned to its more centrally located HSs. Two aspects related to this carcass are of particular interest.

First, despite the location of the site near or within a neighboring pack's territory, and despite at least two howling interactions between the packs, the JP pack adults and yearlings left the pups unattended at the site after one day, apparently because most of the edible portions of the moose had been consumed. The pups remained alone (as determined by radio-signals and howling replies) for another day before they too retraced their route.

Second, a marked change in HS attendance occurred after the visit to the carcass site. In the two weeks before moving to the site, only one of the three radioed subordinates was usually located during each visit to the HS. After the pack returned from the moose, however, an average of two individuals usually attended the HS. This increased attendance rate continued for approximately two weeks before tapering off and returning to pre-carcass levels (Figure 8.10). Such a pattern suggests that HS attendance was reinforced by the exploitation of the carcass. Possibly animals returned to the HS more often in "anticipation" of a similar find occurring. By being at the HS when such a find is made, an individual can assure itself a reasonable chance of obtaining a portion of the remains. However, after several weeks without a second find, the effect seemed to dissipate, as HS attendance returned to pre-carcass levels.

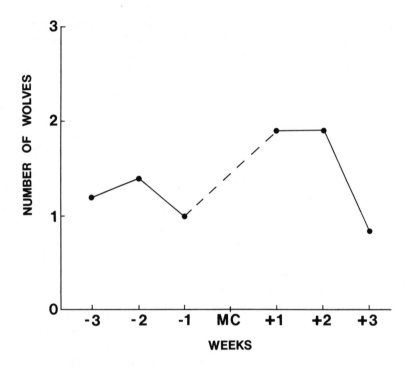

Figure 8.10: The effects of finding an adult moose carcass on the homesite attendance of Jackpine pack subordinate wolves. The mean number of radio-collared wolves (maximum = 3) located at the HS during each spot sample is plotted for the three weeks prior to, and after, finding the moose carcass (MC).

DISCUSSION

Our most unexpected finding was the pattern of nocturnal attendance by adult and yearling wolves at HSs. Most previous studies of HS use have indicated a pattern of day attendance, dusk departure, night absence and early morning arrival (Murie, 1944; Banfield, 1954; Joslin, 1967; Kolenosky and Johnston, 1967; Theberge and Pimlott, 1969; Carbyn, 1974; Haber, 1977; Peterson, 1977). For example, all 11 of the departures witnessed by Murie (1944) in Mt. McKinley National Park occurred between 1600 and 2130 hours. However, most of the observations of HS attendance made during these studies were not made systematically; in many cases they could have reflected primarily the times of day or night the observers were present. The one study that systematically investigated HS attendance for a long period produced the same results as ours (Chapman, 1977).

There are several possible explanations for any real differences between our results and reported HS attendance patterns. Most of the previous studies involves wolves in tundra (Murie, 1944; Banfield, 1954; Clark, 1971; Haber, 1977). In such areas, daytime summer temperatures are relatively high, but more importantly, they cannot be easily escaped due to lack of cover. In Minnesota, however, similar summer daytime temperatures can be more easily avoided in the dense forest, permitting wolves to be more active during daylight.

Second, because of Minnesota's lower latitude, there are two major twilight periods separated by eight to nine hours of darkness. These twilight periods may have a very important effect on wolf activity patterns through their effect on prey activity rhythms. Both white-tailed deer and moose are active primarily around dusk and dawn, and beavers are especially active around dusk although they may remain active throughout the night (Montgomery, 1959; Thomas, 1966; Marchinton and Jeter, 1966; Tibbs, 1967; Nelson, 1979 for deer; Peterson, 1955 for moose; Tevis, 1950 for beavers). The fact that our wolves appeared to be most active (as indicated by arrivals and departures) during dusk and dawn suggests they were taking advantage of prey activity.

We propose that the wolves in our study remained near prey areas during the day to: 1) take advantage of any unusual prey activity during mid-day, and 2) be nearby as prey became active toward evening. Presumably, the wolves continued to hunt throughout the evening and, when successful, returned to the HS where they might feed the pups. Because prey are generally inactive during darkness, the wolves would then remain at the HS throughout the remainder of the night. Then around dawn, they would depart to take advantage of the morning prey activity. Kolenosky and Johnston (1967) found that in an Ontario forest during summer, wolves moved between 0600 and 0800 hours, and again between 2100 and 2300 hours. At least the latter period seemed to be characterized by active hunting for beavers. In addition, the wolves often moved during mid-day between the areas they utilized heavily.

If there is any difference between the results of the Ontario study and ours, it is probably in the greater morning activity of our wolves. This difference could be related to differences in hunting success. Kolenosky and Johnston (1967) found that beavers were plentiful and felt that hunting success was high. Thus, wolves might only need to hunt during the evening. On the other hand, our wolf population was food-stressed during the study (Mech, 1977c), probably forcing the animals to hunt during both prey activity periods and perhaps during the day as well. Other studies have also noted daytime travel, probably related to hunting (Murie, 1944; Clark, 1971).

If the above is true, then, as evening progresses more and more adults and yearlings should attend the HS and just before the dawn departure, the most pack members should be present. The individual HS attendance probabilities of our wolves indicated that the greatest number of animals would be at the HS around 0600. Wolves in Jasper National Park, Canada, were most active at 0400-0800 hours at the HS (Carbyn, 1974). Between 0400 and 0600 hours, 60% of all chorus howling occurred in the Canadian packs studied, suggesting that socially, "this time period was important in the activity patterns of the pack" (Carbyn, 1974:56). We found a peak in chorus howling for both the HL and JP packs during the same period (Harrington and Mech, 1978). A similarly active period has been noted in captive wolves (Zimen, 1971) and coyotes *(C. latrans)* (Harrington, unpub.) around dawn. It therefore appears that dawn is a very important period in wolf social life. Possibly the dynamics of wolf sociality might be best revealed by observations at that time.

The seasonal patterns of HS attendance noted during this study were similar to those found elsewhere (Murie, 1944; Clark, 1971; Carbyn, 1974; Haber, 1977), particularly with regard to the alpha female (Murie, 1944; Clark, 1971; Carbyn, 1974; Haber, 1977; Fritts and Mech, 1981). She generally remains close to the HS prior to parturition, and then for about the next month, while the pups are still nursing. As the pups are weaned during the next few weeks, however, she begins to spend more time away from the HS, presumably hunting. While the alpha female is still nursing, apparently much of her food is provided by other wolves which would explain why alpha male 2499 spent little time at the HS, even while the pups were being nursed.

Thus, the patterns of HS attendance for both the alpha male and female apparently reflect their foraging activities. First, the alpha male, and later the alpha female (once freed of nursing), spend 60-90% of their time away from the HS, presumably to meet the growing food demands of the developing pups. This conclusion is supported by both the alpha male's patterns of departures and arrivals, and the duration of his absences. Typically, he departed at the same time each day, whereas he arrived at random throughout the day, as one would expect of an animal that hunted until it secured prey, and then immediately returned to the HS to feed the pups. The alpha male's variable, and often lengthy, absences from the HS apparently reflected the vagaries of hunting under relatively poor prey availability.

The patterns of the subordinate wolves, however, were quite different from those of the adults. None showed a significant monthly change in attendance once the pups were born, even at the time of weaning. Also, each individual spent a considerable amount of time at the HS, usually close to 50%. Finally, one yearling, studied in sufficient detail, showed a pattern of arrivals and departures and HS attendance that strongly suggested a regular, 24 hour cycle. This evidence indicates that compared with the alpha male and alpha female, these subordinates were not playing a significant role in foraging for the pups. They spent too much time at the HS, and their patterns of attendance were too regular to be dominated by the vagaries of hunting.

A popular explanation for the observation that subordinate wolves often spend considerable time at the HS is the "pup-tending hypothesis." This hypothesis assumes that animals remain at the HS to protect the pups in case of danger (e.g. from bears [*Ursus arctos, Ursus americanus*], strange wolves and humans) (Murie, 1944; Haber, 1977). A usual assumption is that the evolutionary basis (ultimate causation, Alcock, 1975) of the behavior is kin selection (e.g. Bulger, 1975; Haber, 1977). Individual yearlings and subordinate adults supposedly sacrifice a portion of their personal fitness (e.g. forego foraging) but increase their kin's fitness (i.e. protect pups) by remaining at the HS and tending pups. From an individual-selection viewpoint, it might be better for them to forage, increase their food intake and maintain or enhance their physical condition so that they might later attain a territory and mate, be able to secure ample prey and, thus, raise their own young. But from a kin-selection viewpoint, it might be better for subordinates to forego some foraging opportunities and not develop their full potential as quickly, if ever, in order to help safeguard their parents' current litter. Because most wolf packs are essentially one-family units (Mech, 1970; Haber, 1977; Peterson, 1977), a kin-selection explanation for extensive HS attendance by subordinates is plausible.

However, there are no quantitative data to directly support the pup-tending hypothesis. The fact that subordinates have been seen defending pups even in the company of dominants (Murie, 1944) could be an occasional result, but not necessarily the ultimate basis for, HS attendance by subordinates. There are equally plausible explanations based solely on individual selection (see below). To help find the most likely explanation, one must analyze the costs and benefits related to HS attendance, determine how they relate to individual and kin components of selection, and then select the simplest explanation that can adequately account for the observed behaviors ("Occam's Razor", Williams, 1966).

One aspect of the pup-tending hypothesis that we can test involves the allocation of time by subordinates at HSs. If subordinates attend HSs to protect the pups, one would predict that the pups would rarely be left alone. Thus, a subordinate should be more likely to stay at a HS if no other adults/yearlings are present in order to ensure continual protection. Without such a mechanism to coordinate

HS attendance, any benefit to the pups through pup-tending could be easily lost during periods when they are left alone. Therefore, natural selection should favor HS attendance by yearlings or adults that is as continuous as possible. Indeed, some accounts imply, without presenting quantitative evidence, that such coordination among adults and yearlings does exist (e.g. Haber, 1977).

However, we found that the opposite was true: Pups were left alone significantly more often than one would expect by chance. It appeared that the adults and yearlings were *avoiding* staying with the pups if no other pack members were present. This was true for both packs, for both types of data, for both years, and for several thousand wolf-hours of monitoring. These findings do not support the pup-tending hypothesis. Another study involving hundreds of hours of direct visual observations of three Alaskan packs indicated that the pups were left unattended by adults and yearlings 40, 50 and 73% of the time, depending on whether the packs contained eight, at least three, or three yearlings and adults, respectively (Chapman, 1977).

A second means of testing the pup-tending hypothesis is to determine just how likely wolves of various ages and social ranks will defend the pups from external threats. The wolves most active in confronting a threat are dominant individuals, quite often alpha males (Murie, 1944; Mech and Frenzel, 1971; Zimen, 1975; Haber, 1977; Harrington and Mech, 1979). Subordinate wolves, on the other hand, usually have shrunk from encounters (Zimen, 1975; Harrington and Mech, 1979). In fact, one low-ranking female that was the primary "babysitter" at a HS observed by Haber (1977:273), appeared to be "so timid it may have been afraid to go off on its own." It is difficult to imagine how such a timid animal would have been of any use in confronting a bear, strange wolf or human. Thus, even if danger occurs at a pack's HS, there is little evidence that yearlings or subordinate adults would protect the pups. Even if they occasionally do, it is probably for other reasons that they attend the HS.

There are any number of possible explanations for HS attendance by subordinate wolves involving roles in pup socialization, alliance formation, practice in parenting, etc., but because the present study cannot provide data on these points, they will not be discussed. However, there are two likely explanations supported by our data. Both involve the importance of the HS as a focus of foraging activities.

Throughout summer, the HS is a center of food exchange, usually between forager and pups. However, while lactating, the alpha female is provisioned by others and subordinates are also sometimes fed by foragers, or they raid foragers' food caches at the HS (Murie, 1944; Haber, 1977). Thus, subordinates could attend the HS to intercept food meant for the pups. Captive yearlings and subordinates, and some dominant individuals, beg food from "foragers," and occasionally succeed (Bennett, 1979; Fentress and Ryon, this volume). Such behavior could be even more prevalent in the wild because of the more varied and extreme nature of prey availability. In times of food shortage, feeding of yearlings or subordinate adults rather than pups might be advantageous to the parents. If pups are unlikely to survive, as was sometimes the case in the present study area (Mech, 1977c), it would be more efficient to invest in individuals that have survived a winter and are closer to the reproductive state. Of course, the "decision" to feed older animals rather than pups would not be simply made. It would probably involve such factors as pup condition, season (early or late summer, etc.), food availability, yearling condition, and others.

This hypothesis is supported by several aspects of HS attendance. First, subordinate HS attendance generally increased as summer progressed, which could have resulted from the occasional reinforcement provided by each "free meal." Second, the subordinate generally arrived around dusk when successful foragers would begin to return with food, and then remained at the HS throughout the night where it would be on hand as foragers returned. Third, the yearling male of the HL pack exhibited a regular 24 hour activity cycle: He left in early morning, usually stayed away for about twelve hours, and returned at approximately the same time each evening. Since it seems unlikely that he was securing prey at such a regular time each day and bringing it back to the HS, a more reasonable explanation is that he was returning to the HS regularly to chance obtaining food there. Indeed, the alpha male, who presumably was hunting the hardest, usually spent longer and much more variable periods away from the HS. Both the regularity of his departures and the randomness of his arrivals suggest that his returns were dependent on his foraging success. These interpretations are merely speculation; direct observations are needed indicating how often each individual returns with or without food, how often they feed others, and how often they attempt to obtain food from others.

A second, and not mutually exclusive, explanation assumes that the HS is a center of information exchange about foraging areas. In many species, colonial or group living allows individuals to use information provided by others to determine where the most profitable foraging areas are (e.g. bank swallows [*Riparia riparia*, Emlen and Demong, 1975]). Similar information exchange among wolf packmates could occur. In the JP pack in 1973, for example, the effects of a large prey find apparently were revealed. HS visitation by subordinates increased after a moose carcass was discovered, and remained high for two weeks, possibly in anticipation of another find.

However, spending too much time at the HS to chance such a discovery could be detrimental in the long run, for many foraging opportunities would be sacrificed. Thus, a balance must be struck between time spent at the HS, and time spent foraging. The optimal allocation of time would depend on the probability of large prey

finds. If they are common, it would pay to spend much time at
the HS.

SUMMARY

This paper presents a detailed, quantitative analysis of homesite
(HS) attendance patterns in two Minnesota wolf packs based on electronic monitoring of radio-collared wolves. Subordinate animals did
not seem to be contributing heavily to the pups' welfare, at least
through foraging or providing protection. Rather, their HS attendance
patterns could be more parsimoniously explained on an individual,
rather than a kin, level of selection, which cautions against carrying
kin-selection arguments further than the evidence warrants. Such
overuse can obscure important, and often radically different, evolutionary considerations at the individual level. For example, yearlings
that spent considerable time at HSs and may occasionally chase off a
bear or decoy away an errant wolf scientist, can hardly be called
"altruists" if the primary reason they are there is to intercept food
intended for the pups. Wolves are complex animals living in a complex society; we can hardly expect the subtleties of wolf behavior to
yield to simplistic analyses.

The population under study was facing a critical food shortage.
Ecological and physiological conditions can have dramatic effects on
the expression of a species' behavior. Thus, the extent to which the
above results are generalizable is subject to question. However, they
at least represent a subset of possible behavioral expressions in the
wolf. To more fully understand the dynamic relationship between
behavior, ecology and population dynamics, we must expand such
studies to as many populations, and under as many conditions, as
possible. What may apply to the Superior National Forest in 1972-
73, may have little relevance for present conditions in Algonquin
Park, Interior Alaska, or even the Superior National Forest in 1984.

Acknowledgements

Many individuals and agencies helped make this study
possible. Funds were provided by the Patuxent Center's
Endangered Wildlife Research Program (U.S. Fish and
Wildlife Service), North Central Forest Experiment Sta-
tion (USDA), World Wildlife Fund, Ober Charitable
Foundation, and an NSF grant to the Psychobiology
Program, State University of New York, Stony Brook.
Charles Walcott of the Psychobiology Program assisted
throughout all stages of the study and, along with Walt
Pfieffer, U.S. Forest Service, and V.B. (Larry) Kuechle,
Bioelectronics Laboratory, University of Minnesota,
contributed expertise on electronic matters, for which

we are grateful. Jeff Renneberg, Todd Fuller, Jim Klitzke and Ted Floyd provided valuable assistance in the field. Funds for the preparation of this paper were provided, in part, by a Mount Saint Vincent University faculty research grant to F.H. Harrington.

Incidence of Disease and
Its Potential Role in the
Population Dynamics of Wolves in
Riding Mountain National Park, Manitoba

Ludwig N. Carbyn

INTRODUCTION

Wolf (*Canis lupus*) packs establish and defend well defined territories (Mech, 1973). Zimen (1976) discussed the theoretical implications of territoriality and suggested that stability in spatial organization depends upon stable food supply. A temporary reduction in food would influence primarily the number of individuals per pack, but protracted changes in food availability would alter the spatial organization of packs. Accordingly, food induced changes in spatial organization of territories have been reported from Isle Royale (Peterson, 1977:88) and Minnesota (Mech, 1977a). However, a territorial displacement has occurred when food was not in short supply (Carbyn, 1981b). This paper presents a case where disease, rather than food, appears to have contributed to pack displacement.

Diseases affecting wolves have been summarized (Mech, 1970) but the affects of epizootics and enzootics on the dynamics of wolf populations have not been well documented. Murie (1944) discussed mange, canine distemper virus and rabies as possible regulating factors in Alaskan wolf populations. It is thought that both distemper and mange may have reduced wolf populations in Jasper during the 1940s (Cowan, pers. comm.). Rausch (1958) and Chapman (1978) documented cases of rabies in wolves. Neiland (1970) reported on serological evidence of rangiferine brucellosis in wolves. He also experimentally infected wolves with brucellosis bacteria extracted from caribou (Neiland, 1975). Choquette and Kuyt (1974) demonstrated the serological presence of viral diseases in wolves of northern Canada. Todd et al. (in press) discussed the prevalence of sarcoptic mange (caused by the mite *Sarcoptes scabie*) in wolves and coyotes *(C. latrans)* in Alberta. Although documentation of the prevalence of disease-related mortality in wild wolves is difficult to obtain, it has become increasingly feasible with the use of radio telemetry.

106

This paper will document case histories of diseased wolves recovered during field studies in Riding Mountain National Park, and discuss the possible effects of disease related mortality on pack stability and territorial displacement.

Study Area

Riding Mountain National Park is located in southwestern Manitoba (Figure 9.1). Park vegetation is predominantly Boreal mixedwood and Aspen-Oak forest with extensive interspersions of grasslands (fescue prairie), sedge meadows and black spruce (*Picea mariana*) bogs (Rowe, 1959). The climate is continental interior characterized by cold winters with moderate snow depths.

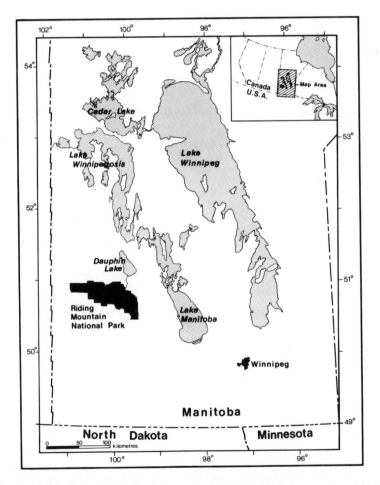

Figure 9.1: Location of Riding Mountain National Park in southwestern Manitoba.

A number of characteristics combine to make the Park a unique biological system (Briscoe et al., 1980). The 2,944 km² wilderness area is completely surrounded by agricultural land, making the Park an "island" where contact between wild and domestic animals is common. Also noteworthy are its diverse faunal composition and periodic eruptions of large mammals.

Prey Populations

The average elk (*Cervus elaphus*) population during the four year study (1975-1978) was 3,500, or 1.2 elk per km² (Carbyn, unpub.). Moose (*Alces alces*) numbers were estimated at 2,300 (0.8 moose per km²). White-tailed deer (*Odocoileus virginianus*) were not counted in regular aerial surveys but were thought to be in moderately high numbers in some parts of the Park, and absent from others. Beaver (*Castor canadensis*) populations were consistently high and evenly distributed throughout the Park.

METHODS

Population estimates of wolves were obtained by (a) locating rendezvous sites and obtaining response indices from nocturnal howling surveys (Pimlott, 1960), and (b) by radio-tracking collared wolves. Radio-tracking resulted in delineation of pack territories and this information was used to calculate wolf densities.

Trapping and radio-collaring were confined to September and October because pups are large enough to be collared, the tourist season is past its peak, and the chance of frozen limbs due to low temperatures is reduced. Trapped wolves were tranquilized using intramuscular injections of phencyclidine hydrochloride in equal proportions (1 mg of each drug per kg of wolf) with promazine hydrochloride (Sparine,™ Wyeth Laboratories). Radio transmitters were mounted on collars of machine belting. Tracking flights employing yagi antennas mounted on a fixed winged aircraft were carried out at variable time intervals, depending on the season. Information on wolf mortality was obtained by recovering radio-collared wolves, obtaining kill records from trappers, and recovering carcasses found during winter ground tracking.

Carcasses were kept frozen until necropsied by pathologists of the Alberta Veterinary Services, Edmonton. The diagnoses reported in this paper are based on gross examination of pathological conditions and clinical testing of results. In instances of suspected tuberculosis, testing for the bovine tuberculosis organism was carried out through isolation and culturing of acid-fast bacteria. Where rabies was suspected, rabies antigen was tested by fluorescent antibody techniques. Canine distemper was diagnosed by the presence of microscopic lesions, pitting of tooth enamel (Figure 9.2) and condition of the organs, particularly the lungs.

Figure 9.2: Portion of the lower jaw of a female wolf pup (D4) captured on 15 November 1978, showing pitting of the enamel at the base of the incisors characteristic of young animals subjected to high body temperatures at early stages of growth and development. (Photo: D. Patriquin)

Special efforts were made to study the Whitewater (WW) pack (Figure 9.3) in detail. However, failing the capture of a member of the WW pack, the movements and predation rates of the most accessible neighboring packs, the Audy Lake (AL) and Baldy Lake (BL) packs, were studied. Prey selection by specific packs was investigated during the winters of 1977/78 and 1978/79.

RESULTS

Incidence of Disease

Twenty-one wolves were captured and radio-collared. Their fates were as follows: (a) eight were trapped, shot or poisoned by humans outside the Park (38%); (b) seven were unaccounted for (34%); (c) three were found dead but the cause of death was unknown (14%); and (d) three died from disease-related causes (14%). In addition, two non-collared wolves were found dead and both deaths were believed to be disease-related. All diseased wolves were between five and eight months of age.

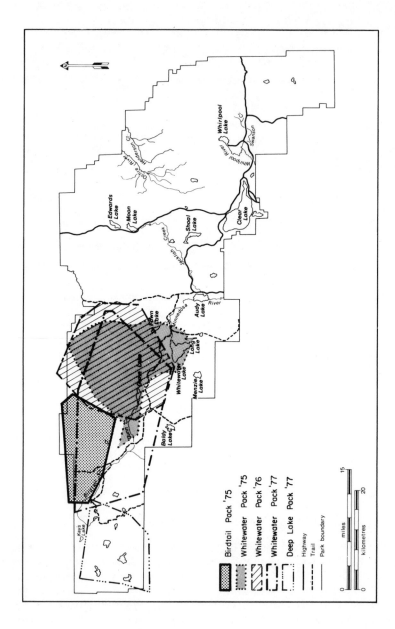

Figure 9.3: Outline of the Whitewater Lake pack territory during 1975, 1976 and 1977, showing the territory extension of 1977. The 1977 territory encompassed most of the area formerly occupied by the Birdtail pack.

Three of the five diseased wolves showed symptoms of canine distemper and two contained lesions of bovine tuberculosis. The latter two were unusually vulnerable to retrapping. One was recaptured four and five days after initial capture, and the second, seven days after initial capture. These two wolves were presumed to be littermates because they were captured within one kilometer of one another. In addition to the diseased animals recovered, one aberrantly behaving wolf was observed in the field and was assumed to be diseased.

Case Histories

D1, a non-collared female pup weighing approximately 29.5 kg, was found curled up, frozen in her bedsite on 22 December 1977. (Exact weights were not always possible because of dehydration of carcasses in winter.) Excessive fluid was present in her lungs but no definite indications as to the cause of this fluid accumulation were readily detectable. Pathologists diagnosed the cause of death as pneumonia precipitated by canine distemper.

D2, a collared female pup weighing approximately 22.7 kg at death, was found on 3 January 1978. She was lying on her side, legs stretched out and lips drawn to the gums, possibly the result of convulsions. Necropsy indicated that she died from pneumonia secondary to canine distemper viral infection. The female, a member of a pack of 10, had been collared on 21 September 1977 when she weighed 27 kg, and her movements were documented until her death. Neither her movements nor behavior appeared abnormal.

D3, a non-collared male pup weighing 24 kg, was found dead along a road south of the Park on 28 September 1978, where he had often been seen during September. He was emaciated and his stomach contained grain (evidence of movements into farmland) and remains of a garter snake (*Thamnophis sp.*). His general condition suggested chronic, not acute, deterioration. Microscopic skin lesions indicated distemper. Fluorescent antibody tests for rabies proved negative. The immediate cause of death was that he had been struck by a vehicle. His debilitated condition was brought on by canine distemper and it is unlikely that he would have survived, even if he had not been run over.

D4, a radio-collared female pup weighing 12.3 kg at death, was recovered on 15 November 1978. She had first been captured and radio-collared on 24 October when she weighed 16.8 kg, and was recaptured and released on 28 and 29 October. At death, she was extremely emaciated and her teeth were discolored and the enamel pitted (Figure 9.2). Numerous focal nodules were found in the lungs and liver. Lymph nodes lining the intestinal tract were markedly enlarged. Porcupine (*Erethizon dorsatum*) quills were present in the head and legs. Death was attributed to emaciation secondary to tuberculosis. The specific strain of tuberculosis was bovine as confirmed from cultures (R. Lewis, pers. comm.).

D5, radio-collared male pup weighing 25.4 kg at death, was recovered on 12 December 1978. The animal had been captured on 1 October when it weighed 22.7 kg, and recaptured on 25 October, and seemed in good health. At death, however, the wolf was emaciated, suggesting acute rather than chronic deterioration. Infected lymph nodes in the abdominal cavity impinged on the intestinal tract. Smears from lymph nodes yielded acid-fast bacteria indicative of tuberculosis. Death was attributed to emaciation secondary to tuberculosis.

D6 was observed by park warden Calvin Allen on 1 May. The unusual behavior which it exhibited may have been disease-related.

> I observed an adult black wolf near the recent washout on the Central Trail. The animal was on the trail, and ran into the bush when I got to within about 200 m. It ran back out again almost instantly from the same spot then proceeded east along a ditch, darting in and out of the edge of the bush several times before disappearing.
>
> When returning west on the Central Trail about 1.5 hours later, I observed the same wolf again on the trail about 4.8 km from its first location. The wolf did not notice me until I was about 50 m away. It was very thin and "ragged" looking. As it ran from the road it hit a clump of shrubs in the area and stumbled around for some time as if blind. The animal then ran almost parallel to the trail until it brushed another shrub where it would turn and run back until it hit some bush and finally turning again, travelled almost back the way it had come. It did this about six times, then seemed to get its bearing and disappeared to the south.

This wolf was in an area where, four months earlier, two wolf carcasses had been collected and diagnosed as having died from canine distemper. The unusual behavior observed may have been due to encephalitic complications brought on by distemper.

Population Decline

The wolf population declined during the four years of this study. Early winter estimates of wolf numbers were as follows: 120 (1975), 73 (1976), 52 (1977) and 63 (1978). These figures represent net change from year to year; they do not consider the number of young produced each spring or their subsequent mortality.

Disease was the second most commonly recorded cause of wolf mortality. Undoubtedly, for every diseased animal discovered, several others went undetected, as only a small percentage of the wolves in the Park were radio-collared. Human harvest accounted for the greatest loss of animals. During the 1975–77 period, a minimum of 38 animals was killed, accounting for 56 percent (38/68) of the known losses.

Territorial Displacement

The displacement of the Birdtail (BT) pack by the WW pack was circumstantially linked to a reduction in pack size by disease. Based on 53, 63 and 21 radio fixes, territory sizes for the WW pack during 1975-76, 1976-77 and 1977-78 were 300, 359 and 544 km^2, respectively (Figure 9.3). In winter 1977-78, the WW pack extended its range across BT pack territory to the edge of the Deep Lake (DL) pack territory. The most dramatic shift in range occurred after 7 December 1977 (Carbyn, unpub.). In early winter 1975 and 1976 (information from tracking in the snow, the pack was not radio-collared that year), the BT pack consisted of six and eight animals, respectively. At least five wolves (four recognized as pups) were present on 1-3 August 1977 at a traditional BT pack rendezvous site, but there was no evidence of any wolves in the area by December. Early winter WW pack size was 16 in 1975, seven in 1976 and 10 in 1977.

DISCUSSION

A wolf pack may be affected by contagious disease in any one of three ways: (1) through loss of experienced adults; (2) through reduced recruitment of young; and (3) through total decimation or permanent disruption of the pack.

This study documented the decline of a wolf population which was not food limited. The decline was at least partially attributable to disease. Although the initial decline in 1976 may have been disease related, no cases of distemper were documented before 1977 and no tuberculosis until 1978, the year after the lowest population count. Trapping of wolves adjacent to the Park may have also contributed significantly to wolf mortality. At the time of the study, there was a demand for long haired fur, with a resultant increase in trapping pressure.

As far as we can determine, the presence of bovine tuberculosis in wolves has never been reported. Although it affects all age classes, it is generally thought to be more deleterious to older animals. This differential effect could be important if older, established pack members, particularly alpha males and females, succumb. Based on observation of one pack in Jasper National Park (Carbyn, unpub.), loss of alpha animals may alter pack prey selection in diverse prey systems, resulting in changed hunting patterns and effects on the prey populations. However, this hypothesis requires further confirmation.

Peters (1979) discussed the concept of "cognitive maps," suggesting that wolves learn routes and familiar landmarks in their territories through experience. As adults, they are more familiar with their territories than the young. When adults are lost, the im-

pact on the pack's habitat use is potentially greater than when pups
are lost. Longevity of alpha animals has been well documented on
Isle Royale (Peterson and Scheidler, 1979) where one pack has had
the same alpha female for eight years and the same alpha male for
five years. Such continuity in pack leadership undoubtedly contributes to stability in spatial organization of territories. The population turnover in Riding Mountain National Park is greater than reported for Isle Royale. Territory instability from year to year therefore can be expected if the Riding Mountain National Park population suffered losses of alpha animals (experienced wolves) and pups
(reduction of pack size and recruitment).

Canine distemper is a disease which directly affects recruitment of young. Because of high densities of wolves and coyotes
in Riding Mountain, and the casual presence of red foxes *(Vulpes
vulpes)* in adjacent areas, outbreaks of the disease are probably
sporadic, dictated by environmental factors. Trainer and Knowlton
(1968) discussed a similar situation involving coyotes in Texas.
They suggested that both canine distemper and infectious canine
hepatitis are enzootic and may be important mortality factors when
compounded by other environmental conditions, such as crowding,
malnutrition and parasitism.

Endoparasite levels are high in Riding Mountain National Park
(Samuel et al., 1978). Factors which contribute to the high parasite
loads may also contribute to an elevated incidence of disease. High
wolf densities in 1975 (one per 28 km^2) and large park sizes (maximum 22 animals — observed in 1974) may have placed social stresses
on the population, possibly increasing vulnerability to potential
disease-causing agents.

Observations indicate that as pack size increases, the effectiveness of predation on larger prey species increases (Carbyn, unpub.).
Above a certain pack size, the amount of food per wolf decreases
and social strife is expected to increase, promoting emigration of
some pack members (Zimen, 1976). However, if pack size is reduced by disease, then emigration of pack members should also decrease as social strife is reduced. However, in three of the five reported cases, the diseased animals appeared to have been separated
from the pack and may have been forced to remain away by their
packs. Animals D4 and D5 (littermates) were separated from their
pack and from each other whenever contacted from the air prior to
their deaths. There is evidence that wolf D3 wandered alone before
being killed. D1 and D2 were littermates and part of a pack in which
no unusual behavioral aberrations were noted. Although diseased
wolves may increase the number of lone wolves in a population, the
effect should be short-lived.

Mange was not detected in Riding Mountain National Park during the period of study, nor is there any historical evidence that it
has ever occurred in the canids of the area. This is curious as the
ectoparasite is widespread in western Canada and has been reported

in wolf populations of Alberta and Saskatchewan (Todd et al., in press). In the areas where it occurs, it may be an important factor in wolf mortality either directly, through exposure to cold temperatures, or indirectly, by weakening the host and increasing susceptibility to disease. Todd et al. (in press) believed that from 1972 to 1977, mange may have been responsible for extensive mortality in wolf pups in Alberta. If mange were to spread to Riding Mountain National Park, the impact of disease on the population, particularly recruitment of young, could increase and compound other disease-related influences.

The responses of neighboring packs to either the complete loss of an adjacent pack, or the disruption of the pack to a point where it ceases to function as an entity, are probably varied. They may depend, in part, on wolf and prey densities and the seasonal movements of prey. For example, when the BT pack disappeared, the WW pack extended its range to include all of the former BT pack's territory. Elk numbers in all seasons were high within the WW pack territory. Similarly, there was no shortage in available beaver colonies. Prey abundance certainly did not appear to have been an obvious factor triggering the expansion of WW pack territory. More likely, this extension was a result of pack crowding and the dissolution of the BT pack in late summer or fall.

Disease was not proven to be the direct cause for the loss of the BT pack, but it was strongly implicated by three observations. First, there were no reports of wolves killed by humans in areas adjacent to the BT pack territory. Secondly, when the BT pack was observed at its rendezvous site in early August there was noticeably less vocalization than recorded in similar situations for other packs. No howling or vocalization of any sort occurred during three days and two nights spent in close proximity to the site. Thirdly, the timing of disease-related deaths in a neighboring pack coincided with the disappearance of the BT pack, suggesting that distemper could have been affecting both packs. In the absence of other evidence, disease is strongly implicated as a contributing factor to this pack's dissolutionment.

Elsewhere, rabies disrupted a pack of 10 wolves in Alaska (Chapman, 1978). Chapman concluded that under low wolf densities, the spread of rabies from pack to pack was unlikely, but that once it occurs, it can destroy entire packs. This clearly documented incident and other reported cases of enzootic viruses in canid populations (Mongeau, 1961; Trainer and Knowlton, 1968; Choquette and Kuyt, 1974), together with observations in Riding Mountain National Park, lead me to believe that disease has not received adequate attention as a mortality factor in wolves. Previous authors have concluded that food abundance by itself, or in combination with social stress, is the main regulatory factor of wolf populations. To these, the role of disease operating under various levels of food abundance and social stress should be included. Undoubtedly, environmental conditions such as climate, soil and alternate hosts are important and will modify the impact of disease from area to area.

Acknowledgements

This study was financed by Parks Canada. I would like to thank C. Allan, B. Briscoe, W. Dolan, T. Hoggins, A. Kennedy, A. MacLean and D. Patriquin for their involvement in various phases of work. T. Anderka (CWS, Ottawa) constructed the radio-collars. Drs. W. Addison, M. Hewitt and R. Lewis necropsied the wolf carcasses and diagnosed the conditions. Dr. E. Broughton commented on the results and critically reviewed the paper. Most especially, I would like to express my thanks to T. Trottier whose dedicated commitment to field work was exemplary.

Part II

Behavior and Ecology
of Wild Wolves in Eurasia

Introduction

Wolves in North America live primarily in remote, uninhabited and often extremely cold northern portions of the continent. We have taken it for granted that all wolves live in such environments and that, given a choice, they would prefer these areas. Eurasian wolves, however, are different, maintaining a lifestyle which varies considerably from preconceived views. Although many Eurasian habitats lack large ungulates, the wolves continue to survive without them. They linger on in Spain, raid garbage dumps not too many kilometers from Rome, wander the deserts of Egypt, Iran and Israel, and survive in scattered pockets in India — all areas where we more readily picture the jackal or even the lion rather than the wolf. Perhaps only in the USSR are conditions similar to those in North America.

The study of Eurasian wolves, many of which are poised on the brink of extinction, may ultimately help ensure survival of the species. Present conditions in parts of Eurasia may foreshadow future changes in less exploited areas of Asia and North America, providing a model of how wolves can adapt to altered habitats. Understanding how they survive may provide insights into those factors which upset the balance between continued existence and extinction.

Dmitri Bibikov opens this section with a review of wolf status in the USSR. As he indicates, there are numerous parallels between Soviet populations and those in North America, as well as a number of aspects unique to Russia. With extensive wild populations, the USSR serves as a reservoir which permits wolves to survive or re-establish in other areas. One region regularly receiving the benefit of healthy Soviet populations is Scandanavia, which long ago reduced or eliminated its wolves. Erkki Pulliainen recounts the flow of

118

migrant wolves to Finland, which has occurred during the past 25 years. Anders Bjarval and Erik Isakson then describe winter and spring movements of three wolves which evidently immigrated to Sweden, via Finland, from the USSR. It is noteworthy that Finland, Norway and Sweden, with climates and habitats similar to those in British Columbia and Alaska, presently harbor only transient wolf populations.

Surprisingly, it is the southern regions of Europe and the Middle East where more permanent wolf populations are located. Luigi Boitani describes the ecology and behavior of the wolves of Italy, and outlines considerations necessary to enact an effective management strategy to ensure their survival. H. Mendelssohn follows with a summary of information available on the habitats and characteristics of wolves in Israel which, in many ways, seems to be the ecological equivalent of the North American coyote. Paul Joslin concludes the section with a description of the wolves of Iran, about which very little is known.

Wolf Ecology and Management
in the USSR[1]

D.I. Bibikov

INTRODUCTION

Diminished wolf *(Canis lupus)* habitats and numbers appear to be inevitable as human population growth and urbanization continue to transform the landscape. This process is normal and perhaps justified since, in developed areas, the wolf's natural role as predator can be assumed by man. Evidence from some European countries demonstrates that abundant and healthy ungulate populations are possible in the wolf's absence. At the same time, however, this highly intelligent, apex predator (in many and diverse natural ecosystems) has the right to exist, where possible, throughout its historic range.

Unfortunately, in North America and western Europe wolf range has been substantially reduced, and some geographical races are now extinct. In the USSR, however, the wolf is still common and widely distributed; as a species it is not endangered. In some areas, it detrimentally affects domestic and/or wild ungulate populations, making wolf control an important and urgent consideration. Because all geographical and ecological forms of the species are still represented, an enlightened stragegy of wolf management is essential. This paper considers important aspects of wolf population biology, behavioral ecology and their relationship to an effective management program.

POPULATION AND DENSITY

The 20th century has seen two peaks of wolf numbers in the USSR, both of which followed years of war and economic difficulties when wolves were not controlled. After World War I, wolf predation on domestic cattle was extreme. For example, in 1924–25 about

[1] A condensed version of this paper was published in *Natural History* 89(6):58–63 (1980).

1,000,000 head of cattle (0.5% of all cattle) were killed by wolves (Dementyev, 1933). The most severe damage occurred in the lower Volga (2.2% lost), Siberia (1.6% lost) and Kazakhstan (1.5% lost). A rapid increase in wolf numbers also occurred after World War II. At the end of the 1940s, wolf numbers were estimated at 200,000, and their depredation on cattle was again severe. Active control was initiated with annual kills of 40,000–50,000 wolves. By the early 1960s, the wolf population was reduced and the annual cull was fixed at 15,000 (Bibikov, 1980).

Between 1968 and 1972, wolf numbers were lower than recorded at any time during the last century. Their range in European USSR became quite discontinuous. In the Baltic region, for example, nearly 50 percent of the area was wolf-free, in Byelorussia and in central Russia the figure was 30 percent, and in the Ukraine, south of the RSFSR, 60–70 percent of the area contained no wolves. Northern populations were more continuous, and their densities were ex-temely low (Bibikov and Filimonov, 1974; Bibikov, 1975). Today, the areas with the highest density ($>$10 wolves/1,000 km^2) are found in Byelorussia (Brest, Gomel, Grodno, Vitebsk and Minsk regions), the Ukraine (Zakarpatie and Ivano-Frankovsk regions), the Pre-Caucasus (Daghestan and Checheno-Ingush Autonomous Republics), the Volga Region (Volgograd, Saratov, Penza regions), the Central Russian Region (Kaluga, Smolensk, Kalinin regions), the Northwest Russian Region (Novgorod, Pskov regions), and also the Urals Region (Kirov, Perm regions and Udmurtia). In 1972, the primary popula-tions, consisting of about 3,000 wolves (40% of the entire west-European plain population), were concentrated in four localities: Smolensk, Kirov-Perm, the Volga Region and the Pre-Caucasus. At the same time, about 3,500 wolves inhabited the Transcaucasus Region. The highest densities reached 20 wolves/1,000 km^2 in Georgia and Azerbaijan, and 10 wolves/1,000 km^2 in Armenia.

Primorie and western and eastern Siberia were sparsely inhabited. Areas with a mean density above 10 wolves/1,000 km^2 occurred in the south of the Altai area, Tuva, Buryatiya, Chita and Amur regions. In the tundra and forest-tundra populations, density did not exceed one wolf/1,000 km^2. The total wolf population for Siberia was about 3,000 in 1972.

In Kazakhstan, a population of nearly 30,000 animals was esti-mated in 1972, including 7,500 in mountain areas, 17,000 in semi-deserts and deserts, 4,000 in the steppe agricultural zone, and 1,500–3,000 in the remainder. High densities on the Kazakhstan plains (10–20 wolves/1,000 km^2) were sympatric with the saiga antelope *(Saiga tatarica)*.

There are no reliable data on wolf distribution and numbers from middle Asia. We estimate that 6,000–8,000 animals were present in 1972. The highest density, above 10 wolves per 1,000 km^2, was recorded in the mountain areas of Kirghizia and Tajikistan.

During the last five years (1973–78), wolf numbers have increased notably (from 50,000 to 66,000–75,000 total wolves). In the Ukraine, in Byelorussia, and in central Russia, wolf populations have nearly doubled. The apparent reasons are slackened control efforts due to reduced wolf density, and a popularization of environmental conservation. Concurrently, however, hunting agencies and societies have increased efforts at controlling wolves. Wolf bounties have been increased. At present, an adult brings a reward of 100 rubles, a female with pups 200 rubles, and a juvenile 50 rubles. Increased control efforts, including shooting from helicopters, ground vehicles and snowmobiles, currently limit wolf population growth. In 1979, more than 32,000 wolves were killed. With continued control efforts, wolf numbers will again decline. Perhaps they will drop below the 50,000 level of the early 1970s. Their range in European USSR will be further reduced, perhaps to half of the former size. However, wolf populations in the mountains of the Caucasus, Central Asia and Siberia are not likely to decrease in the near future. These areas have low human populations, are relatively inaccessible, and effective hunting techniques are not available. Conversely, wolf populations are vulnerable in open landscapes—tundra, steppes and desert—where hunters use aircraft. Data on wolf control in open landscapes indicate that two of seven subspecies in the USSR—the tundra wolf *(Canis l. albus)* and the desert wolf *(Canis l. desertorum)*—are vulnerable and could vanish.

The vulnerability of open landscape populations, especially to aerial hunting, must be considered when developing a wolf management strategy. It is important to bear in mind that complete elimination of a subspecies is not the only concern; a sharp reduction in numbers could also promote emigration to less affected populations of neighboring subspecies, resulting in the potential loss of unique genetic variation through interbreeding. Such a phenomenon occurred at the end of the 1960s when only 30 or so wolves *(C. l. albus)* survived in the tundra of European USSR. A number of these survivors migrated to the forest-tundra and taiga, the range of *C. l. lupus*, possibly to escape shooting from helicopters.

Because of the enormous area and great environmental diversity of the USSR, wolf ecology varies extensively on a geographical basis. A primary difference involves wolf population density and preferred prey. Because the wolf is typically territorial in both the Soviet Union and in North America (Mech, 1970, 1973), population density rarely exceeds 40 wolves per 1,000 km², except in some ungulate wintering yards and wildlife reserves. For example, about 100 wolves dwell in the Caucasian reserve (2,500 km²). The 15 resident packs there remain in permanent home ranges of 60–150 km². Territoriality evidently limits wolf numbers in this reserve to a maximum of 150 animals (Kudaktin, 1979).

In the past, wolves were abundant in the forest-steppe zone of central Russia, the Ukraine, Povolzhie and the western portion of Siberia. But now, due to systematic human persecution, densities of 10-20 wolves per 1,000 km^2 have only been recorded in densely wooded and sparsely populated areas, primarily along the borders between regions. In the tundra and desert, the wolf has always been scarce, with mean densities of one to two wolves per 1,000 km^2. However, in the Aktyubinsk region of Kazakhstan, where the saiga antelope have recently recovered from near extinction, wolf densities of up to 20 per 1,000 km^2 have been recorded. Up to 60 wolves per 1,000 km^2 concentrate in the wintering area of these antelope.

More detailed data are available on wolf populations in nature reserves (Bibikov and Filonov, 1980). Densities fluctuate between 10 and 40 wolves per 1,000 km^2 in winter. In recent years, a minimum of 50 wolves has been recorded in the Altai and Berezina reserves, and approximately 15 or so in the Darwin, Khopyor, Oka, Central-forest, Borzhomi, Chatkal and Sikhote-Alin reserves. The highest density of wolves is found in the Kzyl-Agach reserve, where 60-70 animals inhabit an area of less than 1,000 km^2, much of which is covered with reed marshes and shallow water. Observations in Kzyl-Agach indicate that at least six litters are produced yearly. The wolves prey on the abundant wild boar *(Sus scrofa)*. By eliminating young boars throughout the year, the wolves apparently maintain the population between 700-800 individuals before yearly recruitment.

Large portions of some pack ranges extend beyond the reserves' boundaries, where they may inflict damage on cattle or other domestic animals. Thus, some of the smaller nature reserves (Oka, Mordovian, Khopyor, Central-forest and others) may be little more than safe zones. Only in the Caucasian, Pechyora-Ilych, Altai and other reserves covering more than 1,000 km^2 do wolves typically live permanently within the reserves. However, along their edges, some wolves specialize in killing the sheep and goats which are driven to alpine pastures in summer.

The pasturing of domestic animals in alpine areas creates another problem. When the sheep and goats are driven to lower elevations in late summer from alpine pastures in the Malyi Caucasus, the wolves concentrate in the Borzhomi reserve, increasing their numbers several-fold. As a result, predation on roe deer *(Capreolus capreolus)* increases in autumn and winter. Under these conditions, it is important to control wolves which specialize in preying upon domestic animals outside the reserves.

FOOD HABITS

The wolf in the USSR is a typical euryphagous animal, feeding mainly on wild ungulates (Bibikov and Rukovskii, 1975). The tundra wolf *(C. l. albus)* feeds on wild and domestic reindeer *(Rangifer tarandus)* and, in the eastern part of its range, on bighorn sheep *(Ovis canadensis)* as well. It often is responsible for severe losses of domestic reindeer. The northern taiga wolf *(C. l. lupus)*, in addition to its typical ungulate prey, feeds on blue hare *(Lepus timidus)* in winter when ungulates are scarce. Predation on cattle is very slight. The southern taiga wolf feeds on moose *(Alces alces)* and, in eastern areas, on red deer *(Cervus elaphus)*, roe deer and musk deer *(Moschus moschiterus)*. Damage to cattle is minimal since the grazing season is short. The western forest-steppe wolf takes roe deer (the Baltic republics, the Ukraine, Moldavia), wild boar (Byelorussia), and red deer (the Voronezh region, the Ukraine) in winter. Moose are relatively unimportant. From the end of summer (August) on, the wolf preys on smaller animals and domestics. Damage to cattle is considerable because of their long grazing period. The bulk of the desert-steppe wolf's diet includes saiga antelope in Kazakhstan and Kalmykiya, roe deer and wild boar in river flood-plains, argali *(Ovis ammon)* in Ustyurt, Persian gazelle *(Gazella subgutturosa)* and also domestic ungulates. Domestic ungulates are important where pasture livestock farming is developed and wild ungulates are scarce. The mountain-forest wolf consumes red deer, wild boar, mountain goat *(Capra caucasica)* (the Caucasus), argali, roe deer and maral *(Cervus elaphus maral)* (middle Asia, Kazakhstan, mountains of southern Siberia). Domestic animals are also important in their diet.

The primary cause of the geographical variability in wolf prey is faunal composition coupled with local differences in abundance. Domestic animals are important prey wherever livestock farming is developed, and also where wild ungulate resources are small. Under such circumstances, predation on domestics occurs for significant portions of the year.

Seasonal variability in wolf feeding is characterized by a reduced proportion of ungulates in summer when small animals are abundant. The intake of carrion also increases substantially in early spring. Feeding specialization in individual packs has also been recorded. For instance, over a number of years the major diet component of a pack in the Caucasian reserve was red deer, whereas that of a neighboring pack was mountain goat and wild boar. Near the edge of the reserve, some groups specialized in killing domestic animals. Most frequently, these latter groups were nonterritorial. It appears that techniques for hunting certain prey were passed on to offspring (Kudaktin, 1979).

The abundance and wide dispersal of wild boar in central Russia in the 1970s has not yet resulted in a change in wolf feeding patterns. They still continue to consume their traditional diet—moose. Recently, however, evidence of limited predation on wild boar has been recorded for some packs.

EFFECTS OF WOLF PREDATION ON PREY POPULATION GROWTH

Studies of wolf relationships with ungulate prey have been the focus of research in the USSR. The diversity of prey species and environmental conditions over vast areas has led to highly contradictory inferences. For example, the wolf's role in culling physically inferior prey is definitely not appreciated in areas where much predation on domestic animals occurs. In these areas, many game management specialists regard the wolf as harmful (Eliseev et al., 1973). A number of publications are available documenting reductions of red deer and roe deer populations by wolves, specifically in areas where these prey have been introduced and are fed by man in winter. Ungulates under such abnormal conditions are more vulnerable to wolves. The selective nature of wolf predation, well known and important in the wild, is lost under such conditions.

Pimlott (1967, 1970), Mech (1970) and other North American zoologists have presented evidence that wolves can control and limit ungulate numbers where ungulate densities are not high. When the predator-prey ratio is more than one wolf/30 moose or 100 deer (Odocoileus virginianus), and ungulate forage is abundant, wolves are not able to limit ungulate population growth. Similar evidence for certain regions has been collected by Soviet scientists, although the traditional view of the wolf's capacity to reduce ungulate populations, irrespective of their numbers, still dominates among game management specialists. Comparing wolf numbers with saiga, moose and wild boar populations over the past 10 years, we have concluded (Bibikov, 1980) that an unprecedented increase in these ungulate populations occurred in the 1940s through 1960s when wolves were abundant in the country. From the end of the 1960s through the early 1970s when wolf numbers were minimal, ungulate populations began to decrease (moose followed by wild boar and saiga). A variety of evidence indicates that human activity is the primary cause of ungulate population changes, although the wolf may exert deleterious effects at various stages and in certain regions. These effects of ungulate control by wolves are well known but have not, unfortunately, been studied in detail.

Analysis of many years of data concerning the influence of wolves on ungulates in nature reserves showed that damage to the ungulate populations was minimal (Bibikov and Filonov, 1980; Filonov, 1980) (Table 2.1).

Natural mortality removed an average of two to seven percent of total ungulate numbers (except in the Bashkir and Caucasian reserves). Approximately half of the natural mortality in the reserves was due to wolves. In some cases, they were believed responsible for nearly 75 percent of such mortality. The figures for ungulates killed by wolves are inexact, however, due to the difficulties in discovering prey remains, especially of young animals. In addition, wolf control efforts varied throughout the period; when wolf numbers declined, predation pressure on ungulates decreased thus influencing the relative effect of wolves versus other natural causes. For example, when the wolf was abundant, the roe deer population in the Ilmen reserve

Table 2.1: Percent of Natural Mortality Caused by Wolves to Ungulates in the Reserves of the RSFSR.
(These were determined from recovered carcasses.)

Reserves	Years of Observation	Prey Species	Average No. Prey	Average No. Wolf	Total No. of Dead Animals	Killed by Wolf Absol.	Killed by Wolf %	Annual (mean) Loss of Population (%) Total	Annual (mean) Portion Caused by Wolf
Darwin	1948–1972	Moose	340	12	467	285	61	5.5	3.2
Oka	1944–1972	Moose	275	12*	147	67	46	1.9	0.8
	1938–1971	Sika deer	60		128	62	48	6.7	3.3
Mordovian	1943–1972	Moose	278		104	7	7	1.2	0.1
	1938–1972	Sika deer	180	6*	347	92	26	5.5	2.0
	1945–1972	Maral	57		29	10	30	1.8	0.6
Voronezh	1933–1952	Red deer	230	10*	204	52	25	1.3	0.6
	1955–1960	Red deer	610		305	14	5	2.5	0.1
Caucasian	1936–1968	Red deer	3,000	50–60	280	128	46	0.3	0.1
Bashkir	1942–1970	Maral	160	12–18	35	11	31	0.7	0.2
(Uzyansky area)	1946–1971	Moose	310		20	10	50	0.2	0.1
	1939–1971	Roe deer	50		58	41	71	3.5	2.5
Ilmen	1942–1970	Roe deer	720	11–15	1,260	937	74	6.3	5.0
	1942–1968	Sika deer	38		30	22	73	2.8	2.0

*According to the number of killed animals. The total number of wolves is unknown but data on killed predators are available. Up to the mid-1960s, wolf control was very intensive. Thus, these figures are similar to actual average wolf numbers in the reserves.

lost an average of 13 percent, while the moose population in the Darwin reserve lost approximately 12 percent. Similarly, in the Mordovian and Oka reserves, predators eliminated four to six percent of the moose populations, and eight to thirteen percent of the sika deer populations. Thus, wolves take about 20-25 percent of potential ungulate increase, or about half of the actual increase observed per annum. In reserves with several prey, wolf predation pressure varies among them, being focused on the most numerous and/or the easiest to attack.

SELECTIVITY OF WOLF PREDATION

The beneficial effects of wolf predation on ungulate populations, demonstrated in several North American studies (summarized by Mech, 1970), have also been shown by Soviet studies in areas where man's influence is minimal. Studies in the Crimean reserve, for example, have documented a decline in health of red deer in an over-populated area free of wolf predation (Kostin, 1970). Similarly, a sharp reduction of wolf numbers was followed by a substantial increase of sick caribou noted on the Taimyr (Michurin, 1970). At the other extreme, Kaletskaya (1973) provided evidence of a healthy, increasing moose population in the Darwin reserve, despite an annual loss of about 50 animals to wolves. Later, following the introduction of wolf control, the Darwin moose herd was reduced by skin disease.

Observations in the Caucasian reserve (Golgofskaya et al., 1979) revealed population growth in red deer, wild boar and mountain goat during a period of intensive wolf control which reduced wolf numbers to between 25 and 30 by 1960. Ungulates eventually exceeded 30,000 and destroyed winter feeding yards. Wolf control was relaxed between 1965 and 1971 and, subsequently, the population increased. Ungulate populations began to decline. From 1971 to 1975, the wolf-ungulate ratio stabilized at 1:300, or 21,800 kg of prey per wolf. Recently, both food resources and the physical condition of the deer have improved. In the area under study, deer six years of age and older (59.2%) and fawns (21.1%) were preferentially killed by wolves, as similarly found in North America (Mech, 1970). Animals taken by hunters were predominantly two to five years of age (79.3%) (Kudaktin, 1978), indicating that wolves have a less deleterious effect on ungulate reproduction than humans, who eliminate the most productive animals.

Selectivity by wolves in killing of deer is evident from their hunting behavior, which was studied by Kudaktin (1978) in the Caucasian reserve. The hunting episodes were reconstructed through tracks. It was found that wolves attacked deer at distances between 10.4 and 200 m, conforming to observations of wolves hunting moose and white-tailed deer in North America (Mech, 1966, 1970) where attacks are attempted following an approach toward the prey.

Wolves often gave up the pursuit if the prey was not captured within 100-200 m. Of the 72 unsuccessful hunts recorded, only 12 were pressed for three to four km. Mountain relief with numerous deep gorges, talus slopes and rocks makes such long pursuits difficult.

The flight of the prey seems to impel the wolf to chase it. When the prey stands its ground as the wolf approaches, the latter usually withdraws. If the prey is located higher on a slope than the wolves, they usually do not attack (n = 16). For instance, on 11 January 1975, an adult female wolf climbing a mountain slope encountered six red deer. The distance between wolf and deer did not exceed 15 m (as interpreted from tracks). However, no attempt to attack the running prey was made. On 24 April 1975, four wolves encountered two deer groups of two to three individuals each. The deer, located 20-25 m higher on the slope, fled upward and the wolves did not follow them. The wolves evidently were hungry, though, for 900 m away they killed a wild boar and ate much of it. Wolves often stopped hunting after the first deer fled upward. The same behavior pattern has been described by Murie (1944) in wolves hunting bighorn sheep *(Ovis canadensis)* and caribou *(Rangifer tarandus)* in Alaska.

The dependence of wolf hunting behavior on the relative position of wolf and prey on mountain slopes is understandable. For prey, the probability of colliding with trees and shrubs and injuring extremities on stony talus increases when animals flee downward in a panic. For example, a deer killed by wolves on 21 May 1974 had dislocated its right foreleg when fleeing downward. A female deer attacked by wolves on 23 September 1973 had bruises on the breast and right shoulder measuring 6 x 3 cm, resulting from collisions with trees and stones. Another female deer, killed 12 February 1972, had fractured legs.

The above observations indicate that wolves have a definite advantage if they pursue prey fleeing downward. When a prey flees upward, the outcome is determined by the strength and endurance of the participants. We speculate that wolves may appraise the endurance of a potential prey and its capacity for defense by whether it flees upward or downward. A healthy and strong animal is capable of fleeing upward, whereas a sick or physically impaired individual is less capable of fleeing up a slope and therefore may be more likely to run downward.

Winter tracking data and evidence from deer hunts during other seasons showed that most kills were made in specific locations. These include stony talus deposits, creeks at the bottom of densely-wooded gorges, narrow passages through rocks, etc. It is possible that prey fleeing under such unfavorable conditions cannot maneuver as successfully, lose speed, and sometimes may be badly injured.

Cases of wolves driving ungulates onto frozen lakes and rivers, into ravines with deep snow, and into other unfavorable places have been described by Formozov (1946) and Sludskii (1962, 1970), as well as by Pimlott et al. (1969) in North America.

The sites of wolf kills have been called "wolf corrals" by Kudaktin (1978). Regular use of such corrals can reveal pack adaptations to peculiar ecological features. In addition to fresh prey remnants, bones of prey turned green and partly decayed with age are found in such places. As a rule, "wolf corrals" are situated near areas of prey concentration, exposed alkali soils, water sources, river fords or regularly used passages through gorges, where numerous remnants are found in a very small area. For example, the remains of 36 prey were found in an area 2 km long and 50-70 m wide in the flood plains of the Alous River. One area of 0.5 ha contained 20 deer killed by wolves over a five year period.

Wolves hunting saiga antelope exhibit a pronounced prey selectivity. During the saiga rut, for instance, wolves most frequently take weakened and exhausted males. Similarly, during fall saiga migrations, wolves follow small antelope groups that lag behind the major herds of thousands. In these small groups, the proportion of inferior and wounded individuals is high (up to 5%). Creating test situations by pursuing these small groups allows the wolves to overtake the inferior individuals lagging behind. Saiga carcasses, common on migration pathways and at human hunting sites, are another source of easy food.

Wolves are not always as selective, however. For instance, when wolves take pregnant saiga and newborn young, it appears to be a matter of chance which will fall prey. However, during calving, the death rate of young saigas is negligible because of their large concentrations. It appears that the concentration of many thousands of calving females in a territory of a single wolf pack is an adaptation to minimize wolf predation on young saigas (Sludskii, 1962). More wolves could probably take prey if saiga females were less concentrated when calving. According to Filimonov (1979), the death rate of newborn saiga approaches 33 percent in the small groups which lag behind the main herds. Such predation of young, because of its rather indiscrimminate nature, may not provide positive (directional) selective pressure, especially for the smaller saiga groups.

Wolf predation on saiga also appears to be nonselective at river crossings. The antelope, apparently aware of the hazards, approach water with great caution, avoiding reeds and other cover. If wolves do attack saigas during a river crossing, losses are typically high: dozens of both sexes are commonly killed.

Other situations occur where wolves kill quite healthy ungulates. For example, deep snow with a frozen crust can handicap deer and moose escape, as described by Mech (1970) and Peterson (1977). Similar nonselective predation occurs with wild boar and Caspian seals *(Phoca caspica)* since wolves largely kill the young. Thus, the wolf-ungulate relationship is a very complicated one; no unambiguous solutions exist. Initial conclusions are often substantially altered as more data are accumulated over many years.

INFLUENCE OF WOLF PREDATION ON UNGULATE REPRODUCTION

The wolf's influence on ungulates appears to go beyond its direct impact on numbers and the structure of populations; wolves do more than simply eliminate the weakened and physically (and perhaps ecologically) maladapted individuals. In some cases, wolves can stimulate ungulate reproduction. Mech (1970) noted that after wolf introduction to Isle Royale, the proportion of moose twins increased from 6 to 38 percent. In order to reveal similar wolf influences on ungulate reproduction, we compared the indices of wolf-prey fecundity in the nature reserves for two distinct periods: the 1940s through the 1950s, when wolf numbers were comparatively high; and the late 1950s to the middle 1960s, when wolf numbers were low (Table 2.2).

Table 2.2: Wolf Numbers and Fecundity Indices of Ungulates in Some Reserves of the RSFSR (from Bibikov and Filonov, 1980).

Reserves	Prey Species	Indices of Ungulate Fecundity* Average No. of Calves/100 Females Having Calves	Females with Twins (%)
Darwin	Moose	135	45
	Moose	125	24
Mordovian	Moose	149	41
	Moose	139	32
Oka	Moose	160	57
	Moose	152	48
Ilmen	Roe deer	206	53
	Roe deer	194	51
Caucasian	Red deer	23**	***
	Red deer	14	—

*Upper figures = period of high wolf numbers; lower figures = period of low wolf numbers.

**There are no data on 100 females having calves; only data on 100 females (with or without calves) are available.

***During the years of high wolf numbers, the rate of females with calves reached 32%; during the years of low numbers, about 21%.

The rise in wolf numbers coincided with good feeding conditions for the ungulates. Thus, high ungulate productivity may be a direct result of increased availability of food. During this early period, despite high wolf numbers, the rate of ungulate population growth was high. Regular wolf control in all the reserves led to even higher ungulate population densities which resulted in exhaustion of feeding resources and intensification of competition. Thus, low productivity may be explained by lowered habitat quality.

If wolf predation is the primary cause of ungulate productivity change, feeding conditions should have little effect. At the beginning of the 1970s, wolf numbers began to increase in some reserves. Ungulate populations were declining because of unstable feeding resources and increasing predator pressure. In the Darwin reserve, the number of single calves per 100 females increased five percent while the twinning rate increased 33 percent. In the Mordovian reserve, where wolves were still scarce, these two ungulate fecundity indices increased 2 and 25 percent, respectively.

Thus, it appears that wolf predation affects ungulate reproduction through a change in relative representation of female age groups. Heavy predation pressure eliminates mainly young, old, sick and defective individuals who either do not reproduce, or reproduce at lower levels. A change in population structure among females results in an increase in productivity if mature and efficient animals have greater survivorship. However, more studies on the relationship between wolf predation and ungulate productivity are needed.

Survival of calves was also affected. During periods of low wolf numbers, the survival of young moose in the Darwin reserve was greater through April, as compared to periods when wolves were abundant. In the Oka and Mordovian reserves, moose calf survival was 16 and 50 percent greater when wolf density was low. Roe deer calf survival in the Illmen reserve was 60 percent higher through the first year when wolves were scarce. Partial restoration of wolf numbers in the Darwin reserve increased moose calf mortality.

The above information indicates that intensive wolf control, which substantially reduces wolf numbers, can promote a number of changes in species ecologically associated with the wolf. In short, reduction or elimination of wolves in reserves promoted growth in ungulate populations which eventually led to undesirable effects on forest vegetation and, in turn, required ungulate control and other human interventions. Wolf elimination resulted in disruption of the biologically-regulated system characteristic of the nature reserves. Therefore, in areas large enough to permit self-regulative ecosystems to function, predators should be preserved in densities which can be maintained through natural processes.

INFLUENCE OF HUMANS ON WOLVES

Changes in wolf ecology and behavior caused by human activities, particularly hunting, have not yet been considered. These are of great interest, especially regarding the wolf's adaptation to these newer conditions.

Although the wolf is now most abundant in sparsely populated areas, its current distribution was caused by human activities which displaced it from densely populated areas. At the beginning of this century, the wolf's optimum range was in the forest-steppe zone where it preyed mostly on domestic animals (wild ungulates having been nearly exterminated). As human population increased in the forest-steppe of European USSR, wolves were driven to less densely populated areas. This expansion in the wolf's range, following human distributional patterns, can best be traced in the plain-taiga zone, which is normally unfavorable for wolves because of its deep snow cover in winter. The expansion of wolf range in the plain-taiga of northern European USSR and west Siberia was fostered by the following: an increase in road networks due to logging, new settlements, and better feeding conditions at the expense of wild and domestic animals. The wolf colonized the arid deserts of Turkmenia and Uzbekistan because of irrigation and the formation of water reservoirs near artesian wells.

Although wolves are generally wary of humans, the opposite behavioral pattern has also been observed. Not infrequently, wolves have adapted to living near man, feeding at dumps and livestock burial sites. Similar patterns have been found in wolves in Italy (Boitani, this volume), Israel (Mendelssohn, this volume) and other areas.

When a local population of wolves is nearly eliminated, the most experienced and careful individuals probably survive. Similar characteristics are acquired by their offspring. In addition to some large-scale migrations and the partial breakdown of pack stability in some areas, a substantial change in the pattern of using killed prey has emerged in recent years. Previously, a pack remained near the prey carcass for several days; now, wolves confine themselves to a single feeding, after which they travel long distances. This behavior precludes the use of bait to keep wolves at a site (e.g., for better hunting).

In regions where populations have been severely reduced through extensive control, another situation develops. Unable to find a mate, lone wolves (both males and females), may mate with domestic dogs, producing hybrid populations.[1] After World War II, wolf-dog crosses were recorded in 14 regions—once in Latvia, Moldavia, Turkmenia and Uzbekistan, in the Ukraine (eight regions), and in Kazakhstan (in many places). Hybrid ranges are localized but are large, and sometimes exist for quite a long time. The hybrids either live with wolves or independently of them; they seldom remain with wild dogs. In all cases, they occupy the wolf's ecological niche.

First and second generation hybrids vary in appearance—coloration, structure of fur cover, size and bodily proportions. This variability is probably due to high genotypic variation (Ryabov, 1973). However, such hybrids, especially the offspring of female wolves, probably resemble wolves because of the dominance of the wolf's traits. Wolves usually cross with mongrel dogs and their offspring often inherit physical defects: small teeth, weak extremities, irregular body proportion and abnormal bite. After subsequent crosses (particularly to wolves if wolf numbers increase), these dog peculiarities disappear and the hybrids show greater resemblance to wolves.

---

[1] The vacant ecological niche of the wolf is also filled by feral dogs. The problem of their control is urgent in the U.S. and in a number of European countries, and is of increasing importance in the USSR (McKnight, 1964; Ryabov, 1979; and others).

In the southern Urals, wolf-dog hybrids prey on roe deer. In contrast to wolves, they form large packs in summer, sometimes vocalize while they pursue prey (like dogs), and sometimes chase roe deer up to four km. Individual roles during hunting are differentiated in packs of hybrids, as they are in wolves. However, positive selective effects of hybrid predation on prey populations does not seem to take place. In the Urals, predation by hybrids between 1971 and 1976 reduced the roe deer population four-fold (from 300 to 75 animals) in an area of 100 km^2 (Danilkin, 1979).

In the Voronezh and Belgorod regions, dogs mated with wolves feeding on carrion. The first pairs were observed during the winter breeding season near livestock burial sites. These hybrids were typically bolder predators than wolves. They fed on carrion, but often killed sheep, pigs, goats, geese and dogs. For some, dogs seemed to be their preferred food. Similar behavioral patterns have been observed in hybrids of the Buryat, USSR.

Behavior of hybrids is quite variable (Ryabov, 1973). Being less afraid of man and sometimes more aggressive than wolves, they appear in daylight near populated settlements where they dig dens or rest in buildings. According to Ryabov (1978), a hybrid litter usually consists of six to seven pups. Maturity occurs at ages typical of wolves if a female wolf mates with a dog. However, when the opposite occurs, maturity occurs sooner. Hybrid females of such crosses may mate at eight and a half months of age, making them similar to dogs; wild wolves typically mature at 22 months of age (Mech, 1970), although exceptions have been noted (Medjo and Mech, 1976).

In recent years, dogs have apparently been displaced from many woodlands as wolf populations have recovered. Under wolf pressure, wild dogs disappeared completely from a number of regions in central Russia. Thus, wolf-dog hybrids, and stray and wild dogs will occupy woodlands as wolf numbers decrease. Their presence depends upon human activities, and can cause great damage to domestic and wild ungulate populations. Their persistence and spread is harmful for economic, scientific and aesthetic reasons.

SUMMARY

The problem of wolf management is thus rather complicated. Its solution requires intensive and extensive studies of wolf ecology, including the roles of hybrid and feral dog populations. In recent years, ecological studies have intensified in the USSR, forming a basis for wolf management measures. These reflect two different concerns: the economic impact of wolves on domestic animals and their rightful role in wild areas. Regional programs of wolf control will reduce the wolf's range in the USSR and will decrease its numbers where necessary. But these programs will also preserve the diversity of ecological and geographical populations, as well as the uniqueness of the wolf's genetic and behavioral adaptations.

Behavior and Structure of an Expanding Wolf Population in Karelia, Northern Europe

Erkki Pulliainen

INTRODUCTION

Karelia is a province of the northwest Soviet Union between the White Sea and the Lakes Onega and Ladoga, and extending into Finnish territory in the west (Figure 3.1).

Figure 3.1: Study areas in northern Europe.

134

This area of typical taiga with coniferous and mixed forests, mires and lakes forms a faunal corridor for terrestrial animals dispersing from the east to the Scandinavian Peninsula and vice versa (Pulliainen, in press a). One such animal is the wolf (*Canis lupus l.*), which was exterminated over the majority of the Scandinavian Peninsula and Finland in the 19th century (Palmen, 1913; Johnsen, 1929), but survived in eastern Europe (Kauri, 1957).

Danilov et al. (1978) report that the wolf occurred only occasionally in northern Soviet Karelia prior to 1940 (Marvin, 1959), inhabiting cultivated areas and the coasts of the White Sea. Virtually no wolves occurred in the tiaga. Since World War II, two notable increases in the wolf population have taken place (Pulliainen, 1965, in press b, d; Danilov et al., 1978). This paper will review the behavior and structure of the expanding wolf population and analyze the factors which have contributed to the increase.

METHODS

The frontier between Finland and the U.S.S.R. runs through the western part of Karelia in an approximately north-south direction. The recording of crossings of this frontier by wolves might thus offer information on the changes and dispersal of the population in the inner part of Karelia. Since 1968, the Finnish Border Patrol Establishment has recorded on their daily patrols every crossing by large predators and has estimated their numbers three times a year. Similar surveys have been done on the Norwegian and Swedish borders. Data for the years 1968-1980 will be presented here. Earlier data were published by the author in 1965.

In Soviet Karelia, the Russians monitor the populations of large predators by skiing along observation lines in different parts of the country and counting the numbers of tracks crossing the lines (Danilov et al., 1978). The results are than expressed as the numbers of tracks per 10 km of observation line. The resulting wolf data for the years 1961-1976 have been published by Danilov et al. (1978).

It is naturally worth comparing the Finnish data with the corresponding Soviet Karelian data in order to test the reliability of the two methods used. Figure 3.2 shows the total numbers of crossings of the frontier made by wolves in Finnish Northern Karelia and the numbers of tracks/10 km in the corresponding Soviet Karelian area in 1968-1976. Increasing trends are clearly visible in both sets of data, and can be explained in terms of an increase in the numbers of wolves in Soviet Karelia from 1971 onward, leading to an expansion into Finnish territory after 1974.

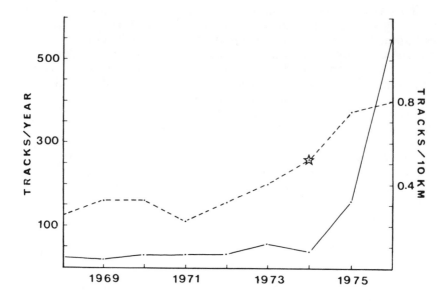

Figure 3.2: The total numbers of crossings of the frontier of Finnish Northern Karelia (left scale; solid line) and the mean numbers of crossings of observation lines by wolves in the adjacent Soviet Karelia (right scale; dotted line) (Danilov et al., 1978). A star indicates the saturation point of the wolf population of this part of Soviet Karelia (see text).

The crossing data can be used to follow relative changes in the movement of wolves (Pulliainen, 1979a), but the numbers of crossings do not express the exact numbers of animals involved, as one individual may cross the frontier several times in one year. Also, it is not reasonable to calculate rates of immigration or emigration from these figures, as only few observations can be made during the snowless season, and it is possible that wolves may move into the U.S.S.R. to breed at that time of the year.

RESULTS

Population Changes

Discussing the wolf of Soviet Karelia in the 1950s, Marvin (1959) reports that:

> It was common in the southern and central parts
> of the area, rare in the northeastern district (Segeza)
> and absent in the northern districts (Belomorsk, Kemi,
> Kalevela and Louhi), although it might sporadically
> at intervals of 5–15 years visit the latter areas.

Between 1959 and 1963, the southern half of Finnish Northern Karelia and the adjacent area in the south experienced an expansion of wolves from the east, 1961 being the peak year (Pulliainen, 1965). The data of Danilov et al. (1978) show that from 1961 to 1966 the wolf population of southern Karelia showed a decreasing trend, while there was an increase in the central part of Soviet Karelia from 1961 to 1965, and then a sharp decrease (Figure 3.3). The Kuhmo-Suomussalmi district in Finland received wolves from this increase in the early 1960s (Pulliainen, 1965:220-221). Intensive hunting contributed a great deal to the decrease in the numbers of wolves in both Soviet Karelia and the adjacent Finnish territory until 1966 (Pulliainen, 1965; Danilov et al., 1978). Numbers of wolves were relatively low in Soviet Karelia (Danilov et al., 1978) (Figure 3.3), and especially in adjacent Finnish territory (Pulliainen, 1973) in the late 1960s.

During 1971-1976, an increase was recorded in the wolf populations of the southern, central and northern parts of Soviet Karelia, the highest density being obtained in the southern part (Danilov et al., 1978) (Figure 3.3). These authors emphasize the threefold increase in the Karelian wolf population from the years 1966-1969 (0.2 tracks/10 km) to 1973-1976 (0.7 tracks/10 km). The highest density was recorded in the areas adjacent to Finnish Northern Karelia and Kuhmo and in the southeastern corner, east of Lake Onega, while there were still very few wolves in the northeastern part of Soviet Karelia (Danilov et al., 1978) (Figure 3.4).

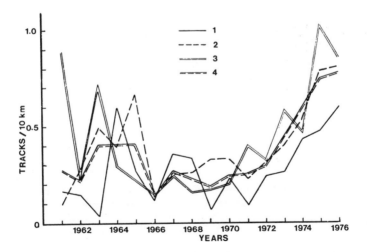

Figure 3.3: The mean numbers of crossings of observation lines by wolves in the different parts of Soviet Karelia in the years 1961-1976 (Danilov et al., 1978). 1 = northern Soviet Karelia; 2 = central Soviet Karelia; 3 = southern Soviet Karelia; 4 = whole Soviet Karelia.

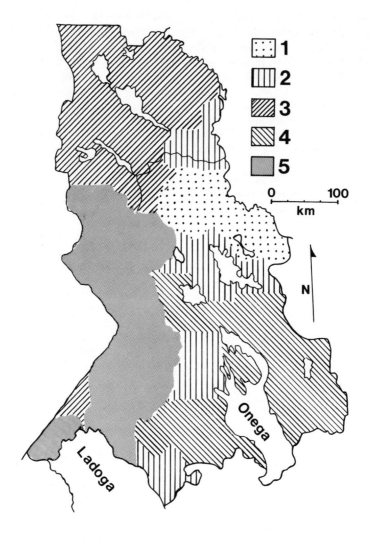

Figure 3.4: Numbers of crossings of observation lines by wolves in Soviet Karelia in winter (Danilov et al., 1978). 1 = 0.19 or less; 2 = 0.20-0.29; 3 = 0.30-0.39; 4 = 0.40-0.49; 5 = 0.50 or more tracks per 10 km of observation line.

Of the 4,656 crossings of the frontier by wolves recorded by the Finnish Border Patrol Establishment in the years 1968-1979, 4,640 took place on the frontier between Finland and the U.S.S.R., 14 on the Norwegian border, and two on the Swedish border. There was a steep increase in the total number of crossings from 1974 to 1977 and a subsequent decrease to 1979 (Figure 3.5). These crossing data and other observations indicate that 1977 may represent the peak year for this expansion of wolves from the east.

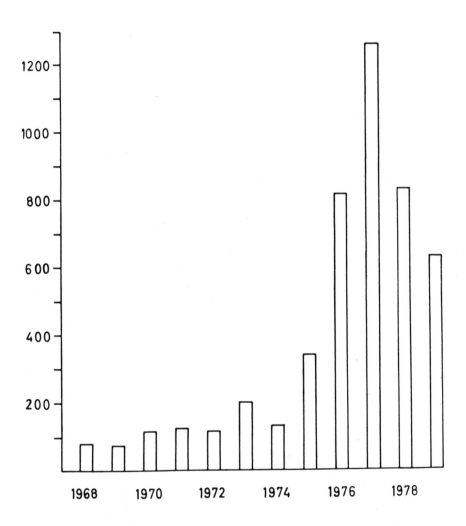

Figure 3.5: The total number of crossings of the Finnish border by wolves recorded by the Finnish Border Patrol Establishment in the years 1968–1979 (see also text).

Figure 3.6 shows that in 1975–1979, the movements of wolves (numbers of crossings of the frontier calculated per 100 km of border) were most pronounced on the frontier of Finnish Northern Karelia where the peak year was recorded in 1977, while in Kuhmo, the adjacent commune to the north, the expansion was still continuing in 1979 (approximately 25 wolves were moving about in the area on 1 January 1980).

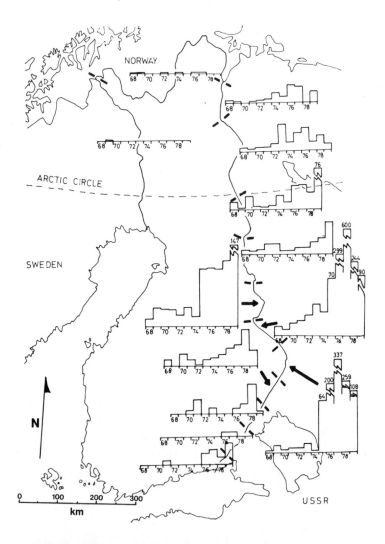

Figure 3.6: The numbers (calculated per 100 km of frontier) of crossings of the different parts of the Finnish frontier by wolves recorded by the Finnish Border Patrol Establishment in the years 1968–1979.

The Finnish border patrols estimated the numbers of wolves within a zone at least 20 km wide adjacent to the frontier three times a year (1 January, 1 June, 1 October). The most reliable results are obtained in January when there is snow on the ground. These show the majority of wolves occurred in the frontier between Finland and the U.S.S.R. (Figure 3.7), the numbers varying between 6-24 in 1969-1975, but increasing thereafter from 1976 to 1978. The total figure reported for January 1978 was between 77-89, but in 1979 and 1980 it was again smaller (between 63-70 and 67-75). The largest packs in the vicinity of the frontier during both expansions consisted of approximately 10 individuals (Pulliainen, 1965, in press d).

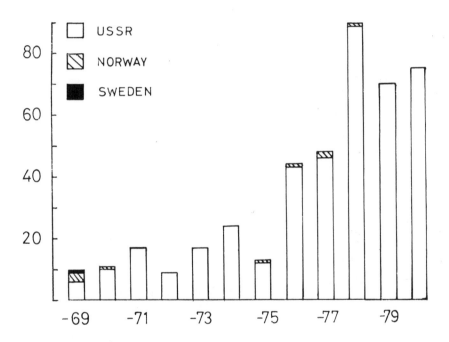

Figure 3.7: The numbers of wolves in the vicinity of the Finnish frontier on the first of January in 1969-1980 according to the estimations of the Finnish Border Patrol Establishment (see also text).

Structure of the Populations

As far as is known, there are no studies on the age and sex ratios of the Soviet Karelian wolf populations, but data do exist on the sexes and, in some cases, the age class of wolves killed or found dead in Finland in 1948-1963 (Pulliainen, 1965) and 1969-1980 (Pulliainen, in press b, d).

The sexing of 154 wolves killed or found dead in Finland in 1969-1980 showed 64.3 percent to be males, a disparity which is statistically highly significant (X^2 = 12.6, p<0.01). In 1969-1975, 38 males and 10 females were studied, the proportion of males being almost 80 percent. The 106 wolves examined in 1976-1980 correspondingly included 61 males (57.5%), still a slight, although statistically nonsignificant, excess of males.

Table 3.1 shows the sex ratios for wolves killed in different parts of Finland (Figure 3.1) in 1948-1963 and 1969-1980. The wolf is known to have bred in the north of Finland (Lapland) in 1948-1963, and at that time there were nine females among the 25 dead individuals. In 1969-1980, almost all the wolves moving in this area were killed by reindeer-owners, and seven of the 10 individuals sexed were males.

Table 3.1: Sex Ratios of Wolves Killed in Finland in 1948-1963*
and 1969-1980

Area	. .1948-1963. . % Males	n	. 1969-1980 . % Males	n
Lapland	64.0	25	70.0	10
Savukoski-Kuhmo District	72.3	47	63.9	72
Northern Karelia	46.9	49	57.8	45
Other parts of Finland	88.9	18	74.1	27

*Pulliainen, 1965

The expansion of the wolf from the east, which reached Finnish Northern Karelia and the southern part of the Savukoski-Kuhmo district in 1959-1963, showed an even sex ratio in the former area but an excess of males in the latter. The Savukoski-Kuhmo district also had a statistically significant excess of males in 1969-1980, while there were five males and one female among the dead wolves encountered in Finnish Northern Karelia in 1972-1975, but a corresponding ratio of 21:18 in 1976-1980. A similar excess of males among the dead wolves is reported in 1948-1963 and 1969-1980 in the other parts of Finland where wandering wolves occurred.

Thirty wolf pups between four and 10 months of age were killed in Finland in 1969-1979, with a slight but statistically nonsignificant excess of males (60%). Altogether, 19 or 20 litters were represented, with pups of 12 to 13 litters taken from communes adjacent to Soviet Karelia.

DISCUSSION

The wolf has a much lighter foot loading (weight-load-on-track 89-114 g/cm^2) (Nasimovich, 1955) than the moose (*Alces alces*) (368-1,204 g/cm^2) (Nasimovich, 1955; Kelsall, 1969; Peterson,

1977), which is its main prey species in eastern Fennoscandia outside the reindeer husbandry area (Pulliainen, 1979b, in press b, c). However, its chest height is so low (approximately 40 cm) that it has difficulty moving in soft snow deeper than 50 cm. Since the old, mature coniferous forests of the northern taiga do not offer enough food for moose populations in winter, it is understandable that wolves have been absent from these habitats in Karelia (Marvin, 1959). This avoidance of old, mature coniferous forests with deep snow cover in the northern taiga has also meant that at times the wolves have failed to occur in areas where there are brown bears (*Ursus arctos*), which spend the wintertime in dormancy, or solitary wolverines (*Gulo gulo*), which have an even lower weight-load-on-track (53 g/cm^2) (Gill, 1978) than wolves. This fact has been poorly appreciated by biologists doing research in mountain areas (Bjarvall and Isakson, this volume) or other environments (e.g. Mech, 1970, 283).

The abundance of wolves in Karelia since the Second World War is, in many respects, a consequence of human impact. An intensive program of clear-cutting in the vast areas of coniferous forest in Soviet Karelia was commenced in the late 1940s, and the conifers were replaced with deciduous trees which offered food for moose populations enabling them to increase markedly (Danilov et al., 1978). After World War II, Finland ceded large areas of Karelia to the U.S.S.R. and most of this land remained neglected. Fields and meadows returned to forest and again provided very suitable environments for the moose and other game (Pulliainen, 1965). At the same time, reindeer husbandry was discontinued in Karelia in the 1950s and the semi-domestic reindeer returned to a wild state, while the forest reindeer (*Rangifer tarandus fennicus*) were no longer hunted (Danilov, 1979). Thus, there was an abundance of food for the wolves, which could use the forest roads, trails of ungulates, etc. when moving from one place to another (Danilov et al., 1978). Thus, the wolf population could expand to the north both in Karelia and in the adjacent Arkangelsk area (Semenov, 1976).

The wolf population increases in Soviet Karelia in the 1950s and 1970s were rapid. In the latter case, the population tripled in less than a decade. This was due to the improved food situation and reduced control during the years when small numbers of wolves were recorded (Danilov et al., 1978). Since Bibikov (1973) mentions that there were 300 wolves in Soviet Karelia in the early 1970s, one might postulate that there may have been a maximum of 800-900 wolves in 1976. In areas where there has been little human impact on wolves, e.g. on Isle Royale, such sharp increases do not seem to occur (Mech, 1970; Peterson, 1977; Allen, 1979).

Wolves from a saturated population disperse in directions where there are few barriers and suitable vacant habitat. In the case of Soviet Karelia, the latter are found in Finland, part of the wolves' former range, because their present range is bordered by the sea in the northeast and east, and a dense wolf population in the southeast and south (Pulliainen, in press d). Bibikov (1973) estimated that the 300 wolves in Soviet Karelia in the early 1970s represented a density of 2.5 wolves per 1,000 km^2. Since expansion could indicate a saturated population and population pressure, recent observations on the increase in the Soviet Karelian wolf population and the commencement of an expansion into Finnish Northern Karelia allow us to estimate (Figure 3.2) that the saturation density of the wolf population under present conditions in Soviet Karelia must be roughly five to seven wolves per 1,000 km^2. Higher densities (10 or more/1,000 km^2) are reached in the wolf populations of the more southerly regions of the European part of the U.S.S.R. (Bibikov, 1973).

Concerning the behavior of a saturated wolf population under present-day conditions in Soviet Karelia, one might speculate that the wolves use of forest roads and the trails of ungulates as pathways in deep, soft snow would lead to increased numbers of encounters between wolf units (single wolves and packs), and thus to increased movement which may be manifested both in an increased number of crossings of the observation line (e.g. the frontier between Finland and the U.S.S.R.) and also in a considerable tendency for dispersal. These aspects have been taken into account in this estimation of the size and saturation point of the Soviet Karelian wolf population.

The majority of the wolves which crossed into Finland from Soviet Karelia in 1959-1963 were killed, and expansion was thus blocked (Pulliainen, 1965). Before and during that expansion, it was found that most of the wandering wolves were males (Pulliainen, 1965), but as the breeding population approached the frontier, the excess of males decreased. The same trend in sex ratios has also been recorded during the recent expansion from Soviet Karelia into Finnish Northern Karelia (Table 3.1). A similar blocking of the expansion is in progress, for at least 104 wolves have been killed in Finland from 1977 to 1980. According to official statistics, 151 wolves were killed in Soviet Karelia in 1978.

The experiences obtained on the expansion of the wolf population in Soviet Karelia since the Second World War are a good example of how man can widen a faunal corridor by virtue of his activities. However, this wolf population is subject to such a high level of control and disturbance by man that normal pack formation and population self-regulation (Zimen, 1976) may not be possible.

SUMMARY

A review is presented of the behavior and structure of the wolf populations of Soviet Karelia and the adjacent Finnish territory from the late 1940s to the 1970s.

The wolf occurred only occasionally in northern Soviet Karelia prior to 1940, inhabiting cultivated areas and the coasts of the White Sea. Practically no wolves occurred in the taiga, characterized by old, mature coniferous forests, deep, soft snow cover and limited potential prey populations. The major increases in the wolf populations of the area have taken place since the Second World War, the first in the late 1950s and early 1960s, and the second in the 1970s. These were, in many respects, due to human impact. The replacement of the coniferous forest with deciduous forest after clear-cutting increased the food available for moose, and thus their populations which, together with other ungulates, made pathways for wolves in the deep, soft snow. The increased prey populations and accessibility of the areas permitted the wolf population to increase. During both increases, wolves from the saturated populations (5–7 ind./1,000 km^2) dispersed into adjacent Finnish territory, males being the first to enter Finland.

Winter Ecology of a Pack
of Three Wolves in Northern Sweden

Anders Bjarvall and Erik Isakson

INTRODUCTION

When the wolf (*Canis lupus*) was granted total protection in Sweden in December 1965, about 10 individuals were assumed to still exist (Haglund, 1968). However, later evidence indicated that this number was probably an overestimation (Haglund, 1975). Following 1970, with the exception of a pair of wolves observed traveling together for a brief period in the winter of 1973-74, there were reports of only scattered individuals (Bjarvall and Nilsson, 1976).

A change occurred in 1977. Late in spring there were several independent reports of a pack of three wolves. The following autumn, tracks of a similar group were found in northern Sweden. Because the wolf remains a controversial but little studied animal in Sweden, we decided to monitor this pack. In addition, we were interested in the possibilities of reproduction, which had apparently not occurred in Sweden since 1964 (Haglund, 1975).

STUDY AREA

The wolves were discovered in Stora Sjofallet National Park and spent the entire winter of 1977/1978 either there or in adjacent parts of Sarek National Park (Figure 4.1). The northern boundary of this area — Lake Akkajaure — is situated about 100 km north of the Arctic Circle and about 420 m above sea level. A narrow zone (less than 100 m) nearest the lakeshore supports coniferous forest that is continuous in the east but gives way to scattered trees in the west. The remaining area supports either mountain birch or alpine vegetation types. Some areas are covered with boulders, making them impassable. Other areas, particularly in the south, are mountainous with average elevations between 1,000-1,500 m. Several peaks extend

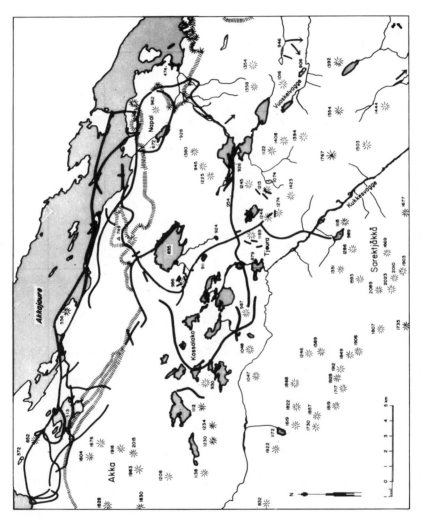

Figure 4.1: Study area and pack travel routes (solid thick lines) between mid-January and early July 1978.

above 2,000 m. There are no roads south of Lake Akkajaure. North of the Lake, a highway leads to a hydroelectric plant constructed in the last few years. Scattered huts, built either for tourism or reindeer management, can be used for lodging. Two trails, one immediately east of the area and the other in the westernmost portion, are extensively used by hikers in summer.

Weather is highly variable. Strong winds and drifting snow frequently sweep all tracks away. During more than two thirds of the 77 field days of this study, the age of the snow in which tracks could be followed was less than one night. On only four days was the snow more than three nights old.

In summer and autumn, the area is an important grazing ground for semi-domesticated reindeer (*Rangifer tarandus*). Calving also takes place here. Late in autumn, the reindeer are herded to forest areas farther east, and in April they are returned. Small groups of stragglers always stay behind and spend the entire winter in the area.

METHODS

It was considered important to try to maintain continuous contact with the wolves so that movements, habitat utilization, mating activities, food habits and relations with other species could be determined. At the same time, however, it was believed equally important not to disturb them. For these reasons, snow tracking seemed to be the only suitable method for monitoring activities. This was done on skis so that details such as vaginal bleeding or predation on rodents could be noted. For safety reasons observers worked in pairs.

Because midwinter days are short and temperatures below -30°C are not unusual, the tracking patrol used the few available huts whenever possible. A snowmobile (Ockelbo 600) was used for transport and to locate wolf tracks after contact had been lost. (In Sweden, motor-driven vehicles may not be used for locating or tracking game. Thus, a special permit was necessary.) On several occasions, a helicopter was used to locate tracks.

While following tracks, the patrol recorded any signs indicative of wolf activity or behavior on prepared forms. Wolf tracks and distances covered were plotted. Other pertinent information about potential prey or human activity was also noted.

A separate form was used for recording data for each animal carcass discovered. Results of carcass examinations and a description of the surroundings were noted, e.g. how the animal was killed and to what extent and by what species it had been utilized. Most carcasses were found along the wolf tracks. However, some were found by searching areas where flocks of ravens (*Corvus corax*) were observed. From sufficiently fresh and complete carcasses, half of the lower jaw and one of the front legs above the tarsus were taken to determine the animal's age and the fat content of the leg bone marrow. These analyses were made at the National Veterinary Institute. Scats were collected along the wolf tracks and their contents were analyzed by Isakson.

RESULTS

Home Range and Movements

Tracks of the wolves were located on 35 separate days from mid-January to early July. Interruptions, mainly the result of bad weather, occurred several times, the longest being 32 days (13 February-17 March). On 12 of the 35 days, only remnants of tracks were located, i.e. footprints close to an old kill. On each of the remaining 23 days, the tracks were followed for one to 19 km (mean 8.2 km) for a total distance of 189 km.

All tracks were in an area approximately 35 x 15 km (Figure 4.1), with the longer axis directed northwest-southeast, parallel to the main river valleys. There were tracks indicating the wolves had been outside this area at least twice, but such trips appeared uncommon. During the observation period, the area utilized by the three wolves was approximately 600 km^2.

The wolves generally avoided areas above 1,000 m (Figure 4.1), traveling primarily along valleys and across plateaus. In deep snow they traveled single-file, but proceeded along individual trails when on top of the crust. Twice, the tracks crossed fairly recent snowmobile tracks but revealed no apparent reaction by the wolves. On one occasion they followed a snowmobile track and, without any detectable reaction, passed within 10 m of 200 liters of paraffin oil stored there less than two days earlier.

Food Habits

On 15-16 April, at least one wolf successfully hunted voles by pouncing through the snow. It made holes approximately 30 cm deep, apparently trapping the rodents with its front paws in much the same manner as the red fox (*Vulpes vulpes*). This behavior has been previously described for wolves (Murie, 1944) but does not seem to have been observed very often (Mech, 1970). No vole remains were found. All other hunting activity involved either moose (*Alces alces*) or semidomestic reindeer.

Moose: Eighteen moose carcasses were found during the study. Six, and possibly seven, of these were wolf kills. Of the six we aged, three were calves and the others were adults of approximately six, eight and nine years. The kills were made in late winter/early spring (Figure 4.2); none were found to be in poor physical condition (Table 4.1). Of the remaining 11 moose, three were possible wolf kills, two were probably killed by other carnivores, five were too old for the cause of death to be determined, and one was killed in an accident.

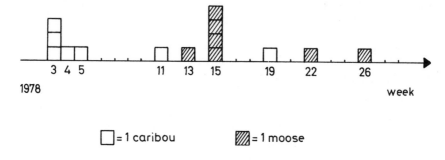

Figure 4.2: Seasonal distribution of 14 known wolf kills.

Table 4.1: Condition, as Judged from Fat Content of Leg Bone Marrow, of Seven Moose and Seven Caribou Killed by Wolves

Condition (fat content)		Moose	Caribou
Good	(70-85%)	2	5
Moderate	(40-70%)	5	2
Poor	($<$40%)	—	—

Information regarding hunting behavior was obtained by interpreting tracks near five of the killed moose. A chase seems to have occurred in only one instance, that of a female calf. Due to drifting snow, tracking conditions were not ideal, but it appeared that she had been browsing in a group of birches (*Betula sp.*). Apparently, the wolves approached at a run for 100 m and chased the moose another 200 m from the trees before killing her. Between the trees and the kill, the moose sunk at least 50 cm into the snow.

The remaining moose were killed without any rush. The wolves apparently located them by scenting and/or tracking and then, without running, approached the prey. In some cases, this approach took place under cover of the birch forest, but in others, the wolves approached in the open at a distance of 100 m or more. There were no indications that these moose attempted to flee. No tracks of running animals were found and the kills were made close to where the moose had been browsing. However, there were signs that the moose had defended themselves. Near two moose, the snow was intensively trodden over areas of about 10 x 10 and 10 x 20 m, respectively, and tufts of moose and wolf hair and blood stains were found. Similar observations were made at a kill discovered on bare ground. Within an area of 3 x 5 m, moss was torn away from both the ground and some boulders and small bushes were broken.

The two kills which were less than one day old when examined, had both been bitten in the lower part of the rump or the upper part of the hind leg. In one hind leg, the muscle was snapped although the hide was not broken. Both moose were also bitten in the nose, but it could not be determined if this occurred prior to death. In addition, one of them (a bull) had been bitten in the shoulders, the other (a cow) in the throat. The throat bite had resulted in extensive bleeding near the trachea.

Wolves are known to locate prey by tracking. They have been observed to distinguish between old and new tracks — following only very fresh ones (Mech, 1970:199). How far they are able to follow such tracks is unknown since, when the behavior has been observed, the wolves located the moose shortly after detecting tracks. In the present study, the three wolves encountered three moose tracks less than four nights old. They followed the tracks for 2 km to an area where the moose separated. Here, the wolves also separated. Unfortunately, our tracking was interrupted and a snowstorm destroyed most of the tracks. However, several days later tracks of a wolf following a moose were found one km further ahead. Thus, the wolves tracked the moose for a minimum of two, and possibly three km.

Even if wolves do follow moose tracks, they do not necessarily catch up with the animals. For example, one night the wolves followed the tracks three moose made the previous night. They ran along the trail for about 50 m to a spot where the moose had several beds and had also been browsing. Here, the wolves apparently lost interest and abandoned the track.

Reindeer: Twenty-six reindeer were found dead during the study. Seven had been killed by wolves. The kills occurred earlier in winter than moose kills (Figure 4.2) and most were in good physical condition, as judged by marrow-fat content (Table 4.1). All seven were adults and the four which were sexed were cows. Fourteen others were killed by unknown carnivores, either wolves, wolverines (*Gulo gulo*) or lynxes (*Lynx lynx*). Thus, the wolf was the responsible predator in all cases in which identity could be established. For the remaining five carcasses, the cause of death could not be determined.

Long chases were much more common for reindeer than moose. In one instance, three reindeer were chased approximately 1.5 km before being killed. The reindeer had been grazing where the chase was initiated and tufts of reindeer hair, one clinging to a small piece of hide, showed there had been direct contact between predator and prey. Blood stains were found along the path leading to the kills. On another occasion a single reindeer broke from a herd of about 100 animals when the wolves began pursuing it. It was killed after a very short rush, but the rest of the herd was followed for an additional 2 km. A fifth reindeer was killed by a wolf after a rush of about 40 m. This reindeer jumped down a precipice about 4 m high. A sixth wolf-killed reindeer was discovered in an area where the wolves forced four small groups of reindeer (6–30 individuals) to run vigorously together for 12 km. Because the wolves followed in the reindeer tracks, it was difficult to establish how far the pursuit continued. However, it was evident that it had lasted for several km. The last of the wolf kills was dead too long to determine the length of the chase, although tracks of running reindeer and a wolf were found in the vicinity of the carcass.

Of the 14 reindeer killed by unknown carnivores, 13 were adults and one a yearling. Of the nine adults sexed, six were cows. No tracks were observed at five carcasses because of ground conditions. Wolf tracks were seen around three of the remaining nine carcasses. However, the wolves may have been scavenging.

In contrast to the successful chases of reindeer described above, there were instances when reindeer escaped without being killed. Both single reindeer and a small group were chased unsuccessfully by lone wolves for 300-400 m. On one occasion, the three wolves discovered a herd of about 30 grazing reindeer after having followed their tracks. A chase was initiated when the distance between the wolves and reindeer was 50-100 m. The reindeer initially left nearly 30 parallel tracks with two tracks slightly separated from the main group. These two tracks were followed by one of the wolves. After 200-300 m, these tracks rejoined the main herd. Several times when the herd curved, the wolves followed short-cuts. Once, going downhill, the herd split slightly but reunited without any reindeer separating. A tuft of reindeer hair, but no blood, was discovered on a frozen lake where the wolves had been running in reindeer tracks. Both species sank less than 20 cm into the snow and were thus probably unaffected by snow conditions. After about 4 km of tracking, weather deteriorated and work had to be interrupted. However, observations later in the spring indicated that no reindeer had been killed. Other unsuccessful chases of between 200 m and about one km were also noted.

Relations with Some Non-Prey Species

One important aspect of wolf ecology is that every large kill provides carrion for scavangers. Visits by the most common scavengers to wolf kills are shown in Table 4.2. In addition, single visits were noted for golden eagle (*Aguila chrysaetos*) (carcass 1–7 days old), white-tailed eagle (*Haliaetus albicilla*) (>7 days old), and common gull (*Larus canus*) (1–3 days old). The arctic fox (*Alopex lagopus*), though occurring in the area, was not observed at any wolf-killed carcass. The hooded crow (*Corvus cornix*) does not winter in this region, however, it was located at carcasses after its return around 15 April. Visits by scavengers to carcasses other than those killed, or probably killed, by wolves are summarized in Table 4.3.

Table 4.2: Observed Visits to Wolf Kills by Important Scavengers
in the Study Area

Age of Carcass (number)	Raven	Hooded Crow	Red Fox	Wolverine
<1 day (2)	1	2	1	—
1-3 days (2)	2	—	1	1
1-7 days (8)	7	1	8	2
>7 days (2)	1	1	1	1

Table 4.3: Observed Visits by Scavengers to Non-Wolf-Killed
Carcasses

Age of Carcass (number)	Raven	Red Fox	Wolverine	Arctic Fox
<1 day (0)	—	—	—	—
1-3 days (9)	1	1	—	1
1-7 days (3)	2	2	—	—
>7 days (13)	8	12	5	2

Additional observations illustrate the relations between wolves and other non-prey species. The three wolves spent two days near a kill, utilizing a number of beds. From three of the beds there were running tracks which approached the track of a single red fox. It appeared that the fox had turned quickly and the three wolves gave up the chase after a short distance.

On several occasions, wolverines followed old wolf tracks but at least one observation indicates that the wolverine may avoid close contact with wolves. A wolverine approached a carcass being utilized by the wolves. It proceeded through the area and then ran away, and did not resume a normal walk until it was about 2 km away. Tracks of three running wolves followed along the wolverine track for approximately 500 m. Twice the wolves crossed lynx tracks, and stopped and examined the tracks before continuing in their own direction. Once, when crossing a mink (*Mustela vison*) track, they exhibited the same behavior.

DISCUSSION

The wolves utilized a range of approximately 600 km during winter and spring, an area larger than previously reported in forested regions, but smaller than reported from tundra. Pimlott et al. (1969) reported home ranges between approximately 100–310 km^2 in Algonquin Provincial Park, Ontario, Canada. Much larger home ranges have been reported from Alaska where Burkholder (1959), using an aircraft, followed a pack which traveled an area of approximately 12,950 km^2 over six weeks. Between these extremes there are reports of ranges around 1,400 km^2 from Alberta (Rowan, 1950), 224 km^2 from Ontario (Kolenosky, 1972), more than 110 km² from Minnesota (Mech et al., 1971b; Mech, 1973, 1974b), 1,050 km² from Finland (Pulliainen, 1965), and 500 km² from Sweden (Haglund, 1968). Mech (1970:164) has pointed out that whereas some variation is the result of differences in methodology and/or the duration of data collection, there are also differences which reflect habitat or prey densities. Additionally, lone wolves are known to cover much larger areas than packs (Mech et al., 1971b), and findings by Kolenosky and Johnston (1967) suggest that home ranges may be smaller in summer than in winter.

From December 1977 to June 1978 the tracks of the three wolves were followed within the Stora Sjofallet National Park. Although Mech (1970:166) has indicated that home range estimates based on tracks are often unreliable, especially because similar sized groups could inhabit the same area, we are confident that only one group of wolves was followed in this study. Every track in sufficient condition to allow detailed study was left by one large and two smaller individuals. Furthermore, the group's behavior was consistent throughout the study; they often reused the same routes (Figure 4.1) and visited previously exploited carcasses.

Large mammals of different species, chiefly ungulates, constitute the main prey of the wolf (Murie, 1944; Stenlund, 1955; Mech, 1970). This was reaffirmed in the present study. Apart from some small rodents, reindeer and moose were the only prey species utilized. However, small rodents appeared to be of little dietary importance.

There are observations of winter interactions between moose and wolf from Alaska (Murie, 1944; Burkholder, 1959; Haber, 1977) and Minnesota (Mech et al., 1971b); however, most information comes from Isle Royale (Mech, 1966). Although wolves are capable of killing moose, many attempts are unsuccessful. The moose may detect the wolves and leave unnoticed, it may hold its ground and defy them, or it may outrun them. Mech found that about a dozen moose were "tested" for every one killed and that no moose that stood its ground was ever killed. Our data indicate a somewhat different relationship. Most of the moose killed apparently stood their ground but with little success. This is difficult to explain but might relate to the size of the attacking pack. On Isle Royale, most moose are attacked by packs of 10 or more wolves. Perhaps these larger packs elicit an escape reaction by the moose more readily than a pack of only three wolves.

The first point of attack is usually the rump or ham area, and the second the nose (Mech, 1966). In our study, the two moose carcasses most thoroughly examined had been bitten in the rump/hind leg area. An attacked moose will defend itself by kicking at the approaching wolves (Mech, 1966). Although direct encounters were not observed in the present study, it was evident from the tracks that struggles had preceded at least three of the moose kills.

Mech (1970:218) stated that lone wolves, pairs and smaller

packs would have more trouble than larger packs killing moose, and would probably rely on the very oldest and weakest moose, killing them in stages. As far as we could determine, the small pack we observed had little difficulty killing moose. Further, there was no evidence that the oldest and weakest moose were the most important prey.

Very little has been written about relations between wolves and reindeer. However, reindeer are very similar to caribou (*Rangifer tarandus*), the principal prey of the wolf in some regions of North America. In many areas, wolves follow caribou on their fall and spring migrations (Kelsall, 1968). Both species live in close contact for extended periods (Mech, 1970) and most predation on adult caribou seems to occur during fall and winter (Murie, 1944).

Wolves normally use three different methods when hunting caribou: ambushing, relay running, or chasing groups from which lagging animals are singled out for attack (Kelsall, 1968). The latter method is probably used most commonly. Since "caribou, except for the incapacitated and very young, can normally outrun single wolves" (Kelsall, 1968), tracks of chases, successful and unsuccessful, are common in areas where caribou are a principal prey. In this respect, there is great similarity between such areas and our study site. In all cases in this study where circumstances prior to a reindeer kill could be reconstructed, the prey had been chased. No evidence of ambushing or relay running was discovered. Hair tufts where a chase was initiated, and blood stains along the trail, might suggest that the wolves concentrated upon individuals less capable of following a fleeing group. Similarity to the wolf-caribou relationship is further supported by reconstructed unsuccessful chases. There were no indications that reindeer attempted to stand and defend themselves against an attacking wolf, as has been reported at least once (Haglund, 1968).

Detailed information of the actual attacks on reindeer is limited primarily because of insufficient remains of kills. The three carcasses examined had bled extensively from the nose, neck, throat and belly supporting reports that wolves focus their attacks on a caribou's front end (Murie, 1944; Burkholder, 1969; Kuyt, 1972).

One or two prey species usually dominate the wolf's diet (Mech, 1970). This was also true in the present study where moose appear to have replaced reindeer as the principal prey toward the end of winter (Figure 4.2). Twice in February, the Lapps gathered straggling reindeer in the area and thus reduced the number of potential caribou prey. However, 100-200 reindeer still remained, compared with an estimated 20-30 moose.

Mech (1970) concluded that when two or more species of large prey inhabit the same region, wolves will concentrate on the smallest or easiest to catch. It is likely the prey vulnerability to wolf predation is drastically affected by snow conditions, as shown for white-tailed deer in Minnesota (Mech et al., 1971a). The late winter change in relative importance of reindeer and moose may have also resulted from a change in vulnerability of at least one of the species. Severe winter weather can diminish a prey's general physical condition, making some individuals highly vulnerable to predation although there was little evidence that this occurred in this study. Changing snow conditions, however, appear to be a primary reason for the switch from reindeer to moose. Above timberline, where reindeer are found, snow conditions rarely influenced wolf mobility. But at the lower elevations where moose were killed, the deep snow, especially near trees, has a high penetrability in mid-winter. A crust strong enough to support a moose rarely develops, whereas wolf mobility gradually improves as spring approaches, making moose more vulnerable.

Mech (1970:205) has suggested that wolves become specialists at killing the important prey in their region. The ability of these three wolves to kill moose suggests they may have dispersed from areas where moose are important prey. One such area is western USSR where wolves primarily depend upon moose (Ivanov, 1967). The number of wolf crossings between the USSR and Finland showed a sharp increase in 1976 and 1977; one-third of all crossings since 1968 occurring during 1977 (Pulliainen, in press b). These wolves roamed widely in Finland and, in December 1976, two wolves crossed from Finland into Sweden (Nyholm, pers. comm.). Thus, the most plausible origin for the pack of wolves we followed seems to be the western USSR.

Remnants of wolf kills are utilized by many species of mammals and birds (Pimlott et al., 1969; Mech, 1970). For some scavengers, such carcasses constitute important winter food; for others, they are rather complementary to prey killed by the animal itself.

This study indicated that ravens and red foxes utilize many wolf kills (Table 4.2). In North America, ravens commonly follow wolf packs or track wolves (Mech, 1970; Harrington, 1978), and the only time one of the wolves in this study was directly observed, it was accompanied by several ravens. There are also reports of red foxes following wolves (Murie, 1944), but our observation of the fox chased by all three wolves indicates that such behavior can be hazardous. In North America, red foxes have been killed by wolves (Mech, 1970).

In Finland, wolves and wolverines compete for food resources and are not sympatric (Pulliainen, 1965). But in North America (Mech, 1970) and northern Sweden they do share the same habitats. Since 1975, wolverines have occupied the area used by the wolves in this study (Bjarvall et al., 1978). However, they were not common and thus were rarely seen at wolf kills or other carcasses. The tracks of a wolverine apparently being chased by all three wolves agrees with similar observations from North America of wolverines being chased (Murie, 1963) or killed (Burkholder, 1962). Thus, the two species can occupy the same area, but the available information suggests that wolves are aggressive to wolverines, which in turn try to avoid direct contact.

The arctic fox population in Fennoscandia has markedly decreased during the first half of the 20th century (Haglund and Nilsson, 1977). Because foxes utilize remnants left by other carnivores (Macpherson, 1969), it has been assumed that the decrease of arctic foxes is related to the decrease of wolves. In the present study, tracks of arctic foxes were observed on eight of 23 days when the wolves were tracked for at least one km. In spite of this, arctic foxes were not found at any of the 13 wolf-killed reindeer or moose, and at only three of the 16 non-wolf-killed carcasses.

Several circumstances could account for the lack of arctic fox tracks at wolf kills: (1) reindeer occupy areas mainly east of arctic fox habitats, (2) moose occupy forest habitats rarely visited by arctic foxes, (3) small rodent prey, e.g. lemmings (*Lemmus lemmus*), were common in 1978, and (4) as tracking conditions around many carcasses were poor, tracks of lighter animals such as foxes might have been more difficult to observe. Although the material is limited, it suggests that arctic foxes did not utilize wolf kills to any appreciable extent.

Acknowledgements

This study was financed by a private contributor. The field work was done by Erik Isakson together with Lennart Arvidsson (16 January-28 February), Tomas Paivio (7 March-7 April) and Thomas Partapuoli (11 April-6 June). L. David Mech reviewed the manuscript and suggested several valuable improvements.

Wolf Management in
Intensively Used Areas of Italy

Luigi Boitani

INTRODUCTION

Italy is an overpopulated country; 56 million people inhabit 320,000 km^2 with an average density of 175 inhabitants per km^2. The relationship of this population to the natural environment is exploitive. Only one-fifth of the country remains forested. There are just four national parks (one in the Appenines) with a total area of only 1,000 km^2. Fifty-eight percent of the country is intensively cultivated, and there are nine million cattle, nine and a half million sheep and nine million swine. Three hundred thousand km of roads traverse Italy, and there is one car for every three inhabitants. There are two million hunters who are generally undisciplined and unregulated for six months of each year. Due to excessive loss of habitat wildlife is seriously endangered; nearly all large herbivores are extinct except for chamois (*Rupicapra rupicapra*), ibex (*Capra ibex*), roe deer (*Capreolus capreolus*) and a few red deer (*Cervus elaphus*) in the alpine areas. The lynx (*Felis lynx*) is extinct and the bear (*Ursus arctos*) has been reduced to a small nucleus of 50–60 animals in Abruzzo. In addition, the science of wildlife management is practically unknown in Italy. The concept of conservation is new, and there is little environmental education. Land management is almost totally absent.

Despite these difficulties free-ranging wolves (*Canis lupus*) still survive in Italy, although their existence is seriously threatened. An urgent, effective management plan to guarantee their conservation must be applied. In 1973, the World Wildlife Fund initiated a wolf protection campaign, and sponsored the first research on status and distribution. From 1974 to 1977 the study was extended to wolf ecology in the Central Appenine areas of central and southern Italy. This work, conducted by E. Zimen and the author, was designed to provide the technical information necessary for the

158

preparation of a wolf conservation plan. The results of the first years of work were published by Zimen and Boitani (1975, 1979), Boitani (1976) and Boitani and Zimen (1979, and in prep.). The present paper discusses data on the distribution and biology of the Italian wolf, necessary to understand the proposed conservation plan. A more general rationalization for the conservation program will be presented later.

Naturally, a conservation plan prepared *ad hoc* for a selected species and a particular area of implementation may have little or no value for other locations. However, I believe that problems associated with conservation of the Italian wolf are similar to those of many other countries, although perhaps more extreme. Therefore, by reviewing the Italian situation, useful guidelines may be derived for other areas. Italy may be at a stage of development comparable to that which other countries will reach in the near future, particularly regarding environmental preservation versus exploitation.

THE ITALIAN WOLF POPULATION

Number and Distribution

A 1973 census determined there were approximately 100 wolves in Italy (Zimen and Boitani, 1975). Since that time, only local censuses and information regarding sightings, killings of wolves and predation on cattle have been accumulated. These data, together with biological information from intensive localized research, indicate the Italian wolf population is relatively stable, although there is a pattern of irregular fluctuation in distribution. The one area where the wolf is consistently present covers about 8,500 km^2 (Figure 5.1), but many other areas occasionally support wolves. The Appenine Mountain Range, potentially suitable for wolves, encompasses approximately 70,000 km^2, though only 12 percent is permanently occupied by wolves. The average density of one wolf per 85 km^2 is much lower than that recorded for North America (one wolf per 25-40 km^2) (Pimlott, 1967; Mech, 1970).

The wolf is unevenly distributed in the mountain ranges of the Appenine chain (Figure 5.1, Table 5.1). Relatively wild and isolated mountains remain the only safe refuge for the species. Among these areas, there is a continuous dispersal of old and young animals. During these movements, the animals are vulnerable as they cross heavily populated areas where there is a danger of being seen and shot. These nomadic individuals may ensure the survival of the species by continually repopulating new areas. In fact, the wolf's present distribution surrounds the zone historically occupied by wolves (Cagnoloro et al., 1974) (Figure 5.1), and changes we observe today always occur within the original zone. Although this zone is still a potentially favorable habitat, wolf presence there is curtailed by increased human pressure.

Figure 5.1: Areas of the Appenine Mountains inhabited
by wolves. Numbers refer to regions noted in Table 5.1.

Table 5.1: Size of the Appenine Regions Inhabited by Wolves
and Approximate Number of Wolves in Each Area

Area	Approx. Area (km^2)	Approx. No. of Wolves
(1) Sibillini — Laga	1,900	8
(2) Altopiano delle Rocche — Velino — Sirente	750	5
(3) Tarquinia — Tolfa — Campagnano — Agro Romano	1,110	12
(4) Maiella — Piano di Cinquemiglia — Parco d'Abruzzo	1,500	22
(5) Matese	400	3
(6) Cervialto — Polveracchio — Terminio	300	8
(7) Alburni	120	4
(8) Sirino — Raparo — Alpi	280	4
(9) Pollino — Catena Costiera	650	12
(10) Sila	1,650	25

The present habitat of the Italian wolf is exclusively Appenine, 800 to 1,500 m above sea level in areas covered by large beech forests interspersed with extensive pasture. The pasture is utilized by domestic sheep and cattle. Cattle and horses are generally left to wild pasture. Sheep breeding is carried out with flocks of up to 3,000 head, under the continuous surveillance of shepherds and trained dogs.

One wolf zone differs markedly from all others. Monti della Tolfa, about 50 km north of Rome, is characterized by hills and low mountains (<600 m), and is typical Mediterranean bush vegetation. The area is wild and infrequently visited. Cattle are left to wild pasture year-around, which contributes to increasing food availability for the wolf.

Food Habits

In Italy, wolves primarily rely upon garbage and domestic animals, determined by analysis of 220 scats collected during winter and summer in Abruzzo (Figure 5.2). In general, samples were placed in categories according to the highest proportional food content. The percentage analysis does not reflect the relative quantity of swallowed foodstuffs (Floyd et al., 1978).

The percentage of identifiable garbage increased from summer (10.7%) to winter (20.6%), while vegetable substances and small mammals decreased from 16.6 to 7.3 percent and from 8.3 to 0.8 percent, respectively. During winter, the percentage of sheep increased from 26.2 to 46.4 percent, contrasting with the number of domestic animals reported killed. There is a need to interpret these percentages carefully, as most sheep and cattle remains found in wolf scats come from the rejects of butchers which are discarded in great quantity in the dumps. If we consider that unidentified substances probably result from food found in dumps, such as fat, meat, bread and spaghetti (found more than once in the Maiella area), then the total percentage of garbage increases to 85 percent in winter, a season during which wolf depredations are limited to a few livestock.

A cautious estimate, however, establishes a variable percentage of between 60-70 percent of garbage in the wolf's diet. These data are supported by blood analysis carried out on trapped wolves. There is a relatively high concentration of nucleic acids in their blood, probably from substances rich in these acids such as liver, spleen and other internal organs (Seal, pers. comm.). These are the rejects most often found in dumps, discarded by butchers as they may be infected by *Echinococcus* and *Fasciola* (liver-worm).

Nearly all available prey species are represented occasionally in scats, including red deer, recently reintroduced (1974-1976) in the Abruzzi National Park area and now totalling about 250 animals. There is no evidence of predation or scavenging on the small popu-

lation of chamois which inhabit the highest mountain tops. The boar
(*Sus scrofa*) is present in Abruzzo but was not found in wolf scats.
However, this species is regularly killed by wolves in Basilicata in
southern Italy.

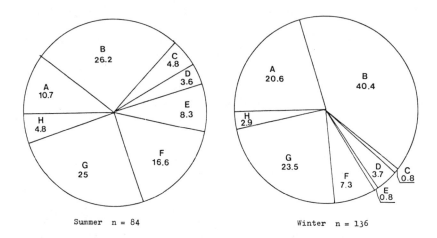

Summer n = 84   Winter n = 136

Figur 5.2: Seasonal distribution of ecologically significant contents
of wolf feces: (A) garbage, (B) sheep/goat, (C) cow/horse, (D) dog,
(E) small mammal, (F) vegetable matter, (G) 100% "matrix" (this
class describes feces that consisted solely of amorphous white
powder whose origin could not be traced, but is likely to be from
garbage dumps), (H) other. Sample sizes: summer (n = 84); winter
(n = 136).

Among domestic animals, sheep are most susceptible to wolf
predation. The traditional method of penning sheep at night with a
cord net, favors wolf attacks. Additionally, modern herding methods
employing numerous herds, and shepherds and dogs unfamiliar with
mountain areas, make the herds vulnerable. Only a few sheep are
normally killed during an attack, although occasionally slaughters
of 200–300 animals occur. Cattle are rarely attacked and only a few
groups of wolves appear capable of regularly killing horses. Therefore, statistics on killing by wolves may reveal little about food
ecology. However, human reactions to these depredations are important and may ultimately decide the wolf's fate.

Movements

Twelve wolves were captured and radio-collared in the region
east of Rome (Figure 5.1) between 1974 and 1976. A full account
of this research is forthcoming (Boitani and Zimen, in prep.). In this
region, the wolf normally moves at night and sleeps during the day,

safely resting in sheltered areas, usually in dense forests. These areas, although not frequented often by humans are not necessarily far from towns. Generally, they are situated higher than the nearby villages (about 1,200-1,500 m) in rocky and inaccessible areas, but they have also been found a few hundred meters from a town's rubbish dump. Each wolf or pack uses more than one of these hiding places, which are visited at varying intervals. In some cases, the animal returns to the same place each morning; in others, the wolf may alternately visit different refuges strategically scattered throughout the territory.

Wolf movements begin at sunset. They move toward urban centers, valleys and areas inhabited by man in search of food. They move surely and carefully between houses, avoiding lights and noises when crossing roads and entering towns. Often, we were able to locate one of our radio-collared wolves in the center of a mountain village by stalking in absolute silence during the night.

The wolf apparently has learned to approach sheep pens without being detected by dogs, and knows where it is most likely to find suitable garbage. At dawn, the wolves return to shelters, arriving before humans are fully active. At times, the wolf is late and does not reach shelter in time, so it spends the day in rock caverns or bushes only 200 meters from a village. In winter, the wolf some-times chooses lower elevation shelters where the snow does not hinder its movements. Sometimes in summer, the routine alters according to seasonal changes and movements of cattle herds. In these cases, the wolf changes its movements and visits potential sources of food nearly every night, often without positive results. However, it is unusual for the wolf to descend below 800 m or to live more than 2,000 m above sea level in this region.

Average home range of the Italian wolf is about 120-150 km². This is smaller than most North American ranges (Mech, 1970), but corresponds to the specific requirements of the Appenine. In many areas, there are two to three villages which, with their rubbish, can amply sustain a pack of wolves. Within the home range, a wolf often follows established routes linked to the topography of the land, human presence, roads (especially in winter when the wolf makes much use of plowed roads) and vegetation. Most movements are in the depth of forests or along the edges of woods, while open pasture land and fields are avoided. In about 50 percent of the ob-servations, the wolves travelled less than one km. They very seldom exceeded 10 km.

Beside daily movements, there were also dispersals. In addition to young animals, one adult female followed by at least one of her four to five month old cubs, also dispersed. These movements, which may reach a total of 60-80 km in a few days, result in the utilization of a new area, at least temporarily.

However, dispersal can be hazardous when the wolf crosses populated areas. In addition, dispersal may lead to areas where survival is difficult. Yet, dispersal is one of the key mechanisms which permits survival of the wolf in Italy because: (1) It permits genetic exchange between neighboring packs and individuals; (2) it allows animals to exploit suitable areas devoid of wolves where, despite the odds against success, there is a potential for establishing new packs; (3) it continually separates young subordinate animals from their natal packs, permitting them to reproduce and thereby increase the population; and (4) it increases the diversity of ecological conditions wolves are exposed to, allowing survival of some local populations while others are being exterminated.

Dispersals cause the sudden appearance of wolf packs in areas devoid of wolves for years. However, the local populace has a more fanciful and, as such, very resistant explanation for this phenomenon: The wolves have been reintroduced from Siberia and Canada by an unidentified authority. Although groundless, this claim is one of the major hurdles to a calm analysis of the problems of wolf conservation in Italy.

Yearly dispersal patterns influence the size of the wolf's range, which should therefore be divided into actual and potential wolf zones. The latter are biologically suited to the species, but persecution by man inevitably leads to its extermination. Nevertheless, these areas remain vitally important buffer zones for the species' survival.

Pack Size

The maximum number of wolves observed in a single pack was seven, but packs often divided into isolated individuals or groups of two or three animals. The adaptative significance of this behavior is clear considering the paucity of large prey. It is more advantageous to be isolated when hunting small animals or eating in rubbish heaps. The typical pack in Italy seems to be composed of one adult pair and their offspring of the year, but there are so many accidents which disrupt packs that it is difficult to make a more precise evaluation. In the Abruzzo National Park, for example, we have never recorded the presence of more than four animals in the same group. There do not appear to be great differences between summer and winter, or among different zones. Pack members often move about alone but associate with pack mates at the refuges. There are also frequent cases of wolves who lead essentially solitary lives.

Natality and Mortality

A rather high reproductive potential (with respect to the whole population) can be expected because of the wolves' tendency to live in small groups and their discontinuous distribution or isolation. Although litter size is average (4-6 cubs) (Mech, 1970), there are relatively more reproductive units. Isolation may also encourage mating with tame or feral dogs. For example, a radio-collared female wolf living alone in a small area bordering the territory of a pack in Abruzzo mated with a dog and produced six hybrid pups. These offspring, three of which were radio-collared, behaved like wolves but their appearance was decidedly hybrid; four were black with a white foot.

The presence of hybrids raises some important considerations: (1) Hybridization with dogs probably has occurred throughout the history of the Italian wolf, which has always been in contact with man and dog. (2) Hybridization may have also facilitated the evolution of a genotype more suitable to the local environment; hybrids may be more successful than wolves in an environment such as the Appenines, if only because they arouse less suspicion and alarm in man, who mistakes them for dogs. (3) However, if the hybrid is more successful, this is because the environment has been modified by man, altering historical selective forces. Conservation philosophy dictates the preservation of the most natural environment and its occupants as is possible, and not just the simple maintenance of current conditions. (4) In similar situations, the wild population of the wildcat (*Felis silvestris*) was able to completely absorb the results of occasional hybridization with domestic cats (Leyhausen, pers. comm.) in a short period. (5) In extremely critical situations of survival, even one hybrid can contribute to saving an otherwise irreversible situation. (6) As far as public opinon and the law are concerned, the presence of hybrids may also have negative effects, as it may reduce the drive for conservation of the wolf. The law protects the wolf and payment is arranged for the damage it causes, but how one must act as far as hybrids are concerned is debatable. These are a few of the considerations concerning the problem of wolf-dog hybridization, and we are far from having agreed upon a course of action.

Wolf mortality is caused almost exclusively by poaching, poisoning and road accidents, although poaching and poisoning are illegal. Shepherds and livestock breeders are not the primary cause of wolf mortality (seven or eight cases come to our attention yearly); hunters, however, do not hesitate to poison or shoot wolves, maintaining that they do so to protect themselves or their dogs. Due to the unpredictable nature of this mortality, it is possible for small groups of wolves to be exterminated within a short period. In the Maiella area, four of six of our radio-collared wolves were killed in a few months. During one month in 1977, a pack of six animals was poisoned in the Monti Simbrvini area. This incidental mortality leaves little opportunity for natural recovery of the species. Of all the dead wolves I examined, I never found a specimen that could be described as old, that is, more than eight years. At the Abruzzo National Park Zoo, the male founder of the group died at the age of 17.

Relationship with Man

The Italian wolf has adjusted to the ecological conditions wrought by man. Perhaps without man, the wolf could not survive. For centuries, the wolf has lived off livestock herds and has recently learned to use the dumps established near villages. It has become almost exclusively nocturnal, moving during daylight only on foggy or hazy days. It has learned the meaning of noises, lights and various human presences. Thus, it now approaches villages and herds, moving even among houses. It has learned the danger of cars and roads, but also the advantages of using roads in winter. It has learned to avoid bait and traps; perhaps conditioned by centuries of hunting and trapping, the wolf does not normally touch meat found in an unusual place. To catch our first wolf, we placed traps around the carcass of a colt; wolves visited it nightly but ate only after 11 days. We never did catch a wolf at that carcass.

The wolf flees from humans at the slightest disturbance, especially when in its isolated refuges. It only hesitates to flee when caught unaware by shepherds during an attack on sheep.

The wolf readily kills and eats most domestic dogs. In fact, many hunting dogs "disappear" in areas inhabited by wolves. However, the large Abruzzi dogs, selected over the centuries by shepherds to guard their flocks and protected by collars studded with iron spikes, are still an effective deterrent against wolf attack. Fights between dog and wolf are unusual; the wolf avoids fighting a pack of aggressive dogs.

The wolf has adjusted both to living in forests and mountains and to living alone and eating what it finds, mainly garbage. It has adjusted well to humans; they cannot adjust to it.

Legal Status

The wolf is protected by various laws covering the Italian national territory. Since 22 November 1976, a decree by the Agriculture Ministry protects the species *in eternum* and bans the use of poison bait for any purpose. The decree was passed following the initial findings of our research, which emphasized the precariousness of the continued survival of the wolf in Italy, and after pressure by the World Wildlife Fund. A separate national regulation controlling hunting, approved by Parliament in December 1977, also totally protects the wolf.

Consequently, many regions have approved laws which, as well as confirming the protection of the wolf, reimburse for damages caused by them. Lazio, Campania, Basilicata and Molise have laws similar to those legislated for the Abruzzo Region in 1974, although not all have worked successfully. Insufficient appropriations and difficulty in ascertaining the extent and cause of livestock damage seriously hinder uniform application. It is not always possible to distinguish between damages caused by wolves and dogs. Moreover, many claims are fradulent. Kills by other agents often are attributed to wolf predation. Facing this difficulty, the authorities frequently refuse reimbursement, thereby causing discontent among the livestock breeders.

On the other hand, requests for indemnity have increased considerably (Table 5.2). In the Abruzzo region, $103,000 was paid in 1974; this figure reached $209,000 in 1977. In the Lazio region, the requests reached $500,000 in 1978, though not all were granted. If we compare these figures with the number of wolves present in these areas, never more than a few dozen, the disparity is obvious.

Table 5.2: Wolf Damages in Abruzzo: Number of Applications for Refund and Number of Kills

Year	Number of Applications		Number of Kills					Amt. Paid in U.S. $
	Presented	Accepted	Sheep	Goats	Cattle	Horses	Pigs	
1974	243	207	1,412	44	43	22	1	103,000
1975	302	258	1,285	62	33	28	7	83,000
1976	277	238	1,082	80	44	51	3	102,000
1977	483	447	1,832	62	52	138	5	209,000
1978	337	289	988	40	47	104	—	113,000

Much of the claimed damage by wolves, perhaps as much as 50 percent, is attributable to stray dogs. A large portion of the remaining claims stem from fraudulent reports. In particular, in the southern regions of Campania and Basilicata, many indemnity requests come from areas where the wolf has been extinct for years, yet breeders pretend they are present. In such situations, a critical revision of the indemnity laws is urgent.

We recently began testing a new method. In cooperation with the Campania region, we prepared a map of the regional territory which was divided into three sectors: (A) areas where the wolf is normally present, (B) areas where the wolf may occasionally be present, and (C) regions where the wolf is never found. When considering indemnity requests, the regional authorities will give precedence to those in the first sector and, if the funds are still sufficient, those in the second sector will be reimbursed. The requests from the third sector will be placed aside. In this manner, we will attempt to eliminate a priori a large part of the fraudulent requests while also trying to cover expenses for damages caused by stray dogs in wolf range. This is to try to control possible private efforts to suppress dogs with methods dangerous to the wolf.

For this type of approach to be complete and widely acceptable, legal protection of the wolf should be abolished in Zone C, and perhaps in all or a part of Zone B. In other words, where damages are not reimbursed, wolves may be caught and killed. Unfortunately, the inhabitants of central and southern Italy do not appear ready for such a proposal. The areas where it would apply are limited, but enforcement is nearly impossible, as it would be easy to illegally carry a dead wolf from a closed zone to a neighboring area where hunting is permitted. Ultimately, wolf survival would rest upon the honesty of the local population and this does not seem dependable, especially in view of what the wolf still represents to many residents. Thus, appropriate wolf management is still evolving in Italy and it will take many years to reach an acceptable solution.

Threats to Wolf Survival

The Italian wolf population receives continuous pressure from man. The limiting effects are reflected in the precarious survival of the species. Threats to survival can be divided into two main groups:

Direct Causes: The consequences are immediate and are sometimes responsible for the total elimination of a wolf population. These are: (1) shooting by indiscriminate hunters, mainly uneducated, and with little respect for the natural environment. The density of hunters in Italy (7 per km^2) makes an encounter between man and wolf highly probable; some kinds of hunting such as organized fox *(Vulpes vulpes)* and boar hunts, increase this probability. About 70 percent of dead wolves recovered, 10 to 12 each year, died as the result of gunshot wounds. (2) Road accidents, the illegal use of poison to control foxes (only seldom is poison used purposely to kill wolves), and the taking of cubs from dens account for 30 percent of wolf mortality. Although road accidents often are unavoidable, the other causes of mortality are all illegal. Better public education could suffice to minimize them.

Indirect Causes: The consequences are slow to ascertain, and their causal link to the survival of the wolf are subtle, but they are certainly the ultimate reason for the animal's uncertain future in Italy: (1) Scarcity of remote, inaccessible areas of security, i.e. the remaining forested areas are small and excessively used, natural food sources are extremely limited, dependance upon human garbage is dangerous, and the isolation of some wolf ranges makes dispersal among neighboring areas dangerous and difficult. (2) The danger of hybridization with domestic dogs. As previously discussed, this process is completely uncontrolled and, at present, we are unable to judge whether its net effect is negative or positive. It seems, however, that if this process were to continue, the Italian wolf-dog could become more suited to the present environment but would also be less of a wolf. On the other hand it seems rather unimportant to worry about genetic deterioration of a species when we do not give the same attention to the deterioration of the environment, which deserves first priority. Species conservation and integrity can be addressed only after a clear environmental policy has been established. (3) Public attitude of those in close contact with the wolf is potentially the most important global threat to the survival of the species (Boitani and Zimen, 1979). Positive opinion, at least in the major cities, is necessary to legislate protective laws, yet it is local opinion which controls, judges and influences application of the law. As long as the local public is hostile and suspicious about the species, it is useless to hope for the peaceful protection of the wolf; breaking the law has little effect if one is pardoned or even encouraged by local public opinion. Indemnity laws may solve the problem of direct conflict of interests, but they do not help convert the wolf's enemies.

Effective management of public opinion therefore plays a primary role in wolf conservation. However, it is a more complex problem than the biological management of the species. It is much easier to reintroduce red deer and roe deer to provide the wolf with natural prey (Boitani, 1976), than to persuade the Appenine Mountain people to protect the wolf.

A CONSERVATION PLAN: GUIDELINES FOR MANAGEMENT

It is obvious, then, that in Italy, as elsewhere in Europe where the wolf and man have lived closely together for thousands of years (Pimlott, 1975; Zimen and Biotani, 1979), wolf conservation directly involves humans and their activities. National parks certainly fulfill the need for protection, but the more integral the ecosystem to be protected, the more limited man's presence and activity must be. For effective conservation of the wolf, protected areas must be extensive to ensure a stable and large population over a long period. The example of the remaining population in Minnesota (Mech, 1977b) is indicative of all North America. In Europe, few such extensive, lightly populated areas exist; only some areas of Scandinavia might compare. But even there it is revealing to watch people's reaction to the natural repopulation of some eastern wolves to understand the problems to be faced to ensure protection (Pulliainen, in press b; Bjarvall, in press).

In Italy, a system for conserving the wolf by setting aside large areas where man's activities are restricted to recreation and tourism is unfeasible. In the not too distant future, similar conditions will exist throughout Europe and perhaps even in North America. The solution may be found if we overcome the old ethic which separates man and civilization from the natural world. The Italian situation offers a clear picture of how this is possible, and could become a model for the future. In Italy, man and wolf have cohabitated since ancient times and have slowly, reciprocally changed. During this slow adaptation, the wolf's behavior has evolved to permit survival under conditions very different from those originally encountered. The wolf now vitally depends upon man. However, the wolf must now be slowly weaned from the more dangerous forms of its dependence.

In Italy today, all traditional human activities in wolf range can be made compatible with the wolf's presence except one: the pasturing of horses. Colts are easily killed or injured when fleeing from the wolf. However, even if all other activities are compatible, one must design a well-defined program of species management which outlines the exact biological, economical and cultural limits which are acceptable. In fact, it is inconceivable to maintain the present situation which allows for easy indemnity claims and furthers a dangerous state of instability.

In Minnesota, authorities believe that "the long range survival of the wolf depends on the preservation of extensive areas where the wolf-human conflict will not be possible" (Mech, 1977b:21). However, in Italy this wolf-human segregation is not possible. Thus a compromise approach must be taken and the best possible integration found.

The Management Plan

A detailed wolf management proposal is in preparation (Boitani, in prep.) and the following is a summary of the principal guidelines, based on the considerations discussed above. The main principle of the plan is to try to reduce wolf-human conflict to an acceptable minimum, so that adequate reimbursement can be provided.

The wolf conservation plan depends upon establishment of defined zones where the species is fully protected by law and by a series of environmental improvements. Other zones where the wolf may be hunted (possibly only by foresters and game-keepers) are also necessary. The total area of the zones of protection should be a minimum of 20,000 km^2. These areas would mainly include all those where wolves already are permanently located (Figure 5.1), but would be enlarged further to include surrounding territory. All such areas must fulfill the following criteria: (1) Sufficient forested and undisturbed refuges where the wolf may retreat. (2) Ecologically suitable habitat for new populations of large prey herbivores (*Cervus elaphus, Capreolus capreolus, Dama dama, Rupicapra rupicapra, Ovis musimmon*). This point, very important in the long run, may also be upgraded by specific management improvements for the herbivores. (3) Wolf distribution ranges which permit frequent exchanges of animals through natural dispersal. Therefore, distance and barriers to dispersal between neighboring areas are important considerations. (4) Freedom from stray dogs, even if owned by villagers. (5) Protection against any form of hunting, at least for an initial five year period. After this, hunting could be closely regulated as regards methods and huntable species. (6) Elimination of garbage dumps and littering of garbage by tourists. (7) Restriction and control of development (roads, houses, hotel, etc.) according to rigid criteria which take into account the needs of wildlife. (8) Limitation of traditional human activities to present levels. Agriculture, pasturing and industry would be permitted, but with certain limitations. The only incompatible activity would be the pasturing of horses. Raising of cows would need to be carried out with care, especially during calving.

These features and provisions, rarely found today, will have to be created. When areas having the first three features are identified, it will then be necessary to initiate the following procedures: (1) elimination of garbage dumps and stray dogs; (2) reintroduction of the most suitable species of herbivores (red and roe deer, in most cases); (3) pasture regulation — in particular, sheep will have to be guarded by a sufficient number of dogs and shepherds and housed in wolf-proof pens at night (made of metal and lighted); (4) artificial feeding locations, possibly moved frequently to different areas of the ranges; (5) enactment of legislation to control hunting and to ensure the total, immediate reimbursement of damages incurred on herders and farmers by wolves; (6) monitoring and research on the ecology of the entire area; technical data are essential both to follow the development of the situation and circumvent undesirable effects, and also to plan and implement new management. (Paradoxically, perhaps this point will be one of the most difficult to achieve in the framework of Italy's traditional approach toward its environment); (7) education and involvement of the public in the various aspects of the operation; the support of the public is vital to achieve this management plan.

Outside these restricted areas the wolf may be removed by authorized game-keepers, and no indemnity would be paid for damages caused by wolves. The protected areas would be chosen according to local socio-economic parameters: traditional activities, socio-economic state of development, development alternatives, attitude of the public and political desire for environmental management.

The last parameter, politics, is the most difficult to resolve in Italy. In fact, even if there is the basis to implement this management proposal from biological, economic and legislative points of view, and perhaps even psychological (in terms of acceptance by the public), a strong political will is still necessary to implement it. Often this is lacking. Implementation of this plan would be difficult, if not impossible, because of a lack of coordination between local and regional authorities, and between the authorities of various regions and the national government.

However, I believe this is the road to follow as dynamically and flexibly as possible, as various political and scientific authorities become involved in the discussion. This conservation plan may surpass its aim of protecting the wolf by instituting a system where human activities are harmoniously integrated into the natural environment. Moreover, it would directly stimulate environmental reassessment and management, areas still in their infancy in Italy.

CONCLUSION

Through the Wolf Specialist Group of the Species Survival Commission of the International Union for the Conservation of Nature (IUCN), research, protection, initiatives and exchanges of information on wolf biology have grown throughout the world, especially in Europe. The Italian Wolf Project benefited early from this situation and has now taken on the role of a model for the conservation of the wolf in Spain, Portugal and Poland, and with regard to some aspects, also in Scandinavia, Yugoslavia, Greece and Turkey. The species probably has special needs in each of these areas which will need to be examined. The more general conservation problems, however, such as the conflict with man and the search for a compromise, are at least partly comparable to the situation in Italy. The experience acquired in Italy may contribute to preventing mistakes in other areas, and may accelerate intervention in similar ecological areas.

In our work we have often had to breach a subject never before encountered in Italy — that the wolf can be saved and not simply killed. Some important steps have been taken but much still remains to be done in order to ensure perpetuation of the wolf in this country. We only hope that our efforts win the race against time and are completed before the last wolf dies.

Acknowledgments

This work is the product of several years of cooperative field work and discussions with my friends Erik Zimen and David Mech, to whom goes my gratitude for their advice and criticism. Many others helped throughout the study, and it would be impossible to mention them all, but a special thanks is due the International Union for the Conservation of Nature (IUCN) and World Wildlife Fund (WWF) staff in Gland and in Rome for their help in all phases of the study. Paolo Barrasso was a perfect assistant in the field. My deepest gratitude goes to the hundreds of aggressive shepherds from the Appenine Mountains who shaped most of the ideas presented here. I also thank the organizers and editors of the Portland Wolf Symposium and two anonymous reviewers who made the manuscript a readable paper.

This work was partly supported by IUCN/WWF project N.989 and 1500 Phase I and II.

Wolves in Israel

H. Mendelssohn

INTRODUCTION

As evidenced by seven quotations in the Bible, the wolf *(Canis lupus)* coexisted with man in the area of Israel for thousands of years and was well known as a predator. Tristram (1885) stated that the wolf is found in every part of Palestine. Wolves still live in over half of Israel, but have disappeared during the last 30 years from the more densely settled areas. It appears, however, that in some areas their population has increased recently due to easily available food from garbage dumps.

The Wild Animals Protection Law of 1954 completely protected almost all wild animals in Israel except, among others, the jackal *(C. aureus)*, which was later given complete protection.

The human population of Israel in 1979 was 3,830,000 in an area of 20,720 km², or 185 people per km². The northern and central part of the country has a much higher human density than the Negev (the southern arid part) and the rift valley (Jordan Valley, Dead Sea depression and Arava Valley), where most of the contemporary wolf population lives. Already in the 1930s, wolves had disappeared from the densely settled areas—the coastal plain between Haifa and Tel Aviv and the mountains between Nablus and Hebron.

Israeli wolves are animals of open areas. They have never inhabited the dense Mediterranean scrub forest that covers about 400 km² in Galilee and on Mt. Carmel. Apparently, the Indian *C. l. pallipes* also do not live in dense forest cover (Shahi, 1977).

Because of Israel's small size, its nature reserves are also small and, thus, are of little use to such wide-ranging animals as wolves. The largest nature reserve in the north, that of Mt. Meron, has an area of about 90 km², which is largely covered by scrub forest and therefore not suitable for wolves.

173

TAXONOMIC POSITION

Ellerman and Morrison-Scott (1951) do not state which wolf sub-species occurs in Palestine[1] but, because they include northern Arabia in the distribution area of *C. l. pallipes* Sykes 1831, it may be concluded that this subspecies also occurs in Palestine. Wolves in Israel display a wide range of size and color differences. Tristram (1885) considered them to be larger and stronger than European specimens.

The information on wolf size and color for the present paper is based on 45 specimens in the Zoological Museum of Tel Aviv University (ZMTAU) and on two from the Beth Ussishkin Museum of Kibbutz Dan. Twelve additional specimens of the ZMTAU could be used only partly because the skulls were incomplete, they were juveniles, or because they appeared to be hybrid. For description of color and quality of summer and winter coats, specimens of a breeding group at the Wildlife Research Centre of Tel Aviv University (WRCTAU) were used as well as 23 skins in the ZMTAU.

As the most reliable indicator of size, the condylobasal length (CBL) of the skull was taken from the anterior extremity of the premaxillary to the rear extremity of the occipital condylus. This measurement was thought to be more reliable than the greatest length of the skull (GtL), measured from the anterior tip of the premaxillary to the tip of the occipital crest, which was more variable. However, because GtL is generally given in literature, it is also used here. CBL, used by Harrison (1964), was preferred to the perhaps better measurement of the basilar length (Zollitsch, 1969) because Harrison (1968) dealt with the wolves of this region and it was considered important to compare our measurements with his.

Other skull measurements are not presented here for the sake of brevity. Neither are body measurements, for they are not sufficiently exact, depending somewhat on the extent to which the specimen has been stretched. Weights, however, are given, notwithstanding that they vary considerably depending upon the physical condition of the specimen and on how full the stomach is.

Israeli wolves are larger and darker in areas of higher rainfall, and smaller and lighter-colored in arid areas. Extremes are represented by a large male (Nr.7425 ZMTAU) from the Golan, with CBL of 226.7 mm and weight of 32.3 kg, and a small female (Nr.5958 ZMTAU) from southern Sinai with CBL of 175.5 mm and weight of 12.3 kg. Both specimens were in good physical condition. The distance between the localities where these two specimens were collected is about 600 km.

Even with specimens of the same sex, size difference is apparent: Nr.6 from near Haifa (ZMTAU), a male with CBL of 218.4 mm and Nr.6320 from the Negev (ZMTAU), a male with CBL of 185.3 mm (Figure 6.1). In this case, the distance between collection localities is only 250 km.

---

[1] Palestine is used in this paper as a zoogeographical term, not as a political one, and includes Israel, the West Bank and the Gaza area.

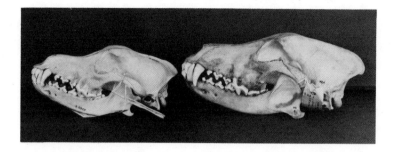

Figure 6.1: Skulls of male *C. l. arabs* (left) and male
C. l. pallipes (right).

Precipitation in Israel is highest in the north and west, and
decreases toward the south and east. The lower Jordan Valley, the
Dead Sea depression, the Negev and the Arava Valley are deserts
with less than 150 mm annual rainfall.

The largest wolves are all from areas with more than 400 mm
annual rainfall and a Mediterranean climate and vegetation, whereas
the smaller wolves inhabit the more arid areas with less than 400 mm
rain (Tables 6.1 and 6.2, Figure 6.2).

There is some size overlap between the wolves of the Mediterranean areas and the arid areas, but not much. One Mediterranean
wolf differed widely from the others: the very large male, Nr.7425
(ZMTAU) with a CBL of 226.7 mm and a weight of 32.3 kg. This
male was not considered when calculating the average in Table 6.1, as
it orginates from a marginal locality in the Golan and may represent
a different population.

Table 6.1: Size and Distribution of Wolves in Israel in Relation to
Rainfall. [Size is expressed by condylo-basal length (CBL) of the
skull in mm. One-way ANOVA: males (F = 39.7, p <0.01);
females (F = 9.7, p <0.01).]

	. . . . . . .Males			. . . . . . Females		
	Average[1] (mm)	Range (mm)	n	Average (mm)	Range (mm)	n
Mediterranean area with >400 mm rain (Mediterranean pallipes)	214.5	206.1–218.4	8	203.8	190.0–211.5	6
Areas with <400 mm rain (without C. l. arabs specimens) (Desert pallipes)	205.1	199.1–208.8	10	193.8	190.2–200.0	5
Small specimens from areas with <50 mm rain (C. l. arabs)	192.9	186.3–196.7	4	181.1	175.5–186.7	2

[1] Excludes one extremely large male from a marginal area that may have represented a different population (see text).

Table 6.2: Size and Distribution of Wolves in Israel in Relation to Rainfall. [Size is expressed by greatest length of skull (GtL). One-way ANOVA: males (F =27.7, p <0.01); females (F = 16.5, p <0.01).]

	. . . . . . .Males			. . . . . . Females		
	Average (mm)	Range (mm)	n	Average (mm)	Range (mm)	n
Mediterranean area with >400 mm rain (Mediterranean pallipes)	228.9	218.0-241.8	8	219.3	210.1-227.0	6
Areas with <400 mm rain (without C. l. arabs specimens) (Desert pallipes)	218.6	210.2-225.0	10	206.4	202.0-212.0	5
Small specimens from areas with <50 mm rain (C. l. arabs)	202.7	195.0-206.2	4	192.1	186.1-198.0	2

In the desert areas (<400 mm rain), two sizes of wolves occur. Most desert wolves are quite uniform in size, but in the most southern area with less than 50 mm rain (Tables 6.1 and 6.2), and in southern Sinai, much smaller wolves occur. They not only have smaller skulls, but also smaller bodies. The males had CBLs of less than 200 mm and the females had CBLs of less than 190 mm. These small wolves were, with one exception, only found in the southern part of the Arava Valley in areas with less than 50 mm rain, and in Sinai (Tables 6.1 and 6.2). Only one old male of this type was found close to the 100 mm isohyet. In the southern Arava Valley, both large and small wolves occur together in the same area.

These small wolves correspond to the description of C. l. arabs Pocock, 1934. Harrison (1968) gives the GtL of C. l. arabs as 184.5-208.0 mm. The type specimen, a female Nr.34.8.4.12 in the British Museum, has a GtL of 198.8 mm and a CBL of 181.8 mm. Five male C. l. arabs in the British Museum and in the Harrison Museum have GtLs of 192.0-206.0 mm and CBLs of 182.0-193.0 mm. Four females in these collections have GtLs between 184.5 and 201.0 mm and CBLs of 169.0-188.9 mm.

These measurements correspond well to those of our local specimens (Tables 6.1 and 6.2). Harrison (1968) included, under C. l. arabs, two considerably larger wolves from Kuwait, but these seem to belong to C. l. pallipes. Perhaps there, as in the southern Arava Valley, both subspecies live in close proximity.

Although the larger wolves in Israel can be divided into two size groups separated by the isohyet of 400 mm (Tables 6.1 and 6.2), they can all be considered C. l. pallipes. Table 6.3 shows measurements of 11 C. l. pallipes from the Middle East and India. The measurements of the wolves of Israel (Tables 6.1 and 6.2) correspond well with these measurements. There is also a considerable difference in the size of the os penis between the local C. l. pallipes and C. l. arabs.

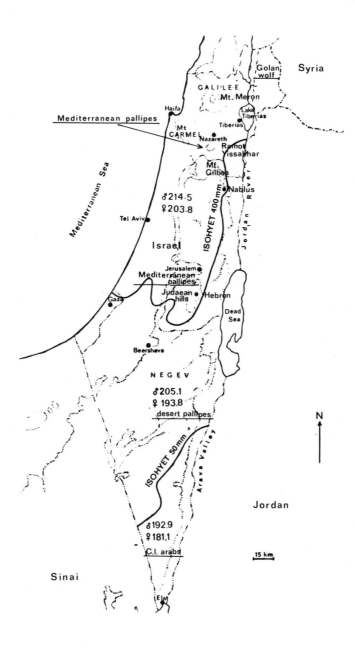

Figure 6.2: Distribution of three different wolf populations in Israel: Two populations of *C. l. pallipes*, differing in size and color, are separated by the isohyet of 400 mm. *C. l. arabs* occur mainly south of the 50 mm isohyet. Size is indicated by the condylo-basal length (CBL) of the skull in mm.

Whereas the lengths of the os penis of three *C. l. pallipes* were 79.7, 81.1 and 81.5 mm, the measurements for two *C. l. arabs* were only 68.0 and 69.3 mm.

Table 6.3: Skull Measurements of 11 *C. l. pallipes* in the British Museum. (Specimens are from the Middle East and India.)

| | . . . . . GtL . . . . . | | . . . . . CBL . . . . . | |
	Mean (mm)	Range (mm)	Mean (mm)	Range (mm)
Males (n = 8)	228	210-238	212	201-222
Females (n = 3)	220	216-225	205	198-214

Wolf size apparently is more influenced by rainfall than by temperature. There is no size difference between wolves from the hot rift valley (mean >23°C) and those from the much cooler Negev Highlands (mean <19°C); rainfall is similar in both areas.

Nothing is known about the relations between *C. l. pallipes* and *C. l. arabs*, that occur together in the southern Arava Valley in areas with less than 50 mm rain. Possibly this area was formerly inhabited only by *C. l. arabs*, which are probably better adapted to extreme desert conditions. Increasing human development of the area improved the conditions for wolves by providing an easily available source of food at garbage dumps, and by stimulating increase of wildlife near areas of irrigated agriculture. These improved conditions may have enabled the penetration of *C. l. pallipes* into this area, perhaps competing with *C. l. arabs* and supplanting it. If this assumption is correct, *C. l. arabs* should disappear from this area in the future. They are now much rarer than *C. l. pallipes*. It is not known if the two subspecies interbreed. Neither is information available to indicate whether the two populations share the same habitat, or whether they are spatially or temporally separated.

A similar case is occurring with two hedgehog species in the coastal plain of Israel, where *Erinaceus europaeus*, following agricultural development, is supplanting *Hemiechinus auritus* (Mendelssohn, unpub.).

There still remains the fact that the wolves of the Mediterranean area of Israel (>400 mm rain) are distinctly larger than those of the more arid areas (50-400 mm rain). The question of whether these two discrete populations should be given separate subspecific status has to remain open until more material from other areas in the Near East can be examined. For the time being, the terms "Mediterranean pallipes" and "Desert pallipes" will be used.

The aforementioned very large male, Nr.7425 (ZMTAU), with a CBL of 226.7 mm, a weight of 32.3 kg and a dense, dark winter fur, is certainly quite different from any *C. l. pallipes* and looks more like a European wolf. Three male European wolves originating from the Carpathian Mountains (#1527, 2581, 2851—ZMTAU) have CBLs of 236.6, 236.6 and 231.7 mm, and weights of 29.2, 30.0 and 32.3 kg.

Their winter fur is quite similar to that of Nr.7425, but more reddish-brown on the sides and legs. A more important difference, however, is that the skulls of the European wolves are not only longer, but also much broader and thicker, especially the jaws. The skull of Nr.7425, although much larger than that of other *C. l. pallipes*, has the same delicate shape when compared to the European wolf skulls. Whereas the CBL/height of the upper jaw behind the canines in Nr.7425 is 6.71 mm, in the three Carpathian wolves it is 5.56 mm, emphasizing their much stronger jaws.

Weights of Israeli wolves are shown in Table 6.4. Altogether, individual variability does not seem very great in these three wolf populations.

Table 6.4: Weight of Israeli Wolves

| | . . . . . .Males . . . . . . | | | . . . . . Females . . . . . | | |
	Mean (kg)	Range (kg)	n	Mean (kg)	Range (kg)	n
Mediterranean pallipes	23.6	22.8-26.3	6	—	—	—
Desert pallipes	20.1	17.4-22.5	7	17.0	15.7-18.9	6
C. l. arabs	18.0	16.3-19.2	3	12.3	—	1

It may seem strange that in such a small country as Israel, only 410 km from north to south, there are three distinct populations of such wide-ranging animals as wolves. There are, however, considerable climatic differences, as mentioned earlier. Perhaps the different populations are well adapted to local climatic conditions. A similar situation is found with the leopard. There formerly occurred in Galilee, and perhaps in other areas in the north, *Panthera pardus tulliana* (now extinct in Israel), one of the largest of the leopard subspecies. However, in the Judean Desert and in the Negev, *P. p. nimr* occurs, one of the smallest subspecies (Harrison, 1968). Perhaps for these two subspecies, the 400 mm isohyet was also the dividing line. The greatly varying environmental conditions over relatively small distances in Israel may stimulate the development of differing populations adapted to special local conditions. Nevo (1969, 1973), and Nevo and Shaw (1972) found in Israel four populations of mole-rat *Spalax ehrenbergi* that differ in size, chromosomes and behavior.

In some European and U.S. zoos, wolves originating from Persia (Iran) are, because of their origin, exhibited as *C. l. pallipes*. According to pictures and a specimen I examined that originated from a pack at Cologne Zoo, these wolves are not *C. l. pallipes*. This specimen, one of the smaller males in the pack (Nr.7332, ZMTAU), had a GtL of 249.3 mm, a CBL of 226.1 mm and a weight of 29.3 kg. Its fur was much greyer and longer than that of *C. l. pallipes*. The specimen had been born and kept in a zoo in Israel, so the long and dense fur did not result from a cold climate. It looked entirely different from local *C. l. pallipes* as well as from Indian specimens, which greatly resemble specimens in Israel, as shown by pictures in Shahi (1977). The specimens in question may be more closely related to *C. l. campestris.*

The fur of *C. l. pallipes* and *C. l. arabs* is very short and thin in summer (Figure 6.3), similar to the short summer fur of *Lycaon pictus.* An exception is the dorsal hair that is somewhat longer in *C. l. pallipes* and *C. l. arabs,* even in summer. Perhaps the longer dorsal hair provides some protection from solar radiation in summer if the animals have to be active during hot summer days. The winter coat is longer, but not as long and dense as that of more northern subspecies. It is rather similar to the summer coat of the northern forms. The differences in both summer and winter coats is conspicuous when comparing local *C. l. pallipes* with European and North American wolves kept in Tel Aviv under the same climatic conditions.

Figure 6.3: Summer coat of *C. l. pallipes.* Note short fur and longer fur on neck and back. Specimen is a dark male from the Arava Valley.

The hair of the summer coat of *C. l. pallipes* and *C. l. arabs* is about 30 mm long on the back and about 10 mm on the sides, but there is much variation. Winter back hairs, particularly from the Mediterranean area, are 45-65 mm long, those of the saddle 70-100 mm, and those of the sides 20-30 mm long. Whereas the summer coat has no wool, or only a little between the longer dorsal hairs, the winter coat has a dense wool layer.

Because of considerable individual variability, it is difficult to describe the colors of Israeli wolves. Harrison (1968) described several specimens from Arabian countries. A color phase quite common in the Arava Valley has a mottled yellowish-grey and black back and tail. Neck, shoulders, sides, thighs, legs, forehead and ears are light yellowish buff with an orange tinge on the sides of some specimens. The lower parts of the legs, their inner surfaces and the ventral surfaces are white. The bridge of the nose is greyish-yellow, muzzle and cheeks are white with some specimens having light spots, some very conspicuous, over the eyes. In other specimens, neck, sides and thighs are light yellowish-grey or light grey, but ears and legs are generally yellowish or buff.

Wolves of the Mediterranean area are darker. Their sides, thighs and ears are a dirty greyish-brown. In another dark phase, the mottled grey and black color of the back extends downward on the sides with the ventral surface being light grey and legs and ears yellowish-grey. The muzzle and forepart of the cheeks are always either white, dirty white or light grey. Dark specimens also occur occasionally in the Negev (Figure 6.4).

Figure 6.4: Winter coat of dark-colored *C. l. pallipes* from Arava Valley. Same animal as in Figure 6.3.

A characteristic feature of many wolves in Israel is that the pads of the third and fourth toes are connected from behind. This connection (Figure 6.5) is conspicuous mainly on the forefeet, but if the pads of the forefeet are connected, those of the hind feet are generally connected too. Under favorable conditions, this connection also shows in the tracks and verifies that the track is from a wolf.

Figure 6.5: Right forefoot of male *C. l. arabs* (Nr.6897 ZMTAU) showing connections of pads of third and fourth toes. Picture taken from a dry museum skin.

Altogether, 23 skins of local wolves were examined. Six were of Mediterranean pallipes and only one had connected pads. Five were of *C. l. arabs* and four showed the connected pads. Twelve skins were of desert pallipes and 11 of them had connected pads. Several skins of *C. l. pallipes* from India also had connected pads, but they were not found in any other wolves, nor in many domestic dogs.

DISTRIBUTION AND POPULATION SIZES

Figures 6.6 and 6.7 show the distribution of wolves from 1935 to 1950, and the more recent distribution from 1970 to 1980. By 1935 there were no wolves in the densely settled areas of the coastal plain and the areas between Hebron and Nablus. Between 1950 and 1970, they disappeared from most of northern Israel and from the areas west of Jerusalem, and they are now (1980) already rare in the areas north of Beersheva. They still occur in about half of Israel in about 70 percent of the area they inhabited before 1950. Figure 6.8 shows exact localities where wolves were observed according to both reliable and not completely reliable records (Ilani, 1979).

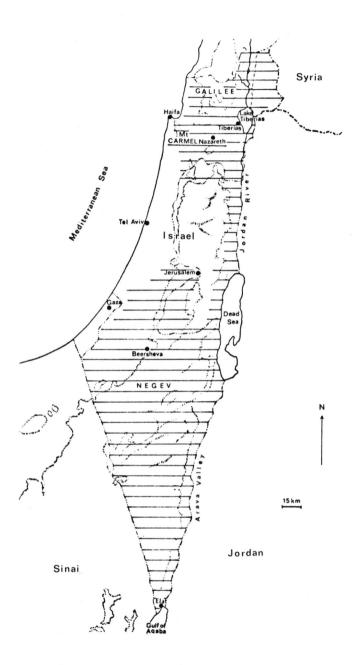

Figure 6.6: Distribution of wolves in Israel between 1935 and 1950, indicated by cross-hatching.

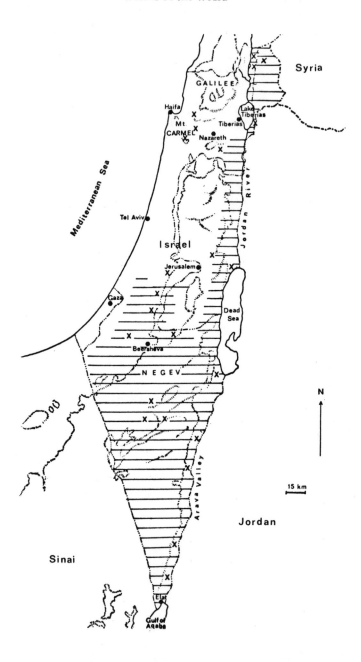

Figure 6.7: Distribution of wolves in Israel between 1970 and 1980.
Crosses indicate areas from which specimens in the Zoological Museum
of Tel Aviv University were collected. One cross may indicate as many
as 10 specimens.

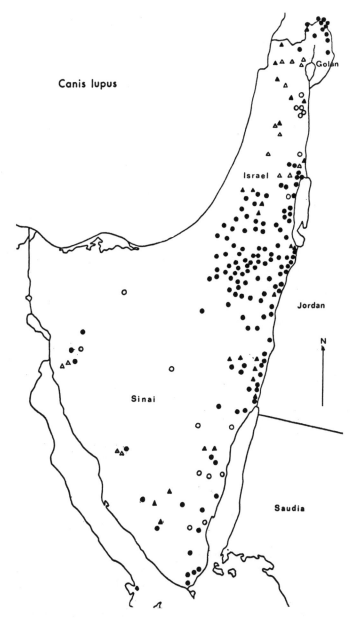

Figure 6.8: Distribution of wolves in Israel, Golan, the West Bank area and Sinai according to Ilani (1979). Records include literature records, reports and observations (courtesy of G. Ilani).
● = verified records during the survey period of 1970-1976.
○ = unverified records during the same period.
▲ = verified records during 1863-1970.
△ = unverified records between 1863 to 1970.

The information on occurrence of wolves before 1970 is largely based on occasional observations, on reports of predation on domestic stock, and on collected specimens. From about 1970 onward, more systematic direct observations were made by the rangers of the Nature Reserves Authority. These observations were summarized by G. Ilani, who kindly contributed the most recent observations of 1978 and 1979 (Table 6.5).

Table 6.5: Wolf Numbers According to Regions Observed During 1978 and 1979. (Observations compiled by G. Ilani.)

Region	Number of Wolves Observed	Date	Remarks
Golan	4	8 Jan. 1978	Shy; harassed by stock-
	2	12 Feb. 1979	breeders; endangered
		12 Apr. 1979	
Mt. Gilboa	3	21 Feb. 1978	Shy; harassed by local Bedouin shepherds
Eastern slopes of Samaria	2	17 Sept. 1978 13 Mar. 1979	
Northern Judaean Desert	4	8 Dec. 1978	Shy; harassed by local Bedouin shepherds
Central and southern Judaean Desert	6 3	4 Jan. 1978 19 Nov. 1978	Very tame; make occasional use of feeding station
Western slopes of Judaean Hills	2 2	5 May 1978 14 Dec. 1978	Shy; harassed by stock- breeders; endangered
Northern Arava Valley	5 3	24 Nov. 1978 2 July 1979 1 May 1979	Tame
Central Negev	3 4 1 7	15 Aug. 1978 7 Dec. 1978 30 Mar. 1979 9 May 1979	Tame
Northwestern Negev	2 1 3 3	6 Jan. 1978 13 Feb. 1978 11 June 1978 22 Apr. 1979	In some areas tame, in others shy
Southwestern Negev	4 2	21 Oct. 1978 30 Mar. 1979	Shy
Southern Arava Valley	5 3	25 Feb. 1978 10 Aug. 1978	Tame

If the largest observations for each area are used, the total minimum estimate is 46 wolves, assuming no movements between areas. Most observations were made during the day, whereas wolves in this area are mostly active at night, even where they are tame. Therefore, G. Ilani (pers. comm.) estimates that the total number of wolves in Israel may be 110-150.

Of the three wolf populations described, only the population of desert pallipes can currently be considered secure. This population subsists, to a considerable degree, by scavenging at garbage dumps. It may well be that the density of this population increased concomitant with the increase of settlements, army camps and garbage dumps in this area.

The Mediterranean pallipes are almost extinct, and urgent measures are necessary to save them. According to G. Ilani, 10-15 of this population survive in the Golan, five to seven on Mt. Gilboa, and five to eight on the western slopes of the Judaean hills, representing three populations now isolated from each other.

The population of *C. l. arabs* in the southern Arava Valley, if it still exists, must also be considered endangered. Their relations with the larger *C. l. pallipes* are not known, but probably the larger subspecies is ecologically dominant under present conditions. Unfortunately, the wolf population in the southern Arava Valley where both subspecies coexist, was badly decimated between 1970 and 1976, despite legal protection. However, by 1980 the populations had recovered quite well.

C. l. arabs also occur in Sinai, but their survival there is precarious. Because little wildlife exists in the badly overgrazed, overbrowsed and overhunted habitats of Sinai, the wolves depend mostly on the goat herds of the local Bedouin and are therefore continually harassed. During the occupation by Israel, the situation improved somewhat as the wolves could scavenge on the rich garbage dumps of tourist resorts and Israeli settlements.

FOOD HABITS

No systematic study on the food of wolves in Israel has yet been made. It is certain that wolves are not dependent on a few species of wildlife, as are many North American wolves (Mech, 1970), nor are they as dependent upon domestic animals as are the *C. l. pallipes* in Bihar, India (Shahi, 1979). According to many occasional observations, Israeli wolves are opportunistic feeders, preying on smaller wildlife, rarely on wild ungulates, occasionally killing domestic animals, but often scavenging on livestock carcasses and at garbage dumps. Garbage dumps are good places to see wolves, especially in the desert, as are the feeding stations run by the Nature Reserves Authority (funded by the World Wildlife Fund) to support populations of scavengers, mainly vultures.

Analysis of 15 wolf stomach contents revealed remnants of the following animals: Jirds (*Meriones* sp. sp.), hares *(Lepus capensis)* and chukar partridges *(Alectoris chukar)*. Hares seem to be a common prey, but many apparently are road-killed hares picked up by wolves. Several wolves that had been killed on roads had undigested pieces of hare in their stomachs.

Gazelles (*Gazella* sp.) are occasionally taken, but there is only one observation of wolves hunting gazelle. In the northern Arava Valley, three wolves were seen one morning chasing a male dorcas gazelle (*G. dorcas* subsp.) which they caught after a chase of about 1 km. As gazelles are diurnal with poor vision at night (Mendelssohn, 1974), they are easily caught at night by wolves.

The possibility that wolves may influence gazelle populations is shown by the comparison of two hill ranges in northern Israel.

They are somewhat similar with regard to climate, vegetation and availability of water. Both are separated by a 5 km broad, intensely farmed plain. Wolves disappeared from one area, Ramot Yissakhar, between 1952 and 1955, but still survive today in the other, Mt. Gilboa. Although gazelles increased over the entire country after they were legally protected (Mendelssohn,1974), their increase was much slower and they did not reach high densities in the area still inhabited by wolves. Several gazelle counts made on Ramot Yissakhar indicated gazelle densities between $15-18/km^2$, whereas a count on Mt. Gilboa revealed only 5.5 gazelle/km^2 (Mendelssohn, 1974). Possibly the wolves helped suppress the increase in the latter area.

In the Judaean desert and in the Negev, wolves inhabit areas in which ibex *(Capra ibex nubiana)* occur, but no cases of predation on ibex are known. Wild pigs *(Sus scrofa)* are very common in northern Israel and in the Golan, but no cases of preying on pigs have been observed so far.

Wolf predation on livestock occurs mainly with the larger Mediterranean wolves, but is not common. A number of cases of predation on sheep were noted among the wolves that formerly inhabited northern Israel. One pack that had its territory southeast of Haifa preyed on sheep, sometimes once or twice a week, but in different places each time; its movements could be followed according to the information received on sheep killing. The killing was carried out by one or two individuals and, generally, in each case only one sheep was killed and dragged away. The killing was done in the late afternoon when the sheep were still grazing or returning to the fold, but sometimes in the morning or at night in the fold where, on one occasion, 16 sheep were killed. Most sheep predation occurred during autumn and winter. Sheep predation has also occurred in recent years in the Golan and southwest of Jerusalem.

Wolf predation on calves is of more concern because of their greater economic value. These cases mainly occurred in autumn and winter southwest of Jerusalem and in the Golan, where beef cattle are kept on the range. Wolves have been observed among the cattle without attacking them until the calves are born in autumn (Z. Choresh, pers. comm.). The wolves then begin to prey on the calves, mostly on the very young ones but occasionally also on larger ones. After the dry summer, cows are in poor physical condition so calf mortality is high and, in a number of cases, wolves probably scavenged on calf carcasses. Here too, kills were generally made by one or two wolves, even though packs of up to eight had been seen there. The largest number of calves killed by one wolf was 17, within two months. After that animal was trapped, calf predation ceased. Both beef cattle and zebu are kept on the range. Zebu calves are only very rarely killed by wolves, as the dams defend their calves effectively, unlike beef cattle. Occasionally adult cows of the small local breed (weight c. 200 kg) were also attacked, mostly by biting at the flanks, but were not killed.

Wolves of the Mediterranean area also feed on small animals, as shown by stomach contents and scats. A female that had been conditioned to people because she pairbonded with a domestic dog, was observed catching and eating a hare, and she was often seen catching and eating voles *(Microtus guentheri)* weighing 25-40 g.

The desert pallipes do not prey on large livestock. Predation on sheep in this area is rare, but the Bedouin consider wolves as predators of their black goats which are smaller than sheep, and in the southern Negev and Sinai, weigh only 12-25 kg. Hairs of the black Bedouin goats have been found in wolf scats collected in this area, but it is unknown whether they were from kills or carrion.

The desert pallipes tend to approach settlements and people more than do the Mediterranean wolves. In a desert kibbutz (communal agricultural settlement), wolves entered the cowsheds at night and moved among cattle and calves without molesting even the youngest calves. However, they entered a hen-house and killed chickens. In another desert kibbutz, the wolves visited the area of the hen-houses at night and caught escaped chickens, but entered a hen-house and killed 10 hens when a door was left open. Their main food at both places, however, was chicken carcasses and offal that they scavenged from the garbage dump. These wolves react eagerly to the cheeping of chicks and were attracted from about one km by these cheeps, both live and tape-recorded (Z. Choresh and G. Ilani, pers. comm.).

Altogether, Israeli wolves do not suffer from lack of food, as almost all specimens that could be examined (n = 28) were in prime physical condition. Two exceptions were noted, both road-killed. A nursing female was killed in May in the Golan, was extremely thin, had eight active teats, and carried a heavy load of ticks (A. Galili, pers. comm.). The other female was a desert wolf killed on 19 July. She had already stopped nursing but had not yet recovered from her poor condition.

RELATIONS WITH OTHER ANIMALS

Wolves and hyenas *(Hyaena hyaena syriaca)* meet quite often at garbage dumps, carcasses and feeding stations. Wolves generally make way for the hyenas which are larger, adults weighing 25-40 kg. In one observation, however, a group of wolves drove a hyena from a carcass.

Wolves feeding on carcasses during daylight may meet vultures. One pair of wolves was feeding on a carcass at a feeding station in the morning. Eight griffon vultures *(Gyps f. fulvus)* from a nearby colony arrived, but did not approach the carcass until the wolves had departed. In another case, a lone wolf fed one morning on a carcass at another feeding station. Seventeen griffon vultures arrived and tried, time and again, to approach the carcass, but were chased away each time by the wolf. They too had to wait until the wolf had departed. In Spain, vultures occasionally succeed in driving away dogs from a carcass (Koenig, 1974).

Jackals *(Canis aureus syriacus)* were common in Israel until 1964 when the Plant Protection Department of the Ministry of Agriculture organized an extermination campaign against them (Mendelssohn et al., 1971). The jackal populations recovered only slowly after this persecution.

Unlike wolves, jackals, which are mainly scavengers, live most commonly in areas with dense human populations where, in some places, they formerly reached a density of about one pair/km^2. In deserts, where wolves are relatively common, jackals occur only in a few localities. Ilani (1979) believed that jackals are more dependent on water since they are found, particularly in the desert, only near human settlements where water is available. It may, however, also be the easy availability of food that attracts jackals to settlements. Desert wolves, on the other hand, have been observed up to 50 km from the nearest water (Ilani, 1979). Possibly they drink only infrequently and husband their body water efficiently. In the few Mediterranean areas where both species occur, wolves are rare and probably cannot influence jackal populations. Cases of direct interactions between wolves and jackals have not been observed, but wolves probably dominate.

Recently, feral dogs have replaced wolves in Israel where wolves have disappeared. These feral dogs are crossbreeds between pariah dogs, which are no longer pure in Israel, and imported European breeds, mainly alsatians. They subsist mainly by scavenging on garbage dumps and killing lambs, sheep and goats. These predations are often ascribed to wolves. They may kill 10, 15 or more animals in one night, mostly by biting them in the throat. Once, three dogs killed 70 kids and goats in one night. Often, feral dogs do not feed on their victims.

Young and Goldman (1944) stated that where wolves are decreasing, they may hybridize with domestic dogs, although few such cases have been reported in North America. Several of the wolves in the ZMTAU may be wolf-dog hybrids (Rohrs, pers. comm.).

One case of a pair bond between a female wolf and a domestic dog in Israel is well documented (M. Shamgar, pers. comm.). The female wolf was one of the last survivors of a pack that lived in the area southeast of Haifa. The case occurred at Mishmar Ha Emeq, a kibbutz 20 km southeast of Haifa. Mr. Shamgar, a farmer on the kibbutz, owned a male alsatian crossbreed that used to spend the day with his master in the fields. One day in February 1963, Mr. Shamgar saw at a distance what he thought was a jackal and set his dog at the animal, which turned out to be a wolf. Both animals sniffed each other, began to "play" and spent most of the day together. The next day the wolf was already in the field when Mr. Shamgar and his dog arrived in the morning. Again wolf and dog spent the day together, and during the following days many interactions, mutual mountings and, eventually, copulations were observed.

Within a month the wolf lost her fear of Mr. Shamgar and approached him to within about 10 m. She disappeared during the last stages of gravidity and returned a few weeks after. During this period, the dog also disappeared occasionally for periods of a few days. The wolf produced five cubs which were found dead, when about six weeks old, in an orchard sprayed with parathion. The wolf was then to be found on most days waiting for the dog and Mr. Shamgar at a certain place in the morning, spent the day with them in the field, and accompanied them in the afternoon back to the kibbutz. She became less and less shy of people and sometimes approached Mr. Shamgar from the rear to a distance of only 1 m, but kept a greater distance when he looked at her directly. She was more shy of other people but began to enter the area of the kibbutz, and sometimes Mr. Shamgar found her in the morning on his doorstep. During her estrus period in 1964, both animals disappeared, returned for some time, but then disappeared for good. According to a film of this pair, the wolf initiated almost all interactions with the dog; the dog's activities were much more influenced by his relations with his master.

PACK SIZE, REPRODUCTION, AGE

Like other information on the life history of wolves in Israel, knowledge of wolf pack size and composition is based on casual observations. Harrison (1968) stated that the wolves of the arid regions of the Arabian peninsula hunt singly or, at most, in pairs. Tristram (1866) stated that in Palestine he never saw two wolves together. In fact, almost all depredations on livestock during the last 45 years have been carried out by single wolves, or a pair.

Any group size, from single specimens to groups of 12, has been seen by reliable observers, with larger groups being seen only rarely. In late summer, autumn and winter when the grown cubs accompany their parents, family groups of up to a pair of adults and five cubs are quite often seen.

Israeli wolves breed in winter and whelp in spring. According to the dentition of young cubs collected in the Arava Valley in summer, the cubs are born there from early to mid-April. A female *C. l. pallipes* originating from that area and kept at the WRCTAU, came into estrus during the second half of January, and whelped between the end of March and the beginning of April. Because she had been kept isolated for several years at the Tel Aviv Zoo, she began to breed only in 1977 at six years of age. She bore four cubs (all males) in 1977, six cubs (4 males, 2 females) in 1978, and a single male cub in 1979. While this paper is being written, in 1980, she is again gravid.

There is only one observation on the time of reproduction in
northern Israel. A female in the last stages of gravidity was shot on
29 April 1952 at the hill range of Ramot Yissakhar. It may be, therefore, that reproduction in the Mediterranean area takes place somewhat later than in the desert.

To obtain an idea of the age composition of the wolf population,
an attempt was made to age the 41 available skulls according to
tooth wear, compared with skulls of known age. This method is not
too exact, as the degree of wear varies between individuals and
among the teeth of a single individual. For example, sometimes the
canines showed the greatest wear, whereas other times the incisors
or molars were more worn. Therefore, the specimens were arranged
in rather large groups (Table 6.6).

Table 6.6: Estimated Ages of Wolves in Israel According to Tooth Wear in 41 Skulls

Estimated Age (arbitrary classes)	25 Males	16 Females
<1 year	6	2
1–2 years	12	4
3–4 years	3	6
5–6 years	2	3
6–8 years	2	—
>8 years	—	1

The estimated age composition of the wolves collected seems to
be quite normal, but the preponderance of young males is conspicuous (Table 6.6). Stenlund (1955), Pulliainen (1965) and Mech
(1975) reported unequal sex ratios in wolf populations. Mech (1975)
found a preponderance of males in a high-density population but,
in low-density populations, an equal sex ratio or a bias toward females. It is possible that the preponderance of males in Israeli wolves
is a real one, but it could also be that the females are more cautious
than the males in relation to such mortality factors as road accidents, poisons and traps. Among eight wolves trapped in steel traps
in the southern Arava Valley, there were five males and three females.
Mech (1975) reported a pair that produced 22 male and four female
pups. A pair of desert pallipes at WRCTAU produced 10 males and
two females in four litters. The seemingly higher mortality rate of
males in Israel could either reflect a preponderance of males in
litters, or greater caution on the part of females.

CONSERVATION AND MORTALITY

The Wild Animals Protection Law of 1954 legally protects the
wolf, but there was no organization to enforce this law until the
Nature Reserves Authority was founded in 1964. Even when this

authority began to prosecute offenders, very few people who killed predators were brought to trial. The management of the Nature Reserves Authority at that time did not view predators as an integral part of the ecosystem that ought to be protected and even gave orders and permits to destroy the interesting wolf population in the southern Arava Valley where *C. l. pallipes* and *C. l. arabs* coexist. Even when people who had illegally killed predators were brought to trial, judges apparently still held prejudices against predators and so imposed small fines. Only one man, who killed a wolf in 1971, was fined I.L.100 (then about $30), and was given an additional suspended fine of I.L.500 for two years.

There were, however, many more wolves killed after the Nature Reserves Authority began its work. In fact, several wolf populations were exterminated or nearly exterminated between 1964 and 1980. In Israel, there is no strong traditional animosity toward wolves, so prevalent in many European countries and in North America. Indeed, the Hebrew word for wolf is often used as a personal name. If, however, wolves preyed upon livestock, the farmers poisoned the wolves, advised by pest control officers who encouraged wholesale application of poisons. This illegal poisoning was one of the main mortality factors that, during the last 30 years, reduced the already small wolf populations in northern and central Israel, and eventually exterminated or nearly exterminated some of them. Poisoning still causes casualties if wolves prey on livestock. However, road accidents seem to be an important, and perhaps the major mortality factor, affecting mainly the relatively dense wolf population in the Negev. Wolves are only occasionally shot in Israel.

Another occasional mortality factor is the ingestion, while scavenging, of poison-contaminated food or of poisoned pests, i.e., rats, feral pigeons, etc., that have been disposed of on garbage dumps. A number of wolves have been found killed in this way.

As the wolves in the Negev seldom prey on livestock, they are not persecuted and their populations seem to be stable.

The wolf situation in Israel shows that legal protection is not always able to protect a species sufficiently. Therefore, the present management of the Nature Reserves Authority is considering compensating stock-breeders who suffer damage from wolves in order to save the last remnants of the wolf populations in northern and central Israel. Hopefully, this consideration will be carried out in time to save these populations.

Although wolves still survive in Israel, with its dense human population, they have disappeared from the most inhabited areas and survive mainly in the less populated regions. The wolves of the Negev are more anthropophilous and less shy than are those of northern and central Israel, but only the future will show how they will adapt to increasing human settlement in this area.

ADDENDUM

After this paper was completed, three more specimens of wolves from the Golan were made available. Specimen Nr.7425, mentioned in this paper, was much bigger than the largest *C. l. pallipes*. The three additional specimens from this area indicate that Nr.7425 was not an exceptionally big specimen, but that all the wolves in this population are larger than the largest *C. l. pallipes*.

The measurements and weights available for these four speci-mens are:

Specimen	CBL (mm)	GtL (mm)	Weight (kg)
Nr.7425, male (c. 6-8 years old), ZMTAU (already mentioned in the paper)	226.7	241.3	32.3
Nr.7511, male (c. 10 months old), ZMTAU	222.9	223.5	29.4
Nr.7510, male (c. 6 months old); ZMTAU (only skull available)	212.8	226.9	—
Nr.7512, female (c. 7 months old), ZMTAU (only skull available)	210.9	219.7	—

It is remarkable that the wolves that lived in the Huleh Valley until a few years ago fall well within the range of the Mediterranean pallipes. The Huleh Valley is only a few km distance from the Golan, but about 1,000 m lower. The amount of rain is about the same in both areas, but the Huleh Valley is much warmer.

All four "Golan wolves" were poisoned by stock owners within the last year after they had caused considerable damage by preying upon livestock, i.e., sheep, goats and calves. The Nature Reserves Authority decided not to pay compensation to stock owners, but to supply food to wolves at feeding stations and to protect livestock with electric fences in order to try to save this wolf population. The data presented here emphasize that the wolves of this population are different from other wolf populations in Israel, and therefore ought to be protected.

Acknowledgements

Many people contributed information and observations on wolves from all over the country. In recent years, particularly the rangers of the Nature Reserves Authority supplied much information and also brought many wolf specimens to the Zoological Museum of Tel Aviv University. Mr. G. Ilani and Mr. Z. Choresh of this Authority provided the author with especially useful information, as well as Mr. Z. Zook-Rimon of the Nature Conservation Research Institute of Tel Aviv University. Mr. E. Hurwitz of the Beth Ussishkin Museum at Kibbutz Dan kindly lent two wolf specimens for this work. A. Arr performed the statistics. Ms. Z. Shariv and Mr. A. Landsmann helped with the museum specimens at the Zoological Museum of Tel Aviv University, and Mr. A. Shoob helped with photographic work. Ms. N. Paz corrected, stylized and typed the manuscript. Their kind help is gratefully acknowledged.

Status, Growth and Other Facets
of the Iranian Wolf

Paul Joslin

INTRODUCTION

Little has been recorded on the Iranian wolf (*Canis lupus pallipes*). This paper reviews limited information regarding status, time of birth, den description, taxonomy, growth, blood analysis, vocalization, prey selection and attacks on humans.

STATUS

According to Harrington (1977), wolves occur in all 24 provinces, inhabiting coastal plain, sandy desert, rocky foothills, steppe and deciduous forest. While their numbers are unknown, it is apparent that they are thinly distributed over most of the desert portions while comparatively common in the Alborz mountains which are blanketed in deciduous forest and form the country's northern boundary.

Not uncommonly, they prey upon domestic sheep and goats, making them unpopular over much of the country, which could bring about their demise. However, they also occur in many reserves and parks, and therefore should not be in any danger of extinction provided these sanctuaries are maintained.

CHARACTERISTICS OF IRANIAN WOLVES

Wolf Confiscation

It was illegal in Iran to keep any wild animal without a permit. On 7 May 1975, Mr. Harrington and I flew to Yazd, in central Iran's more extreme desert regions, to pick up four wolf pups confiscated by the Iranian Department of Environment. The five to six week old pups, 2.5 kg each, were emaciated and suffering from diarrhea. They were brought to Teheran and, on a diet of milk, dog pellets and tinned dog food, quickly recovered.

Den Site

On 17 May, my assistants visited the den site from which the pups had originally been taken. The site consisted of an easily accessible cave on the side of a low hill about 3,000 m from the village of Nir. The cave was approximately 0.5 m in height, 4.8 m in length and varied in width between 1.2 and 4 m. Its narrowest width was at the entrance. The floor was dirt while the ceiling was a large flat stone. Outside were the remains of hooves and parts of leg bones from a donkey and a cow. Nearby were several smaller caves, smaller than a man's width in diameter, which did not appear to have been used by the wolves. The den was discovered by a local shepherd after a pack of three wolves had made several attacks on the village sheep and goats. He then blocked the entrance with stones, leaving a small opening where he set an illegal jaw trap. No bait was used.

An adult male wolf was trapped, bound and carried to the village where several people beat it to death with stones, sticks and a shovel. The shepherd was rewarded with money by the villagers, as was the custom.

Five wolf pups were removed from the den and kept by the village for about a week. They were then confiscated by the Department of Environment. The wolf pups were approximately three to four weeks old when captured. They had been fed meat but no milk and, as a result, one died and the others nearly did before they were confiscated.

Skull

The skull of the adult wolf was collected. Its condylobasal length measured 235 mm. According to Novikov (1956), Russian wolf skulls measure 200 mm or greater in condylobasal length, while jackals' measure 190 mm or less. Harrison (1968), who studied skulls of Arabian wolves, reported that they measured 184.5 mm or greater in length, while jackals measured 176 mm or less.

Novikov recognized two other differences. In the wolf, the anterior incisure of the nasal bones does not have a median protrusion, while in the jackal it does (Figure 7.1). The skull from Nir confirmed his observation. He also noted that in the wolf the cingulum on the external edge of the first upper molar (m^1) was slightly expressed, while in the jackal it was broad and distinctly marked over the entire length. The upper first molar of the wolf, however, was intermediate in appearance, if not tending toward jackal (Figure 7.2).

According to Harrison (1968), *C.l. pallipes* has a skull length exceeding 216 mm based on four specimens, all clearly exceeding the largest adult males of *C.l. arabs* among the great many specimens that he measured.

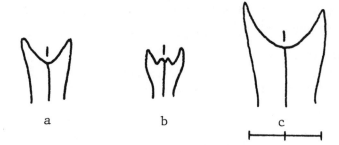

Figure 7.1: Anterior part of nasal bones, indicating differences in medial protuberance. Wolf and jackal are indicated by 'a,' and 'b,' respectively, and are adapted from Novikov (1956). They are not to scale. The last is taken from the Nir wolf. Reference scale is in centimeters.

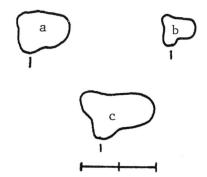

Figure 7.2: First upper molar, indicating differences in cingulum development. Wolf and jackal are indicated by 'a' and 'b,' respectively, and are adapted from Novikov (1956). They are not to scale. The last is taken from the Nir wolf and appears to have an intermediate cingulum development. Reference scale is in centimeters.

Blood Analysis

On 23 May, a blood sample was taken from each of four wolf pups, and cholesterol, uric acid and urea nitrogen levels determined (Table 7.1). Data on gray wolves would appear to differ significantly in terms of urea nitrogen levels. The difference may be due to the Iranian wolf pups receiving milk, puppy food and only a modest amount of meat.

Growth

Between the time of birth and 40 days, the pups averaged an increase substantially less than 50 g per day. After coming under our care at 40 days, they averaged an increase of 119 g per day over the next five months (Figure 7.3, Table 7.2).

Table 7.1: Cholesterol, Uric Acid and Urea Nitrogen Levels in Blood Samples Taken from Four Pups

Wolves	Sex	Cholesterol (mg %)	Uric Acid (mg %)	Urea Nitrogen (mg %)
Iranian				
Dana	F	263	1.7	15.0
Ochi	F	200	0.9	3.3
Kushkush	M	213	1.4	7.1
Mortimer	M	249	1.0	5.7
Mean		231	1.2	7.8
Gray*		203 (n = 13)	0.8 (n = 10)	27.0 (n = 13)

*Source: International Species Inventory System

Table 7.2: Growth Measurements Correlated with Time for Four Wolf Pups

Measurement	Correlation* Coefficient	Linear Regression	Average Growth Rate per Day
Weight	0.99	Y = 0.12x – 2.87	119 g/day
Total length	0.90	Y = 0.67x + 36.36	6.6 mm/day
Body length	0.79	Y = 0.39x + 34.89	3.9 mm/day
Shoulder height	0.93	Y = 0.32x + 12.55	3.2 mm/day
Tail length	0.97	Y = 0.28x + 1.75	2.7 mm/day
Girth	0.91	Y = 0.26x + 19.16	2.2 mm/day
Right ear length	0.94	Y = 0.08x + 3.07	0.8 mm/day
Right foot front length	0.81	Y = 0.02x + 3.22	0.2 mm/day

$$*r = \frac{N\Sigma xy - (\Sigma x)(\Sigma y)}{\sqrt{(N\Sigma x^2 - [\Sigma x]^2)(Ny^2 - [\Sigma y]^2)}}$$

Where: N = number of cases, x = age, Y = growth in weight, total length, etc.

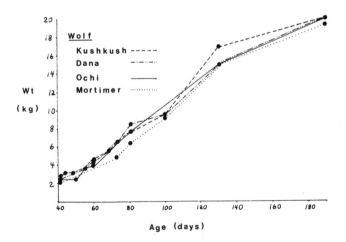

Figure 7.3: Gain in weight in relation to age.

    Body dimensions were recorded only until the pups were approximately 83 days of age (Figure 7.4), by which time their increased size and their playful activity made it difficult to take further measures.

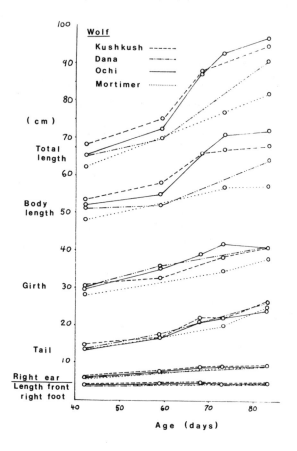

Figure 7.4: Body dimensions in relation to age.

    Tail length was the most accurate of all the linear measurements, having the highest correlation coefficient of any measurement (Table 7.2). It was taken from clearly identifiable locations on the body, i.e., from the base of the underside of the tail to the tip of the distal most caudal vertebra. The measurement was also straight, substantial in length, and the pups were not difficult to steady during the recording.

    Body length, while clearly definable as a measure of the body from the base of the tail to the tip of the nose, was not straight and varied noticeably with changes in posture. The pups' activity also interfered with the measurement. Consequently, it had the lowest correlation coefficient of any measurement. Total length was the addition of body length and tail length.

Ear length was measured from the inside notch to the distal tip. Although the amount of growth during the time was small, it was measurable as reflected in the high correlation coefficient.

The length of the right front foot was measured on the underside from the proximal end of the hind pad to the distal end of the two middle toe pads. Variation in length resulted from changes in the position of the foot and how it was held. Detecting measurable changes was also hampered by its slow growth rate. Not suprisingly, the correlation coefficient was not particularly high.

One wolf (Mortimer) was very ill when first obtained and lagged behind the others in most growth parameters. However, it is interesting that the tail length was not retarded.

Body Temperature

A rectal temperature of 38.8°C was recorded for Mortimer on June 30th. It was assumed to be normal.

BEHAVIOR OF IRANIAN WOLVES

Vocalization

The four pups, which were kept in captivity in Iran until approximately six months of age, never howled, even though I sometimes attempted to induce them to do so with human imitations. Subsequently, they were sent to the Wild Canid Survival & Research Center in St. Louis, Missouri, and housed in a large enclosure next to captive North American wolves. The Center reported that on occasion the Iranian wolves would howl, but only when initiated by the neighboring wolves or by human imitations. They never vocalized on their own.

This disinclination to howl correlated with my field observations. During two and a half years in Iran, often in wolf inhabited areas, I never heard a wolf pack vocalize. This included nights when I howled myself on perhaps 100-150 occasions. Others have said they have heard Iranian wolves howl in the wild, but they too agree it is a rare event.

These data contrast sharply with North American wolves, who begin howling when approximately one month of age (Mech, 1970). Earlier studies in Algonquin Park, Canada (Joslin, 1967) have shown that in the wild, wolf pups will howl when adult members of the pack are not present at about three to four months of age. Thereafter, for at least the next month or two, they can be induced to howl at almost any time of day or night.

One possible explanation as to why Iranian wolves behave differently is that howling may be of little value to them. In Algonquin Park, whenever conditions became windy (a rare event at night), there was a noticeable and statistically significant reduction in howling activity in response to human imitations, even though I was usually close enough to be easily heard (Joslin, 1967). In Iran, nights are more often windy than not. If long-distance communication is the objective of howling, then howling on windy nights is largely a wasted effort as the total area over which wolves can be heard may be greatly reduced under such conditions.

Wolf Predation on Domestic Stock

While predation by wolves on domestic livestock occurs widely over Iran, it is a rare event for any individual shepherd to experience. Probably for this reason, the responsibility of tending sheep and goat flocks is commonly left to children. By contrast, there are a few places where wolves do kill a great number of domestic stock. One such area is Arasbaran Protected Region, an area of approximately 340 km² in northwestern Iran.

The only wildlife in the region of sufficient size and number to be of importance in the wolves' diet were ibex *(Capra ibex ibex)*, a species of wild goat. However, they numbered only a few hundred and inhabited higher mountains largely inaccessible even to wolves.

By contrast, between 50,000 and 100,000 domestic stock, mostly sheep and goats, were estimated to inhabit the area (Shah-savan, pers. comm.). They formed the wolves' primary prey.

All 12 of the villages I visited in the protected region had received recent wolf attacks on their stock (42 cases could be verified). Typically, shepherds reported that they saw wolves near their stock about once every 10 days. On such occasions, it was common for wolves to attack, disappear for an hour or two, then try again and perhaps even a third time sometime later. During my 10 day investigation, I witnessed an unsuccessful wolf attack, the second upon a shepherd's flock in as many hours.

The shepherds managed to save some animals which had sustained injuries from wolf attacks. Two such animals, both goats, were bitten at the throat.

Only adult men or boys in their late teens were seen herding flocks. Two or more large dogs, kept specifically as a defense against wolves, accompanied them. (Because firearms were illegal, none of the shepherds were armed.) Sometimes a dog chased a wolf pack too far from the security of both the shepherds and the other dogs and was turned on and killed by the wolves. These dogs were reportedly eaten. To the herdsman, this was an economic loss; the average dog costs twice as much as a sheep. However, the occasional loss of a dog was preferable to losing several sheep.

Purported Attacks upon Humans

Every winter the two major Teheran newspapers published grisly accounts of people being attacked by wolves, some of which reputedly occurred in the northwestern part of the country. As well as following up on reported livestock losses in Arasbaran, I also investigated possible attacks on humans. Most reports involved villagers travelling alone who, upon seeing a wolf pack for the first time in their lives, had fled believing they had escaped from a wolf attack. This was in contrast to shepherds, who saw wolves frequently and were not afraid of them.

I followed up on the only report of a shepherd having been attacked and supposedly killed by wolves. Eventually, I located a shepherd who had witnessed the man's death. Both had been attending a flock of sheep when about a dozen wolves appeared. One shepherd worked at bunching the flock while the other, with the aid of three dogs, attempted to drive off the wolves. The dogs pulled down one wolf and the shepherd clubbed it to death with his cane. Meanwhile, a boy who also witnessed the attack ran for help. Several men from the local village arrived and helped drive the remaining wolves off. At this point, the shepherd with the dogs sat down, coughed and died. The cause of his death was unknown, but it certainly was not a wolf kill.

Acknowledgements

I wish to thank Mossafas Sharifi, Amir Taimouri, Mohammed Nowruzi, Amir Movassagh and other members of the Iranian Department of Environment who ably assisted me. Special thanks goes to Fred A. Harrington, Brian Mortimer, Raol Valdez and Leticia Valdez.

Part III

Behavior of Wolves in Captivity

Introduction

Many pertinent details of wolf social behavior cannot, at present, be satisfactorily answered by the study of free-ranging wolves. Typically, contact with wild wolves is limited to beeps on a receiver or to brief vocal or visual encounters. Captive studies alleviate several difficulties inherent to field investigations. Researchers are able to limit the context in which the animals perform, reducing the number of variables which may influence their activity, thus allowing for easier interpretation of data. The animals are always accessible, doing away with time spent in locating them and allowing for extended periods of observation. Habituation of the wolves to the presence of observers minimizes experimenter-produced disruptions. Captive studies also facilitate identification of individual animals as well as elucidation of social and biological relationships.

Studies of captive wolves have been criticized, with questions being raised as to the applicability of derived information to wild populations. It is argued that, in the absence of important social factors such as hunting activity, freedom of movement and various environmental influences such as topography, captive wolves should be expected to display little but pathological behavior. However, we simply cannot assume that pathology is inevitable; instead, we must document its existence, and this has yet to be done.

Claims that captivity is stressful to wolves remain undefined and unquantified. Is not being able to travel freely for long distances or to hunt prey stressful? Or, is having to travel long distances and interact with dangerous prey stressful? Is daily exposure to humans stressful? Or, is periodic exposure to humans through their scent, traps and bullets stressful? Although captivity and wild conditions differ, both contain elements which might be considered stressful. It might be argued that wild wolves face natural stresses

206

for which they are adapted, but — is human persecution to be considered a natural stress?

There is no doubt confinement does have its effects. Primary among these is that normal dispersal from the group is prevented. Thus, some behaviors may become exaggerated, such as aggression during the breeding season. Animals which might normally avoid aggression by emigration cannot escape. Other behaviors, especially those related to hunting, may rarely, if ever, occur. However, the problem of confinement must be placed in perspective. A wild wolf pup is confined to a limited area, the homesite, during its first four to six months. Yearlings and adults spend a considerable amount of time at these sites as well, sometimes up to 80 percent of their time. When pups, yearlings and some adults are separated from the pack, they often return to old, abandoned homesites where they may remain for a week or more. As far as we are aware, none of these "confinements" cause, or are the result of, pathological behavior. When the wolves finally move on, they do so because they have become strong enough to travel with the pack, because they have relocated their packmates, or because they are hungry. It is doubtful they travel because of an "innate travel drive," which must be periodically "burned off" lest psychological trauma develop. Indeed, it would be interesting to experimentally provision a wild pack at one location in winter to determine how much wolves voluntarily restrict their movements.

Most assuredly, captivity and the wild are dissimilar in many respects. Captive studies cannot provide insights into all aspects of wolf behavior, and may even misinform. But as long as we remain aware of the limitations, information from captive studies should not be misleading.

Captive studies allow us to clarify aspects of behavior about which wild studies can provide only sketchy and incomplete data. Harrington et al. present a review of wolf mating systems, based on information derived from both wild and captive studies. This paper is presented here, rather than with the wild studies, because the captive studies provided much clearer insights into the dynamics of the system. Wild studies painted a general picture of mating activities, which hinted that there was something more than was obvious with only the brief glances permitted in the wild; the captive studies indicated what that something was.

Captive studies also allow us to rigorously describe and quantify behavior, especially with regard to individual variability and the influence of relatedness, significant factors which must be ignored in most wild studies. Paquet, et al. describe communal denning and nursing, behaviors of theoretical importance largely unobservable in the wild. John Fentress and Jenny Ryon outline the network of distributed feedings that develops each spring and summer in a large captive pack. The opportunity to study a pack over several seasons helps reveal the dynamic nature of wolf behavior. Both studies underline the importance of considering individuals in our attempt to understand behavior.

Finally, captive studies permit us to readily control the wolf's environment, so that careful and controlled experimental studies become possible. In search for causation, the researcher of wild wolves must collect data on every variable which might influence the behavior of interest. A correlational analysis may indicate how the system operates, but until an experiment is performed, conclusions remain untested. In captivity, direct tests can be performed, repeated and modified, so that the variables concerned can be separated, combined and recombined until the details are firmly isolated. That is at least the potential of captive research, though yet to be realized. Lyons et al. and Carl Cheney present promising research strategies for unlocking details of cooperation and foraging strategy. Both studies utilize operant conditioning, which allow researchers to "ask" questions of the subjects, which are then answered by the animals' behavior.

Zimen concludes the section with a detailed model of wolf pack social dynamics, based on data gathered during ten years of continuous observation of a single pack. The value of such a model is that it provides a theoretical framework which can be continuously refined and expanded with the regular addition of new information.

Monogamy in Wolves:
A Review of the Evidence

Fred H. Harrington, Paul C. Paquet, Jenny Ryon
and John C. Fentress

INTRODUCTION

A species' mating system "refers to the general behavioral strategy employed in obtaining mates. It encompasses such features as: (i) the number of mates acquired, (ii) the manner of mate acquisition, (iii) the presence and characteristics of any pair bonds, and (iv) the patterns of parental care provided by each sex" (Emlen and Oring, 1977:222). Attempts to classify mating systems are difficult due to lack of precise terminology (Emlen and Oring, 1977), unequivocal evidence of mating patterns (Kleiman, 1977), or both. Monogamous mating systems have been characterized by the presence of exclusive sexual bonds between two individuals resulting in mating exclusivity (Kleiman, 1977). However, proof of mating exclusivity is difficult to obtain without exhaustive behavioral observations or genetic comparisons between adults and young. Thus, less rigorous evidence typically has been employed: (1) close association of a pair not limited solely to breeding periods, (2) mating preferences, (3) absence of other adults in the pair's home range, and (4) breeding by only one pair in a family group (Kleiman, 1977). None of the above is definitive proof of monogamy, although the first two criteria provide stronger circumstantial evidence than the others, especially the last.

Classification of the wolf's *(Canis lupus)* mating system has been hampered by the animal's elusiveness, wide-ranging movements, and tendency to live in relatively large social groups, and the frequent use of imprecise behavioral terminology by researchers. Most field studies, of necessity, have relied upon aerial observation conducted during the winter breeding season (Mech, 1966; Jordan et al., 1967; Pimlott et al., 1969; Mech and Frenzel, 1971; Wolfe and Allen, 1973; Haber, 1977; Peterson, 1977). However, under these conditions, (1) wolves are difficult to sex, age or individually identify, (2) packs are difficult to follow for extended periods on a daily basis, (3) even if daily contact is possible, the portion of the pack's daily activity cycle potentially observable is limited by short mid-winter days and (4) close observation of the details of behavior is normally precluded even during available observation periods. Thus, observations of mating activity are infrequent, and observations of actual copulations are fortuitous. Determining the number of females which have littered is also difficult due to lack of snow for tracking by early spring, and the seasonal increase in vegetative cover. Furthermore, when telemetry is used, lack of radio-collars on all pack members reduces the chance of locating dens. The frequent combining of litters by mothers soon after parturition (Murie, 1944; Clark, 1977) compounds these difficulties. Nevertheless, the general impression reached by many researchers is that wolves are typically monogamous, although exceptions to this rule are usually noted (Murie, 1944; Mech, 1970; Haber, 1977; Peterson, 1977).

The purpose of the present paper is to re-examine evidence concerning wolf mating systems. Data from wild and captive wolves will be examined separately. This brief review will indicate that at present the data that do exist largely contradict the notion of wolf monogamy. We will offer a classification system which appears to fit the pattern observed in wolves more closely.

MATING PATTERNS

Problems in Observing Mating Exclusivity

The fundamental characteristic of monogamy is mating exclusivity. In wolves, this can be inferred by either the pattern of courtship activity and copulations, or the number of litters produced by a pack. The former provides direct evidence; the latter is indirect and inferential. Both measures, however, depend upon pack composition, which corresponds with the number of mature individuals of each sex. For example, if packs contain only one sexually mature male and female, then mating exclusivity may be the result of demographic patterns (Mech, 1970) rather than direct behavioral strategy (sensu Emlen and Oring, 1977). Thus, the exclusive and reciprocal courtship pattern between the sole mature pair of a pack seems best described as "de facto" monogamy. Evidence for what might be termed "preference" monogamy would be obtained if a similar exclusive and reciprocal mating pattern occurred in packs with more than one mature adult of both sexes. If *such* exclusivity still occurred despite the potential for polygamy or promiscuity, then it would be more legitimate to conclude that monogamy accurately characterizes the mating system.

To distinguish between de facto and preference monogamy, pack compositions must be known. Unfortunately, authors often do not distinguish packs on the basis of their composition, largely because they lack sufficient evidence. The problem is further compounded since smaller packs are easier to characterize but are more likely to contain only a single adult pair, whereas large packs are more difficult to characterize but are more likely to contain several adult members of both sexes. Thus, data from larger packs, necessary to distinguish de facto from preference monogamy, are usually too equivocal to permit firm conclusions.

Once pack composition is known, behavioral observations in principle can establish mating patterns. However, the descriptions will necessarily be limited for many reasons. Search time, fuel capacities, weather and light conditions, and vegetation and topography will jointly determine the quantity and quality of observations. Only a small portion of the breeding season can be covered effectively. Copulations have been observed during only 17 of 52 pack-years of study in Isle Royale and Mt. McKinley National Parks (Mech, 1966; Jordan et al., 1967; Wolfe and Allen, 1973; Peterson and Allen, 1976; Haber, 1977; Peterson, 1977; Peterson and Scheidler, 1977, 1978, 1979). In addition, pack splitting during the breeding season (Jordan et al., 1967; Wolfe and Allen, 1973; Haber, 1977; Peterson, 1977; Harrington and Mech, 1979) further dilutes observational effectiveness, especially on a per-pack-basis. Finally, intrasexual interference (summarized by Rabb et al., 1967; Mech, 1970; Zimen, 1975, 1976; Packard, 1980; Packard and Mech, 1980) may reduce mating activity by subordinate animals in the presence of more dominant ones. Subordinates may restrict overt courtship and mating behavior to periods when dominant animals are otherwise occupied (see below), or visually isolated, making observations of courtship and mating among subordinates unlikely. These factors must be considered in any attempt to clarify wolf mating patterns.

Finally, it should be noted that summary terms such as "courtship", "subordinance", etc. employed by most researchers, and used by us below, are useful precisely because they classify into single categories a range of behavioral details that appear to form common *sets* of action. This can also be a danger when: (1) different investigators combine different details into categories with the same name, (2) the heterogeneity of actual performance details is forgotten through the use of unitary higher-order labels, and (3) terms are used as if they are descriptors when in fact they involve inferences about either (a) causal antecedents to behavioral performance or (b) consequences (functions) of that performance (e.g. Hinde, 1970; Moran et al., in press). However, in this review we must often take terms used by investigators at face value since the precise criteria used to define them are rarely available for independent scrutiny. Sometimes we have also found it necessary to use qualifiers such as "appeared to", "apparently", etc. when the data (observations) leading to conclusions by various authors are not published in detail.

Observations of Courtship and Mating in the Wild

The following observations are based solely on packs known or suspected to have two or more adults of both sexes, although in some instances definitive proof is lacking for certain years.

Isle Royale: Mech (1970) reported a minimum of three adult females in a pack of 15 to 16 wolves between 1959 and 1961. However, only one female closely associated with a male during the breeding season. Attempted copulations between this pair were noted in 1959 and 1960. One copulation observed in 1959 apparently did not involve this pair. Other copulations were observed in 1959 and 1960, but identity of the pairs was not known. At least three different pairs were observed in mating activity, typically involving attempted copulations by the male that were thwarted when the female sat with tail between her legs. In both 1959 and 1960, several males courted the three known females, although no apparent courtship patterns were established. Few observations were reported in 1961.

In 1962, 1963 and 1965, little mating activity was observed. This was interpreted as being due to the wolves' wariness of the spotting aircraft (Jordan et al., 1967).

There was apparently only one breeding pack (Big Pack) on the Island from 1968 to 1970 (Wolfe and Allen, 1973) following changes in the population structure during 1967 and 1968. Courtship was observed between a single pair all three years, and copulation by this same pair was seen in 1968. During all three years, a second male remained in close physical proximity to the courting pair.

Peterson (1977) found that courtship (primarily greeting, mounting, play-soliciting, genital examination) was confined to either the dominant (alpha) pair (n = 49; as measured by courtship behaviors) or to subordinates (pairs?) within the pack (n = 20) between 1971 and 1974; only twice did subordinate males direct courtship behavior to the alpha female. For both main packs (East Pack, West Pack), courtship (and in some years, copulations) was observed for the alpha pairs. However, the East Pack probably contained only one adult pair in 1971 and 1972. In 1972 and 1974, a "subordinate pair" was also observed copulating in the West Pack. In addition to the mutual courtship activity of the alpha pair and the one subordinate pair, "one-sided" courtship was frequently observed, typically involving males attempting to copulate with females which rebuffed the mating attempts.

In 1976, two females reportedly came into estrus in two of the four packs followed (Peterson and Allen, 1976). The East Pack alpha male courted both females, although the subordinate female was unreceptive. No copulations were observed in any of the packs.

In 1977, the alpha male of a pack of four wolves copulated with a subordinate female. During the tie, the male defended his partner from attack by the alpha female. After mating, the subordinate female was chased by the pack for several kilometers over the ice between Isle Royale and Ontario, Canada. However, within two days the subordinate female returned to the pack and the alpha male continued to court her. It was not determined whether the alpha male also mated with the alpha female, or which of the females produced pups, although at least one litter was suspected (Peterson and Scheidler, 1977; Peterson, 1979).

A limited number of recognizable wolves made observations of courtship difficult in 1978, although at least one female came into estrus in each of the four packs (Peterson and Scheidler, 1978). In the East Pack, a subordinate female also came into estrus and was courted by a subordinate male. This pair was frequently "punished" by the alpha pair following courtship activity. No copulations were observed in this pack.

There were apparent changes in the East Pack in 1979, which had the same dominant female since its inception in 1971. During that period, the alpha female had three different mates of two, one and five years' duration, respectively (Peterson and Scheidler, 1979). These mate changes were apparently necessitated by the disappearance of the previous male, although the circumstances surrounding these losses are unknown (Peterson, 1977). Until 1979, mating activity was observed only in the dominant female (although the alpha male courted an apparently unreceptive subordinate female in 1976). In 1979, however, two females were seen urine marking[1] – behavior usually associated only with the dominant female (Woolpy, 1968; Peters and Mech, 1975; Packard, 1980; Harrington, 1981), and the alpha male courted the subordinate female more "actively" than he did the alpha female. Relations between the two females were described as "amicable" (Peterson and Scheidler, 1979). The only copulation observed was between the alpha male and the subordinate female.

Mt. McKinley National Park: Haber (1977) reported that courtship and mating in the Savage Pack occurred predominantly between the same alpha pair in 1970 and 1971. However, a second male (beta male) remained in close proximity to the pair both years and occasionally attempted to mount the alpha female, although he was usually prevented from approaching the alpha female by the alpha male, which often attacked and pinned the beta male to the ground. On one occasion in 1971, the beta male mounted the alpha female twice while the alpha male was feeding 6–12 m away. The alpha female stood receptively both times, and the second attempt was interrupted after 20–30 seconds when the alpha male returned and stood between the two. Two days later the alpha pair copulated.

During 1972, both males showed interest in the new alpha female. The alpha male was usually close by and often physically

---

[1] Urine marking is usually distinguished from other urinations based on several of the following criteria: frequent occurrence, deposition on conspicuous targets, brief duration, and use of raised-leg posture.

touching the female; the beta male remained 3–5 m away. The alpha male was aggressive to the beta male, especially on those occasions when the beta approached closer to the female. Copulation was observed between the alpha pair. In addition, a second female conceived and gave birth that year. Her mate was assumed to be another subordinate male since courtship activity between the two had been observed. Their mating activity was often interrupted by both high-ranking males.

In 1973 the beta male was absent, and the only overt courtship behavior observed was between the same pair as in 1972, which were seen to copulate once. No courtship activity was noted in 1974, when the alpha male disappeared from the pack prior to the breeding season.

Between 1970 and 1973, the same alpha male mated with two different females. Each female was judged to occupy the alpha rank at the time of mating.

Few observations of mating activity were collected for the larger Toklat Pack, although it was known that two females mated in 1973. Unfortunately, the identity of only one of the females' mates was known. Several females evidently mated during a number of years, as judged by the prevalence of multiple litters in this pack (see section on multiple litters, below).

Summary: Few other field observations of comparable detail exist for free ranging wolves. The present evidence suggests the following: (1) Courtship activity frequently appears to be confined to pairs (usually the dominant or "alpha pair"). The second ranking male (commonly termed "beta") sometimes associates closely with the primary pair and is evidently sexually interested in the female. During all, or most, of the female's estrus, however, the beta male is kept away from her by the alpha male, which also prevents approaches by all other males. Thus, the number of mates for the alpha female appears to be partly limited by male-male aggression. In one case where an alpha male relaxed his vigilance briefly to feed, the beta male mounted the alpha female twice; her receptiveness to this male's advances suggests she would have readily mated with him (Haber, 1977). In these instances, male-male aggression may be more important in producing exclusive mating patterns than female preference. (2) Alpha males have sometimes courted and copulated with subordinate females while still courting the alpha female. The alpha female usually is aggressive toward other females during the breeding season, although in one case the relationship seemed amicable. Here, restricted or exclusive mating by the alpha male may be determined more by the alpha female's suppression of mating activities in other females than by the alpha male's commitment to monogamy. (3) More than one male and female in a pack may copulate.

Observations of Courtship and Mating in Captivity

Although it has been argued reasonably that captive observations are "more likely... anomalous, reflecting the greater stress of captivity" (Haber, 1977:240), the precise influence of various environmental conditions upon courtship, mating, and other patterns of behavior in wolves has not been examined systematically at the present time. Certainly the close observations permitted by captive studies on groups of animals (1) with individual histories that are known in detail and (2) that live under environmental conditions that can be controlled, are potential advantages. Ideally, of course, data obtained in captivity should be compared with that found in the field, for in this way the strengths of various constraints on behavior can be assessed directly.

Our survey of the literature does suggest two areas of contrast between most captive and wild conditions that need to be accounted for. (1) The manner of establishing a captive pack (e.g. from a same-age group of siblings [Rabb et al., 1967]) may differ from that of most wild packs (e.g. from a pair of unrelated lone wolves [Mech, 1970]), introducing a confound which may bias the expression of behavior within the captive group. For example, incest avoidance mechanisms (if such exist) could influence mating patterns in the group, possibly for many years. However, it is also important to maintain the distinction between evolutionary (genetic) biases in behavior and individual developmental histories—an area where captive studies can be most useful. (2) The normal expression of some forms of behavior (e.g. foraging, dispersal) is limited, possibly distorted, or even eliminated. This may have effects on the "normal" expression of various other forms of behavior. The potential influence of any given variable upon entire networks of behavior, both directly and indirectly, is again most readily assessed through close and controlled observations such as permitted by captivity. These detailed observations can then be compared to data available in the wild as a test of their generality. Until this is done, it is dangerous to assume *either* similarities or differences of behavior under different circumstances.

In this survey we necessarily omit some of the desirable refinements of captive/field comparisons, and focus upon those data that do exist which are relevant to the broad theme of observed monogamy (and its limitations). While captive and field studies cannot be simply superimposed we do find apparent similarities in mating that invite more detailed comparisons in the future.

Brookfield Pack: Rabb et al. (1967) observed a pack in a 0.3 ha enclosure. The group initially consisted of five pups from two litters sired by a common father. Between 1960 and 1963, courtship patterns were fairly restricted: each of three females *primarily* courted a single male while one male courted a single female and the other courted all three, although with unequal frequency. Only the dominant female copulated, except once when a subordinate female mated while the dominant female was tied.

Mating was not as restricted in later years. The dominant female had two and three mates in 1963 and 1964, respectively. Another female had two mates in 1965. In 1966, individual females had three, two, two and no mates, while males had three, two, one, one and no mates, respectively. Throughout the study, individuals demonstrated long-term sexual preferences for certain other individuals. However, these individual preferences were not always predictive of copulation patterns. For example, the original dominant female courted and solicited the dominant male preferentially, but he did not reciprocate and they never mated. In addition, these preferences were usually not exclusive. For example, one female courted two, two, four and two males between 1963 and 1966. In turn, two, three, five and three males courted her in the same years. Her only observed copulations were with two males in 1966.

Bavarian Forest Pack: Zimen (1975, 1976) observed a group in a much larger enclosure (approximately 6 ha). The only adult female in the original group of four littermates did not mate with her brothers in 1971 or 1972, although she was courted by two lower ranking brothers. An unrelated female pup, introduced in 1971, became the dominant female in 1973. She was courted by all the males except a littermate and the alpha male (the same as in 1971 and 1972). She courted the alpha male, which did not reciprocate. She eventually copulated with the beta male.

In 1974, the same alpha female was courted by several males. For a period of 12 days, she mated with her mate of the previous year who, at that time, was the dominant male. She then copulated for three days with a two year old male who, based on the date of parturition, evidently sired the litter.

In 1975, the alpha female maintained her status. Early in estrus, she mated with her original mate from 1974. She then copulated with another two year old male. Finally, during the last portion of her estrus, she mated with her second mate from 1974, who apparently once again fathered the litter. No other females mated during the study because (1) most were excluded from the pack by the dominant female, (2) those that remained with the group were not courted by males, or (3) their courtship activities were obstructed by the dominant female.

Shubenacadie Pack: Fentress and co-workers (Fentress and Ryon, this volume, and unpublished data) have observed a group of wolves in a heavily-wooded 3.4 ha enclosure since 1974. Initially, there was an adult pair, their six pups (two male/four female), and two female yearlings (unrelated to the other animals). Two pups (male/female) were removed at 11 days of age and reunited with the pack six months later. The female pup was killed the following mating season, several months after reintroduction.

Prior to the pups' reintroduction, the 11 year old adult male died. In 1975, no mating activity was observed. Mating resumed in 1976 when the pups were 22 months old. From then until 1980, the reintroduced male was observed to copulate, although he was socially subordinate to his brother as determined by outcome of aggressive interactions between the two males. He copulated with his mother in 1976, 1978, and 1979, and with at least two of his sisters in 1978, 1979, and 1980. In addition, several other females also mated and produced pups each year, but their mates were unknown. Other males in the group were not observed to copulate until 1980, when the dominant male from the 1974 litter, and another born in 1976, mated with the two females born in 1977.

All six females which mated and whelped between 1976 and 1980, courted the reintroduced male more frequently than any of the other males. Aspects of female courtship included initiation of body contact (T-formation [Golani and Mendelssohn, 1971], muzzling, body rubbing) and tail aversion.

Connecticut Pack: Schotté et al. (1977) and Schotté and Ginsburg (1978) detailed the mating patterns of a pack of wolves descended from an unrelated pair, placed together in a < 0.1 ha enclosure. In the second breeding season, several of the 10 month old males exhibited interest in their mother (e.g., investigated genital area, or placed chin or feet on her back), but all observed mounting and copulation was done by the dominant adult male. The following season, one of the then 22 month old males courted his mother more persistently than did the adult male, and both males copulated with her. The other 22 month old males also showed interest in the female although, evidently, none mated with her. In the next breeding season, three of five males actively courted the dominant female although only the original, dominant male was seen to copulate with her. The dominant female did not appear actively to discourage courtship by her offspring, although she was less likely to stand during their attempted copulations (12% of offspring's mounts versus 84% of the alpha male's). High ranking males interfered with mating attempts by subordinate males. Less interest was exhibited by the males toward the subordinate females ($\bar{x}$ = 116 courtship actions to subordinate females versus 925 to the dominant female). It was suggested that these younger females (two and three years old) may not have reached behavioral sexual maturity, "perhaps due to the presence of an older dominant female who is their mother" (Schotté and Ginsburg, 1978:12).

Summary: Captive studies indicate that mating patterns can involve a complex web of interaction among many individuals. Common to these studies is a low incidence of mating exclusivity. Mating preferences do occur, as measured by the direction and frequency of courtship between various individuals, but these preferences are relative rather than absolute. Females and males that mate often have two or more mates during the same breeding season. In cases where mating is restricted to a single pair, intrasexual aggression appears to be a common factor limiting breeding opportunities. When socially-imposed restrictions are absent, a pattern of polygamy/promiscuity occurs. Since promiscuity implies the lack of long-term interindividual bonds and indiscriminate mating (Selander, 1972), polygamy better characterizes captive wolf mating systems, with its often long-term individual preferences.

NUMBER OF LITTERS PER PACK

A further characteristic of a mating system is the number of individuals producing each year's young. Not all females that breed necessarily produce young. Wild packs are typically characterized as producing single litters (Van Ballenberghe and Mech, 1975; Haber, 1977; Peterson, 1977); multiple litters (Murie, 1944; Clark, 1971) are thought to be exceptional. However, since pack compositions are often poorly known, single litters might indicate, in at least certain cases, that packs contain only a single breeding female. Thus, pack composition must be considered when calculating the ratio of multiple to single litters in packs.

When only data from packs known to have at least two adult females are considered, the frequency of multiple litters is substantial (Table 2.1).

Table 2.1: The Prevalence of Multiple Litters in Wild and Captive Wolf Packs Containing at Least Two Adult Females

| | ... Number of Pack-Years With ... | | | |
Location	Single Litter	Multiple Litters	Unknown	Reference
Wild Packs				
Alaska	1*	3*	1	Rausch, 1967
Mt. McKinley National Park	3	1	0	Murie, 1944
Mt. McKinley National Park				
Savage Pack	4	1	1	Haber, 1977
Toklat Pack	3-8	0-5**	1	Haber, 1977
Baffin Island	2	1	0	Clark, 1971
Captive Packs				
Brookfield	14	8	0	Rabb et al., 1967, pers. comm.
Bavarian Forest	3	0	0	Zimen, 1975, 1976
Shubenacadie	0	5	0	Fentress & Ryon (this volume)
Carlos Avery	2	1	0	Packard, 1980
Wolf Park	8	1	0	Klinghammer, per. comm.
Altmann's Pack	1	4	0	Altmann, 1974
Barrow Colony	2	2	0	Lentfer & Sanders, 1973
Connecticut	2	0	0	Schotté and Ginsberg, 1978
Washington Park	2	1	0	Paquet et al., this volume

Multiple litters (%) - wild packs - 22–41% (n = 27 pack-years)
                        - captive packs - 39% (n = 57 pack-years)

*Determined from postmortem examination of female reproductive tracts from entire packs collected during the breeding season. The number of multiple litters may be overestimated if losses occur *in utero.*

**"Probably" two litters in three years; "possibly" two litters in one year, and between one and three litters in a fifth year.

In the wild, frequency of multiple litters may range between 20 and 40 percent. These estimates are probably minimal figures since early mortality, combining of litters (Murie, 1944; Clark, 1971), and observational difficulties reduce the chances of locating all litters. All the wild data come from relatively open habitats. Finding dens and observing litters in forested habitats is extremely difficult and rarely accomplished (Mech, 1966; Peterson, 1977). Thus, whether forest dwelling wolves have multiple litters at rates comparable to tundra wolves is unknown.

The frequency of multiple litters is as high in wild packs as it is in captivity (Table 2.1), suggesting that the patterns of mating behavior observed in captivity have their counterparts in the wild.

DISCUSSION

This review of wolf mating systems was not intended to be exhaustive. Rather, we have selected representative long-term and detailed field and captive studies to examine two primary questions: (1) can wolf mating patterns be legitimately termed monogamous, and (2) do packs typically produce only one litter per year? We focused on these aspects because both scientific and popular writing frequently allude to them, and because, as we have shown, both appear to be erroneous.

The frequency of multiple litters in the wild is much greater than a cursory reading of the literature might suggest. Therefore, rather than dismissing multiple litters as exceptional, or attempting to place statistical limits on their frequency, we suggest that future studies concentrate on identifying factors which influence their expression. Neither multiple nor single litters should be viewed as anomalies; both should eventually become predictable results of a definable set of interacting social and ecological factors. It will, however, be important that these factors, as well as the rules by which they interact, be specified with as much clarity as possible.

Several factors leading to multiple litters have been suggested. Temporary pack splitting (Mech, 1966; Jordan et al., 1967; Haber, 1977) during the breeding season is thought to facilitate multiple litters by freeing subordinate wolves of social suppression by dominant pack members (Zimen, 1975; Haber, 1977), although subordinate females have mated in the presence of the alpha female (Peterson, 1979; Peterson and Scheidler, 1977, 1979). In one pack, a subordinate female came into estrus and mated with no apparent antagonism from the dominant female, which at the time was at least nine years old (Peterson and Scheidler, 1979). The loss of the former alpha female may also facilitate multiple litters; the only case of multiple litters observed in one Alaskan pack occurred after the alpha female disappeared prior to the breeding season (Haber, 1977). Kinship patterns (which apparently can influence mating patterns in a pack [Zimen, 1976]), female reproductive value and various ecological factors (prey type, abundance, etc.) might also influence the number of litters in a pack.

One issue that has confronted us in our attempts to evaluate critically the existing behavioral data related to monogamy is that details of behavioral performance which might help clarify both phenomenology and mechanism are frequently lacking. It is not always clear, for example, whether one investigator's use of a term such as "courtship" corresponds in detail with the observations similarly classified by another (or even by the same observer on a different occasion, for a different animal, etc.). Certainly it is important to bear in mind that summary terms such as courtship reflect a diversity of particular actions that may fluctuate more or less independently. The problem is compounded when terms that imply description also incorporate inferences about either causation or function (see Hinde, 1970).

Even apparently straightforward measures of "amount of", "intensity", etc. applied to a single measure of behavior must be viewed cautiously if the details of measurement are not provided. Thus, frequency, duration, bout composition, persistence, and the like are often taken as equivalent measures of behavioral tendencies when they can in fact vary independently (Fentress, 1973). Two additional considerations limit the precision of conclusions available at the present time. The first is the frequent use of terms of intention (e.g. this wolf meant to do that) with no clear reference to overt action. The second is a common tendency to treat individually defined actions (behavioral classes) as if they are totally independent in their expression when in fact tendencies to perform one set of activities almost invariably influence tendencies to express descriptively quite different action classes. This, combined with the non-unitary nature of behavioral categories (above), makes the interpretation of direct versus indirect influences of any given variable in behavior a problem for analysis rather than assumption. Here future comparisons between field and captive studies, if pursued rigorously, could make a substantial contribution to our knowledge.

Resolving the question of monogamy also depends, in part, upon how it is defined. True or "preference" monogamy implies that a pair develops a social bond which precludes the formation of similar bonds with other individuals, thus limiting courtship and mating to one partner. By these criteria, wolves are not monogamous, since exceptions to mate exclusivity have been observed in many captive packs and have been indicated in several wild studies. It could be argued that relative, rather than absolute preference should be used to characterize monogamy; as long as one partner is "typically" preferred, the label of monogamy can still be employed. Loosening the criteria, however, introduces the necessity of establishing limits. At what point does "monogamy" cease and "polygamy" begin? In addition, social constraints limiting courtship must be considered.

How much of the observed pattern reflects preferences of A for B, and what proportion reflects restriction imposed upon both A and B by intrasexual competition.

Additional qualifications such as "de facto" or "forced" monogamy may be useful as they indicate the dependence of mating systems on the availability of mates in both a spatial and social context. Both reflect situations commonly encountered by wolf packs. De facto monogamy occurs whenever packs contain only one adult pair, which typically would occur for several years after the founding of a pack. In later years, de facto monogamy will recur whenever mortality or emigration again reduces the pack to a single adult pair. "Forced" monogamy occurs when there is a potential for polygamy in the pack which is not realized because of intrasexual dominance and aggression. With both de facto and forced monogamy, an individual has no choice as to a mating partner. Because options for mating are limited by factors other than choice, it may not be appropriate to label these patterns monogamous.

Kleiman (1977:39) distinguished two general forms of monogamy. Type I, or facultative monogamy, with "males and females being so spaced that only a single member of the opposite sex is available for mating," is analogous to de facto monogamy. Type II, or obligate monogamy, in which "a solitary female cannot rear a litter without aid from conspecifics," encompasses both de facto and forced monogamy and, if it occurs, preference monogamy. However, it must be emphasized that the aid given the female need not depend on a monogamous bond. The wolf may be an obligate social species, but its mating system need not be obligate monogamy. Type I and II monogamy characterize only a subset of mating patterns found in wolves, and are thus inadequate for characterizing wolf mating systems.

A classification scheme more relevant to wolves has been provided by Emlen and Oring (1977:217). Monogamy, where "neither sex has the opportunity of monopolizing additional members of the opposite sex," fits most, or all, new and many smaller packs characterized by de facto monogamy. Larger packs, containing more than one adult of each sex, would be characterized by two forms of polygamy. The male mating patterns resemble those of *female defense polygyny* in which "males control access to females *directly*, usually by virtue of female gregariousness" (Emlen and Oring, 1977: 217). But unlike many mammalian systems in which dominant males have many mates and subordinates few or none, dominant male wolves may have only one mate while subordinates have none. The results are nevertheless similar: dominant males have greater genetic representation because of their apparent ability to restrict subordinates' access to females. The female mating patterns resemble those of *female access polyandry*, in which "females do not defend resources essential to males but, through interactions among themselves, may limit access to males" (Emlen and Oring, 1977:217).

Again, some females have greater genetic representation because they prevent other females from mating. This classification not only appears to fit most observations of wolf mating activity, but also has the advantage of being dynamically tied to the composition of adults in a pack. Changes in adult composition, or *operational sex ratios* (Emlen and Oring, 1977) will have important effects on courtship and mating patterns. Finally, because packs are normally composed of related individuals (Mech, 1970), kinship relationships should influence the pattern and frequency of intrasexual and intersexual reproductive activity. Here, of course, it is important to retain the distinction between evolutionary (genetic) and developmental (ontogenetic) precursors to behavior—a point where both field and captive studies offer potentially important, but incompletely explored, perspectives.

SUMMARY

A review of mating activity in wild and captive packs indicates monogamy is only one form that wolf mating systems will follow. Three forms of monogamy are distinguished. De facto monogamy occurs when only one adult pair is present, as in new and some small packs. Forced monogamy occurs when intrasexual aggression limits courtship and mating to one partner, despite the presence of other potential mates. Preference monogamy occurs when individuals, by choice, limit their courtship and mating to one partner. Unequivocal evidence for the latter form does not exist at present, and the former two forms characterize only a subset of observed mating patterns. Other evidence suggests that wolf mating systems are often polygamous, with courtship and mating patterns influenced strongly by intrasexual competition and aggression. Males exhibit a *female defense polygyny* and females exhibit a *female access polyandry*. The need for more precise measures of behavioral profiles as well as their distal and proximal antecedents is emphasized.

Cooperative Rearing of
Simultaneous Litters in Captive Wolves

Paul C. Paquet, Susan Bragdon and Stephen McCusker

INTRODUCTION

Two concurrent pregnancies were recorded during 1977 at the Washington Park Zoo, Portland, Oregon. The resulting offspring were successfully cared for in a communal den, and were nursed without regard to relatedness by both mothers. Communal denning has rarely been reported in wolves (*Canis lupus*) and has never, to our knowledge, been closely observed. There are no published records of communal nursing.

Multiple litters are believed to occur infrequently. They are thought to be limited, in part, by suppression of mating activity in low status wolves (Schenkel, 1947; Rabb et al., 1967; Fox, 1971; Mech, 1970; Zimen, 1975, 1976; Sullivan, 1978, 1979; Haber, 1977; Packard and Mech, 1980), strong mate preferences, and delayed sexual maturity (Rabb et al., 1967; Mech, 1970; Fox, 1971; Packard and Mech, 1980). As a consequence, a single litter is usually born to a pack each year, most often the result of a mating between dominant wolves.

However, evidence concerning the frequency of multiple litters suggests they are relatively common. Surveys of captive packs containing two or more adult females indicate a multiple litter rate of 28 to 39 percent (Harrington et al., this volume; Packard, 1980). Several captive packs have produced only single litters, whereas others have consistently displayed a high rate of multiple litters. For example, Altmann (1974) reported multiple litters in four of five breeding seasons, and Fentress and Ryon (this volume) in each of five years. In contrast, Zimen (1976) and Schotte and Ginsburg (1978) reported only single litters in each of four mating seasons.

Multiple litters also occur in the wild. Sixteen percent of all Alaskan packs studied (n = 62) had more than one litter (Packard, 1980). When only packs known to have at least two adult females were considered the rate of multiple litters was between 22 and 41 percent (Harrington, et al., this volume). Other observations of courtship behavior, estral bleeding and copulation by more than one female per pack (Peterson, 1977, 1979; Haber, 1977; Jordan in Packard, 1980) suggest that multiple litters may be relatively common in the wild.

Most field records of communal denning originate from Alaska. Several reports are the result of direct observation, whereas others are somewhat speculative, based primarily upon observations of abnormally large litters presumed to be the progeny of more than one female. Stephenson (pers. comm.) discovered a den in Alaska with a minimum of 11 pups attended by at least three adults. Another den containing 14 pups was found by Eskimo hunters in Alaska (Stephenson, pers. comm.). Although conclusive observations were lacking in both cases, it appeared that the pups were being cared for communally. In Mt. Mckinley National Park, Alaska, Murie (1944) observed two litters totalling 10 pups being brought together by their mothers. The pups were subsequently cared for by both females in a communal den. In the same area, Haber (1977), reported communal denning in one pack on three separate occasions. In addition, he recorded communal denning in other packs within the same general area.

Communal denning has also been observed in captivity (Fentress and Ryon, this volume; Klinghammer, pers. comm.; Packard, 1980; Rabb et al., 1967, pers. comm.; Traverso, unpub.). Commonly, however, the pups have not all survived, in some cases being killed by one of the mothers (Fentress and Ryon, pers. comm.; Klinghammer, pers. comm.; Rabb et al., 1967; Packard, 1980). Pups were killed in 54 percent of the multiple litters (n = 13) reviewed by Packard (1980). However, not all of these pups were being cared for in a communal den.

There are no records of wolf pups being voluntarily nursed by females other than their own mother. This is possibly due to a lack of den viewing opportunities, rather than the uniqueness of the phenomenon. In one case three litters and their mothers were placed together and a one week old pup was observed attempting to nurse from another litter's mother (Fox, 1971). In another captive pack, both the beta and gamma females nursed the alpha female's pups whenever she was absent from the den, although they had no pups of their own (Klinghammer, pers. comm.).

Although cooperative behavior is generally acknowledged to exist among wolves, there remains a lack of systematically collected data confirming the extent of its development. Accordingly, Fentress and Ryon (this volume) and Harrington and Mech (this volume) have pointed to the need for long-term, detailed observations documenting the role of social structure and the degree of individual participation in cooperative pack activities. Individual cooperative strategies can then be associated with age, sex, and social position and critically compared with analagous behaviors in other canine species.

This paper presents information gathered from close observation of two communally denning female wolves and their offspring. It is compared to similar observations of individually denning mothers and their pups, collected during two previous years. The paper quantifies the division of maternal care and describes social relationships among females and their offspring. Feeding behavior of pack members during denning is also described.

MATERIALS AND METHODS

Site Description

The wolves were kept in a cement grotto which consisted of three tiered, semi-circular platforms leading down into a dry moat. A small gravelled area, a pool, and a cement shelter were provided. The back of the enclosure was surrounded by high cement walls. No dirt or grass was available. Total area was approximately .03 ha.

Contiguous with the grotto were four adjacent concrete holding rooms. Each room was approximately 2 x 4 x 3 m. Three of the four rooms were provided with sliding guillotine-style doors opening into the main enclosure. Three similar doors connected the four rooms. At the rear of each holding cage was an access door. Each door was provided with a slanted feeding funnel and viewing window (12 x 30 cm). Water troughs, fillable from the outside, were in each of the four cages. Illumination for each holding cage was provided by a single 60 watt light bulb controlled by the keepers. Indirect light was available from the keeper's hallway which ran along the rear of the holding cages. There was no natural illumination.

Daily, except Sunday, the wolves were fed a commercially prepared meat substitute (Zu Preem™) supplemented by dry dog kibbles (Blue Mountain™). Frozen salmon was provided once weekly when in season. Water was available *ad lib.*

The Wolves

All wolves were of the subspecies *C.l. pambasileus.* None were socialized to humans, although they quickly habituated to observers and zoo staff. At the outset of the study in December 1975 the pack consisted of two wolves, a 15 month old male (Zane) and a 15 month old female (Taquish) (Figure 3.1) Subsequently, pack membership varied from two to thirteen wolves. Prior to the 1977 breeding season, when the multiple litters were born, the pack consisted of five wolves: a four year old male, Zane; a 22 month old male, Mowgli; a three year old female, Juneau; a 22 month old female, Sitka; and a nine month old female, Nahanni. Zane and Juneau were wild born in interior Alaska and captured as pups. The two

yearlings, Sitka and Mowgli, were littermates, the result of a 1975
mating between Zane and a wild born female Taquish. The pup,
Nahanni, was also sired by Zane in a 1976 mating with Juneau.

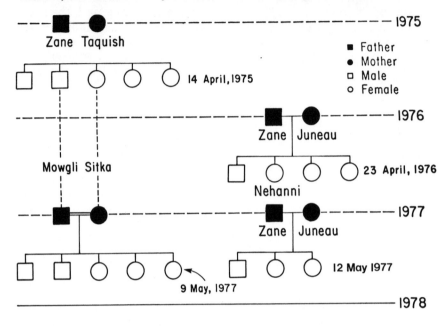

Figure 3.1: Graphic geneology of wolf pack at the Washington
Park Zoo, Portland, Oregon. Not shown is an additional unrelated female (Laska) who was captured as a pup near Mt. McKinley,
Alaska. She was approximately the same age as Juneau and was
introduced to the pack when Juneau was. She remained at the
Washington Park Zoo until December 1976 when she was surplused
to Utah State University, Logan, Utah.

Dominance Relationships

Dominance shifts occurred among the females in both the
1975-76 and 1976-77 breeding seasons (Table 3.1). Two year old
Juneau became alpha female in June of 1976 after killing Taquish,
the previous alpha female (Sullivan and Paquet, 1977). She retained the position until the end of January, 1977, when she
was replaced by Taquish's 22 month old daughter, Sitka (Figure
3.2).

The social order of the adult males shifted only in 1977 (Table
3.1). Four year old Zane, dominant since the pack's inception in
1973, was displaced by 22 month old son, Mowgli, during the
second week of April 1977 (Figure 3.3). At this time Zane also
became subordinate to Sitka who, supported by Mowgli, would
aggressively attack him.

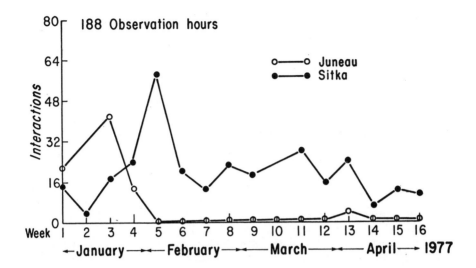

Figure 3.2: Each graph represents the total number of observed interactions in which a designated female was dominant. A dominance shift is indicated during the fourth week of January 1977.

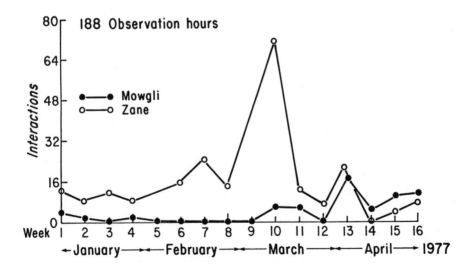

Figure 3.3: Each graph represents the total number of observed interactions in which a designated male was dominant. A dominance shift is indicated during the first week of April 1977.

Table 3.1: Dominance Relationships During Breeding Seasons, 1975-77

| Months | . . . . Males . . . . | | . . . . Females . . . . | | |
	Zane	Mowgli	Juneau	Sitka	Taquish
Jan-April 1975	Alpha	—	Yearling	—	Alpha
Jan-April 1976	Alpha	Yearling	Alpha	Yearling	*
January 1977	Alpha	Beta	Alpha	Beta	—
February 1977	Alpha	Beta	Beta	Alpha	—
March 1977	Alpha	Beta	Beta	Alpha	—
April 1977	Beta	Alpha	Beta	Alpha	—

*Isolated from pack December 1975-May 1976, during which time Juneau mated with Zane

Mating Behavior

Courtship and breeding involved only the wolves 22 months and older. Ten month old animals showed no physical or behavioral signs of mating activity. Although the litters sired by Zane in 1975 and 1976 did not have the same mothers, all observed mating during each breeding season was restricted to single pairs. In 1975 Zane was observed mounting Taquish 19 times and tying with her once. In 1976 he was observed mounting Juneau 48 times and tying three times. In 1977, although solicited by his daughter Sitka, Zane was only observed courting and breeding Juneau, his mate of the previous year. He aggressively rebuffed Sitka's courtship attempts. He was observed mounting Juneau 71 times, and tying with her successfully 5 times (9-10 March, 1977). Mowgli was initially unsuccessful in attempts to court his sister Sitka as she exhibited a strong preference for her father Zane. Sitka eventually became receptive to Mowgli, who was observed mounting her 47 times and tying with her three times (4-6 March, 1977).

Techniques

Behavioral observations were recorded in field notebooks and data sheets and later transcribed onto computer cards. Grotto observations were generally made from a maintenance ramp overlooking the entire grotto. Holding cage observations were made through viewing windows in the keeper access doors. Binoculars (9 x 36 or 8 x 35) were used as needed.

Individual relationships were assessed by quantifying the relative frequencies of designated behaviors, primarily body postures (Schenkel, 1947), during dyadic or group interactions. Other behaviors recognized as typical of dominant animals were also taken into account (Mech, 1970; Zimen, 1971, 1975). Total observation time of denning behavior was 225 hours (Table 3.2). Observation of general pack behavior was in excess of 2,200 hours. Interobserver reliability averaged over 94 percent as determined twice weekly throughout the study.

Table 3.2: Denning Observation Hours

Year	Female	Observation Hours	Dates
1975	Taquish	51	14 April–29 April
1976	Juneau	47	23 April–7 May
1977	Isolated pregnant females	13	5 May–9 May
1977	Sitka with solitary litter	15	9 May–12 May
1977	Communally denning females	99	12 May–25 May

RESULTS

Denning

Four days prior to the appearance of the first litter, zoo management isolated the pregnant females. Sitka delivered five pups (2 males, 3 females) on 9 May 1977. Three hours after delivering she was reunited with Juneau. With the exception of occasional sniffing there was little interaction between the two females. Juneau was never observed to approach the pups although Sitka made no apparent attempts to prevent her from doing so.

Within 24 hours of parturition it was observed that one noticeably smaller pup, crowded out by its four littermates, was unable to nurse. Sitka repeatedly carried it to the rear of the denning area and nursed it separately. This pup later disappeared and apparently died and was consumed by one of the adult wolves.

Juneau delivered three pups (1 male, 2 females) on 12 May. All pups were cleaned and nursed and then carried one at a time to Sitka, the dominant female. No signs of submissiveness were observed as she presented the pups to Sitka. Sitka received the pups and immediately commenced to clean, groom and nurse the new pups together with her own litter. Later the same morning Juneau was observed nursing both litters while Sitka slept nearby.

Both females delivered their pups in the same area of the den, constructing a ring with a raised perimeter from the straw covering the floor. One day prior to Juneau's delivery, Sitka moved her pups to the rear of a more isolated den.

Division of Maternal Care

Both litters displayed a preference for Sitka, based on the speed with which the pups approached Sitka, without apparent prompting, after a period of absence. Conversely, Juneau always had to physically retrieve individual pups and place them in a position to nurse. This pattern became evident three days after Juneau delivered.

In 1977, Sitka spent significantly more time in the den nursing, than did Juneau (Table 3.3, $t = 1.99$, $p < .05$). Although nursing was defined as total suckling time, it was impossible to be certain that

each pup was constantly suckling. We therefore may have miscalcu-
lated active milk exchange. However, because Sitka typically nursed
a greater number of pups (5 versus 3), we are confident her total
contribution to nutrition was greater than Juneau's. Juneau spent
significantly more time in care and grooming of the pups (Table 3.3;
t = 2.52, p <.05). In 1976, when Juneau alone littered (1 male, 2
females), neither percent time nursing nor percent time in care
and grooming of pups differed significantly from her figures in 1977
(t = .77, d.f. = 47 and t = .89, d.f. = 47). Due to den configuration in
1975 we were unable to observe Taquish with her pups.

Table 3.3: A Comparison of Care Behavior of Denning Mothers

	1976 Juneau	. . . . .1977. . . . . Sitka	Juneau
Time spent			
Nursing	14.0%	33.0%	14.0%
Care and grooming	3.2%	1.8%	4.9%
Total hours observed in den	22	53.4	52.2

Time Budgets

A regular periodicity in the mothers' care of the pups was established within two days after commencement of communal denning
and was maintained until approximately the 14th day after Sitka
whelped. The mothers were seldom with the pups between 1900–
2000 hours, but during all other times, one was in regular attendance. Both mothers had preferred hours. There was a weak inverse
relationship between the periods when the two mothers were in attendance (r = -.436, p <.05) (Figure 3.4).

To obtain an independent estimate of the care schedule, zoo
staff made spot checks and recordings at random times. To avoid
bias the observations were done when regular data collectors were
absent. The checks corresponded well with the time schedule generated from observer data (Sitka; r = .737, p <.001 and Juneau;
r = .932, p <.001). Although preferred hours of attendance had
been evident in both 1975 and 1976, no verifiable patterns were
established.

In 1977, Sitka and Juneau each spent about half of the total
observed time with the pups (Table 3.4). These data do not accurately reflect real time because the 24 one-hour observation
periods were biased toward Sitka's preferred hours. We compen-
sated for this bias by calculating mean time each female was with
the pups during each one hour period. This method revealed that
although Juneau spent more time with the pups than did Sitka
(Table 3.4) the difference was not significant (t = .979, d.f. = 98).

The time each female was alone with the pups was equivalent using
both the raw data (Sitka 34.6% vs Juneau 33.5%) and the adjusted

data (Sitka 28.3% versus Juneau 40.1%; t = .801, d.f. = 98). Data from the other two females in 1975 and 1976 were similar (Table 3.4).

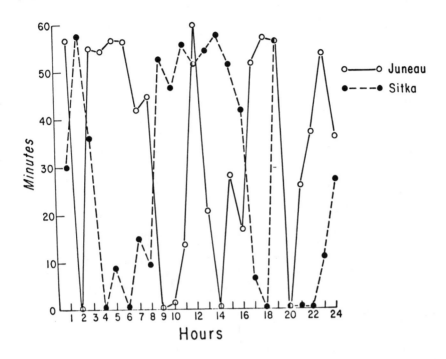

Figure 3.4: This graph depicts the time periods when communally denning mothers were in attendance (1977).

Table 3.4: Percentage of Time Females Present in Den with Pups
(Includes Observed Time and Adjusted Time)

	1975 Taquish	1976 Juneau	. . . . .1977. . . . . Sitka	Juneau
Time spent				
With pups	51.0%	47.3%	53.9%	52.7%
(adjusted time)			(46.5%)	(57.4%)
Alone with pups	51.0%	47.3%	34.6%	33.5%
(adjusted time)			(28.3%)	(40.1%)
Total hours observed	51	47	99	99

The pups were alone significantly less often in 1977 than in either 1975 (t = 14.7, p <.001) or 1976 (t = 15.1, p <.001) (Table 3.5). There was no significant difference between 1975 and 1976 (t = .66, d.f. = 47). In 1975, Taquish did not allow any other wolves into the den, so the pups were always alone whenever she was away. In 1976, Juneau allowed one other female access to her pups and that female "pup-sat" 14.8 percent of the time Juneau was absent. During 1977 Sitka would not allow other pack members to approach the pups. Therefore, the pups were cared for by only the two mothers. However, even without additional help, communal care significantly minimized the time pups were alone.

Table 3.5: Time Pups Left Alone

	Dates	Percent Total Observed Time Pups Alone	Percent Total Observed Time with Pups
Taquish	14 April-29 April 1975	49.0	51.0
Juneau	23 April– 6 May 1976	37.9	47.3*
Sitka	10 May– 12 May 1977	19.6	80.4**
Juneau and Sitka	12 May– 24 May 1977	12.7	87.3

*In 1976, a second nonparous female was in attendance 14.8% of the time Juneau was absent.

**Sitka's high attendance rate is due to the fact that the time frame displayed in the chart represents only the first three days after parturition. The other females showed a similar high attendance rate for equivalent periods. However, none of the females alone exceeded the attendance rate of the two parous females caring for pups cooperatively.

Feeding Behavior

Juneau was the only mother observed to cache food in 1977. She also frequently regurgitated food for Sitka. Sitka was never seen to reciprocate. The only aggressive encounter recorded while the mothers denned together occurred when Sitka attacked Juneau, pinned her to the ground and secured food from her mouth. Mowgli provided whole and regurgitated food to both mothers and the pups. When access to the den was restricted, Mowgli forced food under the door which separated him from the mothers. Zane and Nahanni both brought food to Juneau but not to Sitka. Nahanni begged food from Zane and Juneau but neither was observed to respond (Table 3.6). In Tables 3.6, 3.7 and 3.8 the numbers indicate observed feedings of one wolf (recipient) by another (provisioner), of either whole or regurgitated food.

Table 3.6: Feeding Distribution During Denning, 9 May - 14 June 1977 (154 Observation Hours)

Provisioners	. . . . Recipients.					
	Zane	Juneau	Mowgli	Sitka	Nahanni*	Pups
Zane	—	4	0	0	0	3
Juneau	0	—	0	19	0	5
Mowgli	0	3	—	7	0	1
Sitka	0	0	0	—	0	7
Nahanni*	0	4	0	0	—	1

*Yearling

In 1976, when Juneau produced the only litter, both Juneau and Laska were observed caching food and regurgitating food for one another and the pups. The yearling female, Sitka, never provided food or begged for food from other wolves. Her brother Mowgli never regurgitated food but brought whole pieces of food to both Juneau and Laska. Zane, the father of the litter, deposited regurgitated food near the entrance of the den, which was consumed by both Juneau and Laska. He also regurgitated food for Mowgli (Table 3.7).

Table 3.7: Feeding Distribution During Denning, 23 April - 7 June 1976 (107 Observation Hours)

| Provisioners | . . . . Recipients . . . . | | | | |
	Zane	Juneau	Laska	Mowgli*	Sitka*	Pups
Zane	—	9	3	6	0	2
Juneau	0	—	5	0	0	7
Laska	0	6	—	0	0	2
Mowgli*	0	3	5	—	0	1
Sitka*	0	0	0	0	—	0

*Yearlings

In 1975, Taquish only accepted food regurgitated by Zane. Both Juneau and Laska regurgitated food at the den entrance but it was ignored by Taquish and usually eaten by Zane. Zane was never observed feeding the yearling females, nor were the females observed feeding one another. After seven weeks of age the pups were recipients of whole and regurgitated food provided by Zane, but not the yearling females (Table 3.8).

Table 3.8: Feeding Distribution During Denning, 14 April - 1 June 1975 (93 Observation Hours)

| Provisioners | . . . . Recipients . . . . | | | |
	Zane	Taquish	Juneau	Laska*	Pups
Zane	—	15	0	0	4
Taquish	0	—	0	0	0
Juneau*	0	3**	—	0	0
Laska*	0	7**	0	—	0

*Yearlings
**Food provided but not eaten

DISCUSSION

Wolf populations seldom realize their reproductive potential (Rausch, 1967; Mech, 1970; Zimen, 1976). This has been attributed, in part, to reproductive limitations imposed by pack social organization which prevent some sexually mature females from breeding (Fox, 1971; Zimen, 1975, 1976; Packard and Mech, 1980). It is conceivable that a breakdown in social structure could disrupt restrictive mechanisms which limit breeding, thus resulting in an increased litter rate.

One common factor prior to the births of 21 captive-born multiple litters we reviewed was an inability on the part of established breeding females to effectively restrict the mating activity of subordinates.* This appeared to be associated with one or both of the following conditions: (1) social changes resulting from loss of status, death or removal of the dominant female, either prior to or immediately following estrus, and (2) large packs containing several sexually mature females.

In several instances, the removal or displacement of a key pack member appeared to have contributed to an increase in pregnancies (Rabb et al., 1967; Traverso, pers. comm.). For example, Rabb et al. (1967) noted that removal of an alpha male and the death of an alpha female promoted changes in pack social order, which they believe resulted in multiple matings and, ultimately, several pregnancies. Similar social changes preceded the double litters at the Washington Park Zoo, Portland, Oregon. Dominance shifts occurred among the females in the three consecutive years preceding the births, and male hierarchy changed immediately following estrus the year multiple litters were born. In contrast to previous years, social inhibition of mating was absent throughout the breeding period. The new alpha female made no apparent attempt to prevent subordinate females from breeding. Similarly, there was no attempt by the dominant male to prevent the subordinate male from mating. The only social restrictions to mating were imposed by the alpha male, who aggressively rebuffed sexual solicitations by the new alpha female. Although changes in the rank order occurred in each of the study years, the shifts in 1975 and 1976 preceded the breeding season by several months. Possibly this permitted the dominant female to become firmly established, allowing the pack to stabilize prior to mating. Woolpy (1968) has suggested that similar social changes could be produced by human exploitation of wolves (i.e. removal by hunting, trapping, poisoning), an idea which has been reiterated by Haber (1977).

In large packs with several sexually capable females, it may simply be beyond the capacity of one female to effectively restrict all breeding. Fentress and Ryon (pers. comm.) have observed that low ranking females will attempt to mate whenever out of contact with the dominant female. With several sexually mature females and topography allowing for easy concealment, a single female would

---

*This information is based upon our interpretation of data collected from personal communication with researchers and a review of pertinent published reports (Rabb et al., 1967, pers. comm.; Lentfer and Sanders, 1973; Packard, 1980; Fentress and Ryon, pers. comm.; Altmann, 1974).

find it difficult to suppress all mating activity. In wild packs, which may temporarily separate into subunits during the fall and winter (Wolfe and Allen, 1973; Peterson, 1977; Harrington and Mech, 1979), females may breed prior to later regrouping. When reunited, the pack might contain several pregnant females, each from a different subgroup.

Evolutionary Aspects

Offspring are an evolutionarily important resource. Their survival increases the probability of future genetic representation for all related individuals. If wolf packs are "expanded family units" (Mech, 1970), cooperative rearing of the young would be expected, as members share a portion of the same genetic complement (Hamilton, 1963). Offspring require a costly investment which individuals must balance with the potential benefits to be derived from the offspring's survival (Alexander, 1974). Consequently, individuals should behave toward pups in a manner which maximizes their own inclusive fitness, even if it results in lowered survival of the young (Bertram, 1976). Individuals cooperating in pup care would most likely be closely related (Bertram, 1976; Hamilton, 1963, 1971). However, the degree of cooperative behavior which they exhibit might vary with changing environmental constraints, social conditions, and pack size (Packard and Mech, 1980).

The probability of more than one litter surviving in a pack would likely depend upon acceptance by the dominant breeding female of the "extra pups". Survival would be enhanced if all the offspring were closely related and were not competing for limited resources. Based on relatedness, current resource availability, and social climate, the dominant female could "determine" whether to permit extra pups to survive and whether she should directly contribute to their care. Field evidence of mothers sharing the same den prior to parturition is not available. However, in two cases (Murie, 1944; Clark, 1971) where pups were born at separate den sites, and subsequently integrated, the mothers shared the den of the dominant female. In one case (Clark, 1971) interactions of the subordinate female with the pups declined following repeated threats by the dominant mother.

When circumstances are favorable, the dominant female might permit subordinate females to raise their litters cooperatively with her own. A cooperative arrangement potentially offers benefits for both mothers. Obviously, it would be advantageous for a subordinate to participate in the care of another litter, as long as it would guarantee her own pups survival. Energetically, the mothers would probably expend no more energy caring for the two litters cooperatively, than caring for one alone, yet the litters would profit from the increased attention of two mothers. If one mother stopped lactating or died, the pups would still be able to obtain milk, although younger pups would be at a competitive disadvantage if there was a shortage of milk. Overall, both mothers could help insure the survival of their own offspring while gaining the added advantage of increased genetic representation via related young.

Pup Preferences

Although wolf pups are thought to be blind, deaf and have little, if any sense of smell during the neonatal period (Scott and Fuller, in Mech, 1970) the pups observed at the Washington Park Zoo displayed a remarkable ability to discriminate between females at distances of up to two meters. It seems unlikely that this could have been accomplished exclusively with taste and tactile senses. Klinghammer (pers. comm.) has tentatively noted that neonatal wolf pups respond positively to auditory stimuli, and we believe the pups may have been reacting to auditory cues.

The onset of auditory function in domesticated dogs (*C. familiaris*) begins four to five days after birth, but is only measurable with intense stimulation. At seven to eight days, action potentials are recognizable at the round window and the entire auditory system responds to normal acoustic stimuli (Pujol and Hilding, 1973). If we presume similar physiological development in wolves, a plausible explanation for the pups' discriminatory behavior is that the older litter responded to auditory signals from Sitka, and the younger pups simply followed their movement. The quality and quantity of Sitka's milk, as a result of an earlier littering time, may have been superior to Juneau's. Thus, the pups would have been positively rewarded by selecting Sitka.

Prediction for Wild Populations

The social behavior of wild wolves is determined, in part, by additional constraints not associated with life in captivity (Kleiman and Brady, 1978). It is believed that wolf density and resource availability influence territory size, and that territoriality limits the number of packs per unit area (Packard and Mech, 1980). Pack size is thought to vary with population density, prey biomass, size of prey, and the capacity for social attachment by individual pack members (Mech, 1970; Rausch, 1967; Zimen, 1976).

Where wolf numbers have been seriously reduced for reasons other than resource limitation (e.g. catastrophic disease, excessive human exploitation), wolves could accelerate reproduction by breeding at younger ages (Medjo and Mech, 1976), increasing litter sizes, increasing the number of females per litter (Mech, 1970; Zimen, 1976), and increasing the number of litters per pack. All of these changes are probably influenced by an interaction of social and ecological factors (Packard and Mech, 1980). The proximate regulating factor is most likely nutrition, which subsequently affects pack size and possibly social structure. Normally, packs which are not nutritionally stressed would exhibit little intrapack strife (Zimen, 1976; Packard and Mech, 1980), resulting in lower mortality among dominant females, less chance for social disruption and therefore low frequency of multiple litters. Packs which are nutritionally stressed but nevertheless retain a stable dominance hierarchy should likewise produce multiple litters at a low frequency. However, packs which have been randomly reduced may not retain an intact social structure, thus creating favorable conditions for multiple matings. Therefore, we would expect multiple litters to occur more frequently in populations which are not resource limited but have been reduced because of a catastrophic event, such as disease or excessive human hunting.

SUMMARY

Multiple litters are believed to occur infrequently in wolf packs. However, an examination of the literature suggests they may be more common than previously acknowledged, both in captivity and in the wild. Multiple litters might occur frequently in conjunction with social disruption.

The successful rearing of more than one litter per pack may depend upon the degree of relatedness among pack members, particularly breeding females, along with prevailing environmental conditions. If wolf packs are "expanded families," then cooperative care of pups should be expected, including communal denning and nursing of multiple litters. Both of these forms of cooperative behavior may occur with regularity, but lack of observational opportunities could limit reports.

Cooperative rearing of young potentially increases time pups are being actively cared for, while concurrently freeing mothers for other endeavors. In addition, it provides a safeguard against untimely death of nursing mothers and the subsequent mortality of dependent pups. Cooperative care may also be energetically more efficient, with mothers systematically arranging attendance time to an optimal level.

Acknowledgements

We gratefully acknowledge the invaluable assistance of Washington Park Zoo staff and volunteers, especially Jill Mellin, Gordon Noyes and Dan Heath. We wish to thank Zoo Director, Warren Iliff, for his unfailing support and encouragement throughout the project. Also, Richard Forbes, Stanely Hillman and Robert Tinnin for help in preparation of the manuscript and their constant support. Lastly, we would like to extend a very special thanks to Hal Markowitz and John Sullivan.

A Long-Term Study of
Distributed Pup Feeding in Captive Wolves

John C. Fentress and Jenny Ryon

INTRODUCTION

For both the biological and behavioral scientist, the intricacies of social organization in wolves (*Canis lupus*) are a major attraction (e.g., Fentress et al., 1978; Moran and Fentress, 1979; Schenkel, 1947, 1967; Zimen, 1976). The basic social unit, the pack, both constrains the range of activities of its individual members and provides important benefits. As packs are generally extended family units (Harrington and Mech, this volume; Mech, 1970; Murie, 1944; Peterson, 1977), a major question of interest is: How do pack members distribute their reproductive and care-giving activities, thus permitting the successful integration of offspring into the family?

Pup survival depends upon food resources provided by others, whether by lactation, regurgitation, or carrying food in the mouth (e.g., Murie, 1944; Rutter and Pimlott, 1968; Mech, 1970; Fentress et al., 1978). The manner in which different pack members contribute to the provisioning of pups is also of broader theoretical interest in terms of evolutionary models of cooperative behavior (e.g., Brown, 1978; Hamilton, 1964; Krebs and Davies, 1978; Maynard Smith, 1964; Trivers, 1972; Wilson, 1975). While cooperative feeding in wolves has often been mentioned in the context of broader studies and reviews (e.g., Allen, 1979; Kleiman, 1977; Kleiman and Brady, 1978; Kleiman and Eisenberg, 1973; Mech, 1970; Murie, 1944; Peterson, 1977; Rabb et al., 1967; Schonberner, 1965; Vehrencamp, 1979; Wittenberger, 1979; Woolpy, 1968; Young and Goldman, 1944), the data available in published reports are surprisingly sparse (cf. Bennett, 1979; Paquet et al., this volume). Recent data on other canidae such as red foxes (*Vulpes vulpes*) (Macdonald, 1979) and black-backed jackals (*Canis mesomelas*) (Moehlman, 1979) suggest important species differences as well as similarities in the various strategies (and possibly consequences) of cooperative pup care.

From the perspective of behavioral analysis, one need is for detailed observations that document individual contributions and total pack structure over a protracted period. In this way, the network of both short-term interactions and longer term relationships can be traced systematically for individual age, sex and social role differences (cf. Harrington and Mech, this volume; Hinde and Stevenson-Hinde, 1976; Lockwood, 1979). We report data here, abstracted from our ongoing program of social behavior research in captive canids (e.g., Fentress et al., 1978; Roper and Ryon, 1977; Ryon, 1977, 1979; Unpub. theses by Bennett, 1979; Field, 1979; Havkin, 1977; Moran, 1978) that may contribute to a more complete picture.

METHODS

The geneology of our animals is summarized in Figure 4.1. The original male (L63♂) was obtained at four weeks of age from the Whipsnade Zoological Park in London and hand-reared (Fentress, 1967). Two females were placed with L63♂: a four week old pup in 1968, and a six month old hand-reared pup (Z69♀) in 1969. By spring 1973, none of our animals had bred. The older female was removed and two pups (Sa73♀ and Sz73♀) were obtained from the Anchorage Zoo in Alaska. They were hand-reared and then placed with L63♂ and Z69♀ at three and a half months of age.

The first litter of pups was born to L63♂ and Z69♀ in 1974. Subsequent litters from sibling-sibling and parent-offspring matings were born in 1976, 1977, 1978 and 1979. Some pups were removed from the group each year. We gradually increased the ratio of males to females from 1:3 (1973, 1975) to 1:1 (1979). The sex and year of birth of the animals that have been retained are indicated after each animal's identifying initial(s).

Reproductive status of individual females for a given year was determined by: (a) observed mating, (b) morphology (i.e., distended abdomen), and (c) presence of female and pups in dens. Abortion, stillbirths and/or loss of neonatal pups was determined by: (a) presence of dead pups in den with female, and/or (b) a female having had a distended abdomen and then appearing with a sunken abdomen and blood under her tail. Females which lost pups will be termed "unsuccessful mothers."

Data for 1973–74 were obtained in a 1,011 m² enclosure which we monitored from an adjacent tower at the BioSocial Research Center, University of Oregon. Since autumn 1974, the animals have been kept in a 3.8 hectare wooded enclosure at the Provincial Wildlife Park, Shubenacadie, Nova Scotia. Seven mounds of bulldozed earth and tree trunks were made around the periphery of the enclosure to provide denning areas. Observations were made from a trailer near a cleared feeding area and from huts near each of the major den sites.

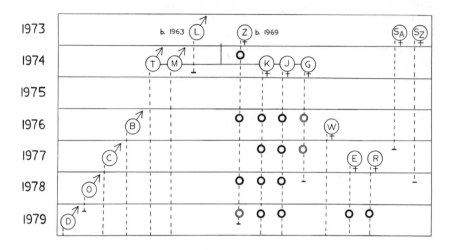

Figure 4.1: Geneology of pack animals. Year of birth is indicated on the margin unless otherwise noted. Death is indicated by termination of dashed line under animal's initial(s). For females, a heavy, double circle represents a mother for the year shown; a light, double circle represents an unsuccessful pregnancy (abortion, stillbirth or loss of neonatal pups); and lack of symbol represents no observed pregnancy. (Format derived from Macdonald, 1979.)

Animals were fed 7 to 30 kg of chicken parts and vitamins plus commercial dog chow or occasional road-killed deer approximately every other day. The food provided at each feeding always exceeded the amount the group ate in a day. Food was placed in a single pile and all animals had equal access to the food.

Our data were collected over 430 days of observation during feeding periods (duration range 30–180 minutes) between 29 August 1973 and 1 January 1980. Only observations made between the first and last feedings of mothers and pups each year (normally April to January) are included. One to three observers (see acknowledgements) recorded behavior continuously during observation periods with time noted each minute (cf. Ryon, 1979). A shorthand consisting of 150 coded behaviors was developed and used throughout most of this study (Ryon, unpub.). Samples of inter-observer agreement in the identification of individual animals and in classification of behavior exceeded 95 percent.

The behavioral activities with which we were primarily concerned here were: (1) feeding, and (2) food solicitation. Feeding of one animal by another can involve: (a) lactation, (b) regurgitation, and (c) carrying plus depositing food. Solicitation of one animal by another involves interpretation as well as description, but we have found almost no disagreement among trained observers. Food solicitation includes muzzle contact plus licking, extension of the body, twisting of the body to the side with the head lowered and/or ears flattened, rapid tail wagging of the lowered tail, bounding, pawing and squeaking. The pattern in older animals is similar to "submissive behavior" (Schenkel, 1947). Feeding plus possible rebuffs, and solicitation are typically closely intermeshed. This makes simple declarative cause-and-effect statements difficult (cf. Moran and Fentress, 1979). Together, these various actions serve the biological function of provisioning food to some pack member(s) by others. While we focus on food provisioning, it is important to recognize that any behavior occurs within the context of other features (Fentress, 1976, 1978). Thus, we provide additional background when it appears to assist interpretation.

RESULTS

We have, to date, observed a total of 1,102 feedings of one pack member by another. We shall present our data in two sections: (1) a detailed chronological summary of cooperative feedings and solicitations from 1973 through December 1979 ("chronology"), and (2) summary cross sections of the data by individual, age, sex and social role ("cross-sections").

Chronology

1973: The two unbred adults, L63♂ and Z69♀, fed two three and a half month old "adopted" female pups (Sa73♀ and Sz73♀) (Figure 4.2). The male fed 111 times versus the female's 40. He was also solicited by the pups twice as frequently as the female (79:34). We observed the female soliciting five times from the male, all unsuccessfully.

No feeding of pups was observed on the day the pups were introduced. On the second day, the pups solicited the adult female persistently by running beside her and poking her muzzle with theirs until she regurgitated. When the pups tried to eat the regurgitant, the adult female growled and ate it herself. During the third feeding session, the pups solicited both adults. The female rebuffed them by growling, but the male regurgitated. Eventually, both adults fed the pups, but the female continued to rebuff them more often.

1974: The first regurgitation to Z69♀ by L63♂ was observed 18 days before the female gave birth. Z69♀ also ate food provided by human caretakers. The male continued to feed her throughout the prenatal period and until postnatal day 24 (Figure 4.3). He fed the yearlings twice as often as he fed Z69♀ during this period (34:16). Z69♀ did not feed the yearlings. After the pups came above ground, L63♂ was observed to feed Z69♀ only once. This contrasts with the

21 times he fed the yearlings and 33 times he fed the pups for the rest of the observed season which ended in October when the animals were moved to Nova Scotia. Z69♀ regurgitated and/or carried food to the pups 26 times, and to the yearlings five times during this period. The yearlings also fed the pups 26 times. Often, the yearlings fed the pups soon after they themselves had consumed regurgitated food, thus acting as a relay in the feeding network.

Z69♀ solicited feeding from L63♂ by lowering her body, and licking his muzzle from below. The yearlings solicited in a similar way (Figure 4.4a) as well, by pawing. The pups would often leap toward the muzzle of older animals (Figure 4.4b). All other animals solicited from, and were fed by L63♂ (Figure 4.4c). His feeding, however, was selective in the sense that the ratio of his feeding to the solicitation of pups was highest (33:12), yearlings intermediate (55:25), and adult female lowest (17:31). The pups solicited from each of the older animals, with the ratio of feedings to solicitations being higher for the two adults (59:22) than for the yearlings (26:24) (Figure 4.3).

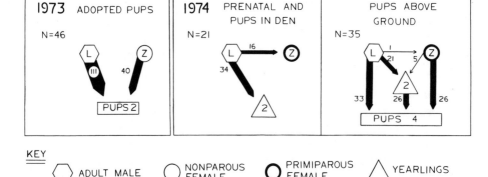

Figure 4.2: Feeding of adopted pups and subsequent litter. 1973 data show feeding by unbred L63♂ and Z69♀ of two pups introduced at 3.5 months, Sa♀ and Sz♀. The number of feedings are indicated within, or next to, arrows indicating caregiver and recipient. In 1974, the first litter was born to the above adults. L63♂ fed both yearlings and Z69♀ during the latter's pregnancy, and early post partum. When pups came above ground (4 weeks), they were fed by the yearlings. L63♂ continued to feed yearlings, but was seen feeding Z69♀ only once more. Note small amount of yearling feeding by Z69♀. In this and subsequent figures, the term *nonparous* is used for nonpregnant females whether or not they have bred previously.

FEEDING   SOLICITING

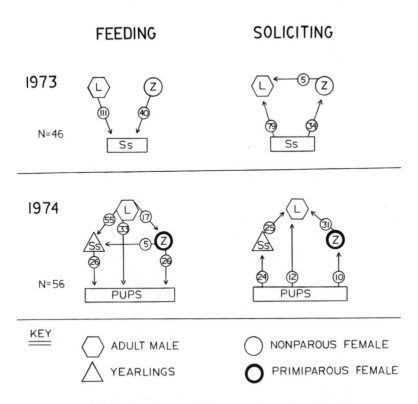

Figure 4.3: Soliciting and feeding, 1973-1974. The numbers and directions of observed food solicitations are compared with the summed feeding scores from the previous figure. Note that while there is an overall positive correlation between numbers of feedings and solicitations, the ratios in individual cases may deviate substantially from 1:1. Pups and yearlings are each treated as single units since they often solicited and were fed together. N = number of observation periods.

1975: No wolves bred in 1975.

1976: The overall network of feeding/solicitations for 1976 is summarized in Figure 4.5. Again, the correspondence between feedings and solicitations is positive, but imperfect. The three pups that remained in the compound until 19 September (when one was removed) solicited, and were fed by, *each* of the older pack members, consisting of three mothers, one unsuccessful mother, two nonbreeding females, and two males. Two of the mothers (Z69♀ and J74♀) were fed by the two adult males and the unsuccessful mother (G74♀); the third mother (K74♀) was not fed, nor did she solicit. Each of the mothers fed pups only, and was solicited by pups only.

Figure 4.4: Feeding and solicitations. (Top) Sa73♀ and Sz73♀ as year-lings soliciting food from L63♂. (Center) 3½ month old pup soliciting from Z69♀. (Bottom) L63♂ regurgitating to five month old pups.

The two females that apparently did not breed (Sa73♀ and Sz73♀) were solicited by mother J74♀, but did not feed her. Most feeding of mothers was done by the males (T74♂ and M74♂) who were solicited by them the most. The pups most frequently solicited two of the three mothers, plus the unsuccessful mother. These females fed the pups most often. The overall correspondence between feeding and solicitation at the level of individual interactions is again less than perfect. For example, G74♀, the unsuccessful mother, fed the pups most often, while Z69♀ was solicited most often (feeding to solicited ratio = 74:28 for G74♀ and 52:62 for Z69♀). Sa73♀ and Sz73♀ were observed feeding the pups only twice each, but, on many occasions, they were chased away from the pups by Z69♀, who was a high-ranking female. Thus, feeding of others must be "allowed" as well as "desired."

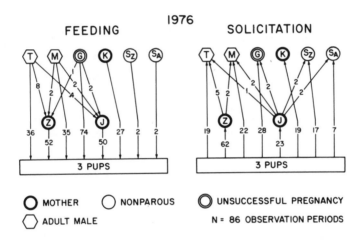

Figure 4.5: Network of feedings and solicitations in 1976. The number and direction of feedings and solicitations are shown by the arrows going from feeder or solicitor to recipient. Three pups remained in the pen until 19 September, when one was removed.

Feeding of mothers prior to partuition was not observed this year, although T74♂ was seen standing, holding food, and squeaking toward Z69♀ several times. Z69♀ appeared to ignore T74♂ on these occasions, and left to chase Sa73♀ each time. The first observed solicitation was by J74♀ on 7 May, 12 days after she whelped, and was directed toward both Sa73♀ and Sz73♀. Sa73♀ responded by leaving, while Sz73♀ pinned J74♀ to the ground. The first observed feedings were regurgitations to Z69♀ by T74♂ and M74♂ after she solicited them on 12 May, 10 days after she whelped. Feedings of mothers continued until 31 May, and ended by the time the first pup feeding was observed.

1977: Feeding/solicitation data are summarized in Figure 4.6. The two mothers (K74♀ and J74♀), the unsuccessful mother (G74♀), and the two yearlings (B76♂ and W76♀) solicited feeding from other pack members and were fed, as were the pups. Z69♀ was not a mother this year and, in contrast to previous years, she neither solicited nor was fed. Z69♀ and G74♀, in many respects, had reversed feeding/solicitation roles from the previous year when Z69♀ was a mother. (Z69♀ appeared to be an "outcast" this year.)

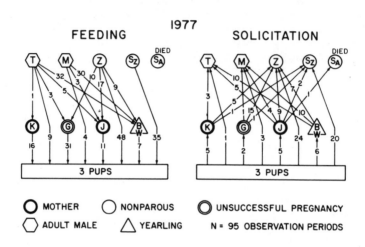

Figure 4.6: Network of feedings and solicitations in 1977. The number and direction of feedings and solicitations are shown by the arrows going from feeder or solicitor to recipient.

As in 1976, J74♀ solicited from the greatest number of other adults in the pack, and at least once from the yearlings. The two nonparous females and the unsuccessful mother fed the pups most frequently. The two barren females in this group were also solicited most frequently by the pups and had the highest rate of pup feeding. One of these (Sz73♀) had a low rate of pup feeding in 1976 which was associated with interference by the higher ranking Z69♀, but increased her pup feeding this year after Z69♀ lost status. The adult males, T74♂ and M74♂, fed pups less than did any other adults. They were solicited, however, by the greatest number of different animals. Sa73♀ died early in the pup rearing season. By then she had been observed being solicited only once, but did not respond.

We observed feeding of the mothers between one and 30 days after the birth of the first litter on 22 April. Z69♀ fed mothers more often than did any other pack member, but she did so preferentially. She fed J74♀ and G74♀, with whom she had shared a den the previous year (Bennett, 1979), but refused K74♀, in spite of five solicitations. K74♀'s solicitation of Z69♀ is interesting since K74♀ had directed considerable aggression toward Z69♀ earlier in the year (Bennett, 1979). Z69♀'s ratio of feeding to being solicited was 17:9 for J74♀ (mother) and 10:15 for G74♀ (unsuccessful mother). The adult males fed the yearlings at a higher rate than their combined feedings of adult females and pups (32 [yearlings] to 18 [pups plus adult females] for T74♂; 30 [yearlings] to 7 [pups plus adult females] for M74♂). As in other years, no selectivity could be discerned in the feeding of individual pups, although considerable selectivity is suggested for the feeding of older animals. The yearlings this year behaved similarly to those in 1974 in that they were both fed by adults, and themselves fed pups. Adult males and adult females which did not become pregnant were not fed, nor did they solicit.

1978: Feeding and solicitations for 1978 are summarized in Figure 4.7. The summer was unusually hot and the general activity level, appetites and overall feeding frequencies among pack members were depressed. In contrast to previous years, only a single pup was maintained in the pen. Two of the three 1978 mothers were observed being fed, but selectively and at low frequencies. The unsuccessful mother (G74♀) was not fed, nor did she solicit. Adult males and nonparous females fed other animals but were not fed themselves, although W76♀ was observed soliciting twice from T74♂. The single pup was fed solid food by each of the three mothers, the three adult males, one other adult female, and two of the three yearlings. The youngest male (B76♂) both solicited from, and was fed by, other adult males.

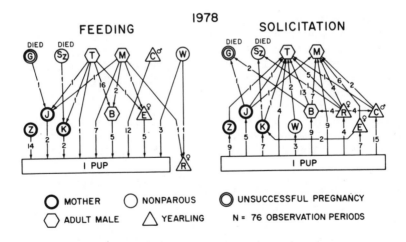

Figure 4.7: Network of feedings and solicitations in 1978. The number and direction of feedings and solicitations are shown by the arrows going from feeder or solicitor to recipient. Only one pup was kept in the compound for the postnatal feeding period.

Each of the three yearlings behaved differently: (a) E77♀ solicited, was fed, and also fed the pup as had yearlings in previous years; (b) R77♀ solicited and was fed, but did not feed the pup; (c) C77♂ was not fed, solicited infrequently, but fed the pup relatively frequently. Thus, while the behavior of the mothers, nonparous adult females, adult males and yearlings showed broad parallels to previous years, some exceptions were also apparent.

Only two of the three mothers (K74♀ and J74♀) were fed at a comparatively low rate, and mostly by the two oldest males (T74♂ and M74♂). Z69♀, who was a mother for the first time since being an "outcast," was not fed at all and solicited (T74♂) only once. The three animals we observed feeding the pup most frequently were Z69♀–a mother (n = 14), C77♂–a yearling (n = 12), and M74♂–an adult male (n = 7). C77♂ was solicited more frequently by the pup than was any other pack member (n = 15); Z69♀ and M74♂ were each solicited nine times, as was B76♂ (the adult male who himself solicited feeding and was fed by the two oldest males). The low level of observed pup feeding plus feeding of two of the three mothers and one of the yearlings (n = 1 each) contrasts with T74♂'s tendency to feed the younger adult, B76♂ (n = 16). T74♂ was also solicited more frequently by B76♂ (n = 13, compared to n = 1 for each of the three mothers and the yearling C77♂, n = 5 for a second yearling, and n = 2 for nonparous W76♀). T74♂ was solicited by the greatest number of animals (n = 6). No other single animal was observed to be solicited by more than two individuals. Given the relatively small number of solicitations and feedings in 1978, a considerable degree of selectivity is suggested. This selectivity in feeding/soliciting was less for the pup than for other pack members.

1979: Data for 1979 are summarized in Figure 4.8. Each of the four mothers was again fed. Also, each adult fed the lone pup. Z69♀ aborted, and later died in 1979. Unlike the occasions when she was a successful mother, she fed a mother by regurgitation in addition to feeding the pup. The other nonparous adult female (W76♀) fed three of the four mothers in addition to the pup. B76♂ was again unusual in that (a) he did not feed any of the mothers (although he was solicited by two of them), (b) he, in turn, solicited feeding from four other adults, including one mother (K74♀), and (c) he was fed by one of the adult males (T74♂). The pup solicited from each of the other pack members.

The feeding of mothers was observed from 21 April to 16 June (seven days before and 39 days after the first litter was born). As in previous years, none of the mothers fed each other. Unlike previous years, T74♂ fed only one mother (K74♀), although he was solicited by each of the mothers. He was twice observed to avoid a soliciting mother (J74♀) on his way to K74♀. This male was also the only animal solicited by K74♀. This contrasts with the other three mothers, each of whom was observed soliciting from four individuals. M74♂ and C77♂ each fed R77♀ preferentially. Both of these males were solicited by three of the mothers. R77♀ solicited the males most frequently. As in the previous year, Z69♀, who died during the period of observation, and C77♂ fed the pup most frequently.

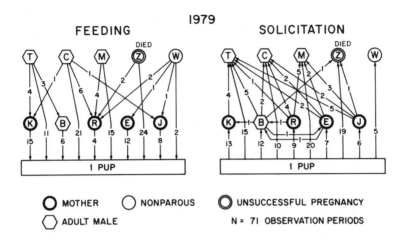

Figure 4.8: Network of feedings and solicitations in 1979. The number and direction of feedings and solicitations are shown by the arrows going from feeder or solicitor to recipient. Only one pup was kept in the compound for the postnatal feeding period.

Cross Sections of Data by Individual, Age, Role and Sex

Figure 4.9 summarizes the feeding behavior (by regurgitation/food dropping) for Z69♀ from 1973 to 1979. She fed adopted pups in 1973 even though (a) she had never bred, (b) she had not, as a pup, been fed solid food by adult wolves, and (c) she had not observed pup feeding by others. When she bore pups in 1974, she was observed regurgitating to them 26 times and to the yearlings only five times (compared to 40 times when they were pups). In 1975, when there was no breeding in the pack, she regurgitated to her own yearlings (an exact count was not kept). In 1976, she fed only her own and others' pups, and no other pack members. This contrasts to 1977 when Z69♀ did not breed. That year, she fed other mothers at least 27 times in addition to yearlings nine times and others' pups 48 times. She bred again in 1978, and was only observed feeding the pup left in the pen (which belonged to J74♀); she was not seen feeding either the yearlings or other mothers. Upon aborting her own litter in 1979, she again fed other mothers (though infrequently), as well as the pup left in the compound. These data suggest that the feeding behavior of a female can vary as a function of whether or not she produces a viable litter.

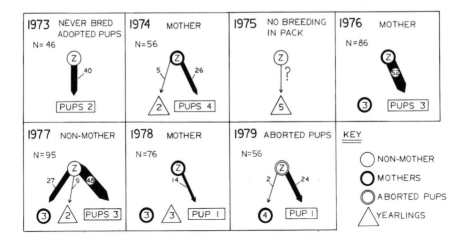

Figure 4.9: Summary of feeding behavior by Z69♀, 1973–79. Numbers within rectangles (pups), triangles (yearlings), and heavy circles (mothers) indicate the number of animals in each category. The data demonstrate altered tendency of same female to feed non-pups as a function of whether or not she is herself a mother that year.

The totaled data on feedings and solicitations each year are summarized for each animal in Table 4.1. (Mothers for a given year are enclosed in heavy solid-line rectangles; unsuccessful mothers are enclosed in broken-line rectangles.) We observed 13 litters. The mothers solicited feeding in 12 cases. In 11 cases, the mother was fed by other pack members. Four instances of abortion, stillbirth or loss of neonatal pups occurred. In one of these instances, the female solicited and was fed. There were nine cases in which adult females (two years plus) did not become pregnant. They were never fed, and were observed to solicit in only two instances.

With the sole exception of B76♂, adult males did not solicit nor were they fed by others. Pups solicited and were fed in every case. All yearlings solicited and, with one exception (C77♂), were observed to be fed by older pack members. Yearlings also characteristically fed pups (R77♀ being the sole exception). Pups and yearlings were treated as single groups in Table 4.1, when we could not distinguish between the behavioral profiles of individuals. As the animals got older, clear profiles emerged on the basis of sex, role (e.g., mother versus non-mother), and individual characteristics.

The feedings (regurgitation plus food dropping) by individual females and males are subdivided in Tables 4.2 and 4.3, respectively. The numbers of observed pup feedings, yearling feedings and adult feedings are indicated. Mothers tend to focus their feeding on pups (which may not be their own), when available.

Table 4.1: Summary of Feeding and Solicitation Data for Each Animal[1]

	...1973...	...1974...	..1976...	...1977...	.1978..	.1979..	..Total...
L63♂	111, 0	105, 0					216, 0
	84, 0	68, 0					152, 0
Z69♀	40, 0	31, 17	52, 11	84, 0	14, 0	26, 0	247, 28
	34, 5	10, 31	62, 7	60, 0	9, 1	21, 0	196, 44
Sa73♀	0, (75.5)	(13), (30)	2, 0	0, *	1, *		15, 105.5
	0, (56.5)	(12), (12.5)	9, 0	1,	2,		22, 69
Sz73♀	0, (75.5)	(13), (30)	2, 0	35, 0	1, *		51, 105.5
	0, (56.5)	(12), (12.5)	19, 0	24, 0	2,		57, 69
T74♂		. (removed).. / . (removed)..	48, 0	50, 0	20, 0	18, 0	136, 0
			25, 0	19, 0	28, 0	32, 0	104, 0
M74♂		0, (21.25)	39, 0	37, 0	12, 0	19, 0	107, 21.25
		0, (11.5)	26, 0	18, 0	29, 0	30, 0	103, 11.5
K74♀		0, (21.25)	27, 0	16, 1	2, 2	15, 5	60, 29.25
		0, (11.5)	19, 0	5, 9	9, 2	14, 4	47, 26.5
J74♀		0, (21.25)	50, 8	11, 25	2, 3	8, 2	71, 59.25
		0, (11.5)	23, 9	5, 22	5, 3	6, 11	39, 56.5
G74♀		0, (21.25)	77, 0	31, 13	1, *		109, 34.25
		0, (11.5)	30, 0	2, 17	3,		35, 28.5
B76♂			0, (139)	(3.5), (35.5)	5, 18	6, 3	14.5, 195.5
			0, (98.5)	(3.5), (13.5)	13, 22	14, 10	30.5, 144

(continued)

Table 4.1: (continued)

	1973	1974	1976	1977	1978	1979	Total
W76♀			0 (139)	(3.5) (35.5)	4 0	6 0	13.5 174.5
			0 (98.5)	(3.5) (13.5)	3 2	5 0	11.5 114
C77♂				0 (53.7)	12 0	29 0	41 53.7
				0 (22)	17 3	20 0	37 25
E77♀				0 (53.7)	5 2	12 1	17 56.7
				0 (22)	7 8	8 7	15 37
R77♀				0 (53.7)	0 2	4 14	4 69.7
				0 (22)	4 16	9 11	13 49
O78♂					0 51		0 51
					0 72		0 72
D79♂						0 118	0 118
						0 116	0 116
Total	151	162	297	271	78	143	1,102
	118	102	213	141	129	159	862
Feed/solicit	[1.28]	[1.59]	[1.39]	[1.92]	[0.61]	[0.90]	[1.28]
Observation days	46	56	86	95	76	71	430

*Died

[1]For each wolf, each year four numbers are entered: (a) number of times animal feeds others—upper left corner; (b) number of times animal is solicited—lower left corner; (c) number of times animal is fed by others—upper right corner; and (d) number of times animal solicits feeding from others—lower right corner. When a female produces pups, her scores are enclosed in heavy solid-line box. When a female has an unsuccessful pregnancy (abortion, stillbirth or loss of neonatal pups), her scores are enclosed in broken-line box. The ratio of feeding to solicitation, and the number of observation days for each year are given at the bottom of the table. Parentheses enclose data averaged for individuals that were not individually distinguished (e.g., pups in a litter).

Table 4.2: Summary of Feedings of Other Animals by Individual Females[1]

	1973	1974	1976	1977	1978	1979	Totals
Z69♀							
Pup	40 (100%)	26 (84%)	52 (100%)	48 (57%)	14 (100%)	24 (92%)	204 (82%)
Yearling	—	5 (16%)	—	9 (11%)	0	—	14 (6%)
Adult	0	0	0	27	0	2 (8%)	29 (12%)
	40	31	52	84	14	26	247
Sa73♀							
Pup		13	2				15 (100%)
Yearling		0	—				0
Adult		0	0				0
		13	2				15
Sz73♀							
Pup		13 (100%)	2 (100%)	35 (100%)	0		50 (98%)
Yearling		0	—	0	0		0
Adult		0	0	0	1 (100%)		1 (2%)
		13	2	35	1		51
K74♀							
Pup			27 (100%)	16 (100%)	2 (100%)	15 (100%)	60 (100%)
Yearling			—	0	0	—	0
Adult			0	0	0	0	0
			27	16	2	15	60
J74♀							
Pup			50 (100%)	11 (100%)	2 (100%)	8 (100%)	71 (100%)
Yearling			—	0	0	—	0
Adult			0	0	0	0	0
			50	11	2	8	71

(continued)

Table 4.2: (continued)

	1973	1974	1976	1977	1978	1979	Totals
G74♀							
Pup			74 (96%)	31 (100%)	0		105 (96%)
Yearling			–	0	0		0
Adult			3 (4%)	0	1 (100%)		4 (4%)
			77	31	1		109
W76♀							
Pup				4 (100%)	3 (75%)	2 (33%)	9 (64%)
Yearling				0	1 (25%)	–	1 (7%)
Adult				0	0	4 (67%)	4 (27%)
				4	4	6	14
E77♀							
Pup					5 (100%)	12 (100%)	17 (100%)
Yearling					0	–	0
Adult					0	0	0
					5	12	17
R77♀							
Pup					0	4	4 (100%)
Yearling					0	–	0
Adult					0	0	0
						4	4

[1]The data are subdivided into observed feedings of pups, yearlings and adults. When no yearlings were present, this fact is indicated by a dash. Heavy solid-line boxes indicate when female was a mother; broken-line boxes indicate when female had an unsuccessful pregnancy.

Table 4.3: Summary of Feedings of Other Animals by Individual Males[1]

	1973	1974	1976	1977	1978	1979	Total
L63♂							
Pup	111 (100%)	33 (31%)					144 (67%)
Yearling	—	55 (52%)					55 (25%)
Adult	—	17					17 (8%)
	111	105					216
T74♂							
Pup			36 (75%)	9 (18%)	1 (5%)	11 (61%)	57 (42%)
Yearling			—	32 (64%)	1 (5%)	—	33 (24%)
Adult			12 (25%)	9 (18%)	18 (90%)	7 (29%)	46 (34%)
			48	50	20	18	136
M74♂							
Pup			35 (90%)	4 (11%)	7 (58%)	15 (79%)	61 (57%)
Yearling			—	30 (81%)	2 (17%)	—	32 (30%)
Adult			4 (10%)	3 (8%)	3 (25%)	4 (21%)	14 (13%)
			39	37	12	19	107
B76♂							
Pup				4 (100%)	5 (100%)	6 (100%)	15 (100%)
Yearling				0	0	0	0
Adult				0	0	0	0
				4	5	6	15
C77♂							
Pup					12 (100%)	21 (75%)	33 (83%)
Yearling					0	—	0
Adult					0	7 (25%)	7 (17%)
					12	28	40

[1]The data are subdivided into observed feedings of pups, yearlings and adults. When no yearlings are present, this fact is indicated by a dash.

Non-mother adult females, including females who become pregnant but lose their pups, may also feed mothers as well as yearlings. Adult males are most likely to feed mothers, yearlings and pups. Yearlings restrict their feeding to pups, and pups do not feed other animals.

Figure 4.10 summarizes the distribution of feeding of other pack members for females and males, each group collapsed over age, role and individual differences. In our total observations, adult females distributed 91 percent of their feedings toward pups, three percent toward yearlings and six percent toward mothers. This contrasts with the males as a group, who distributed their feedings more evenly: 60 percent toward pups, 23 percent toward yearlings and 17 percent toward adults (usually, but not exclusively, mothers).

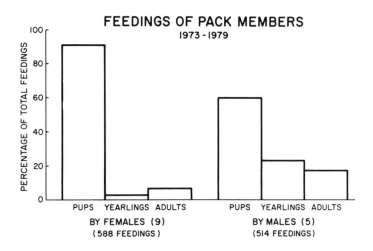

Figure 4.10: Summary of feedings by females and males, 1973-79. These summarized data are broken into detail by individual, age and social role in Tables 4.2 and 4.3.

DISCUSSION

We have summarized our observations on pup feeding for six seasons in a wolf pack whose chronology continues to be investigated. As emphasized by Allen (1979), even relatively long time samples may reflect momentary trends that will subsequently alter. We thus offer our present generalizations in necessarily tentative terms. A second obvious caution is that we have obtained our data from captive animals. While the possible influence of captive conditions upon caregiving behavior is poorly understood (e.g., Kleiman and Malcolm, 1979), we have been fortunate to obtain supporting information from field workers (Harrington and Mech, pers. comm.). We have been impressed by individual differences among the 16 animals observed, plus the possible alterations that may result from pack composition, shifts in individual status, etc.

Although we expect that careful future study in both captive and wild conditions will generate rules of regularity that encompass such differences, nonetheless, as pointed out by Harrington and Mech (this volume), detailed information has long been needed about which wolf feeds which (and which solicits from which). Our long-term observations provide the first such data base. Further, our study suggests a strong and biologically based predisposition of wolves toward communal care under a variety of environmental circumstances. For example, a non-mated (and otherwise inexperienced) adult pair readily adopted two 3.5 month old pups and, in each of the breeding seasons, mothers solicited and were fed by others, *in spite of* the provision of ample food supplies in the enclosure.

General Picture of Communal Feeding

Every adult wolf feeds other wolves during breeding and pup care seasons. Pups are fed by, but never feed, other animals. Yearlings tend to be intermediate in that they are both fed by adults, and also feed pups. Mothers almost always solicit, and are fed by, other adults. In contrast, barren adult females were never fed, and rarely solicited feeding. Only occasionally do females who have unsuccessful pregnancies solicit or get fed. With the exception of one animal, we have never observed adult males to be fed.

Pattern of Feeding Behavior

If we look at the patterns of feeding more closely, several additional trends emerge. Yearlings (n = 13) have *only* been observed to feed pups, never adults or other yearlings. We have *never* observed mothers to feed other mothers. This contrasts to six instances where non-mother adult females did feed mothers: (a) one case where the same female (Z69♀) was a mother in some years, but not in others, (b) two cases where the females never bred (Sz73♀, W76♀), and (c) three cases where the female had an unsuccessful pregnancy (Z69♀ one year, and G74♀ two years).

Adult females may be more selective in their feeding of individual mothers than are adult males, but our base for comparison here is small. The additional suggestion, that this selectivity can be related to alliance groups, etc. (cf. Bennett, 1979), deserves further examination. The feeding of adult females is most commonly performed by adult males; adult males also devote a greater proportion of their feeding to yearlings than do adult females.

Part of the feeding pattern appears to be due to the disinclination of a mother to have some other adults feed her pups; thus, potential female helpers may be actively prevented by mothers from participating in pup care (e.g., the chasing away of non-mated Sa73♀ and Sz73♀ by Z69♀ in 1976). In addition, while the correspondence between solicitations and feeding is strong, it is imperfect. This suggests that both individual styles and relationships among animals may be important (cf. Hinde and Stevenson-Hinde, 1976). In contrast to the complex patterns of feeding of adult animals by other adults, the feeding of pups appears to be nondiscriminant. This is true even where some females are selectively associated with other females before the birth of pups and while pups are still in separate dens (Bennett, 1979).

Some Literature Comparisons on Distributed Care

The general issue of distributed care of offspring has recently been the subject of considerable speculation and some thoughtful review (e.g., Barash, 1977; Brown, 1978; Kleiman, 1977; Kleiman and Eisenberg, 1973; Kleiman and Malcolm, 1979; Krebs and Davies, 1978; Kruuk, 1972; Mason, 1979; Pianka, 1970; Trivers, 1972; Vehrencamp, 1979; Wilson, 1975; Wittenberger, 1979). It would go beyond our scope to do other than emphasize that distributed care is widespread among social vertebrates, that our comparative information base is, at present, inadequate, and that further documentation is of direct relevance to models one might propose for underlying networks of selection and biological adaptation (cf. Bertram, 1976; Hamilton, 1964; Harrington and Mech, this volume; Maynard Smith, 1964, 1977, plus above references).

Three recent canid studies are particularly relevant to the results we report here. Camenzind (1978) reported that coyotes (C. latrans) he observed on the National Elk Refuge in Jackson, Wyoming frequently live in packs and den together. Further, he found that the pups may be nursed communally by more than one mother, and subsequently fed by more than one adult male. Similarly, Macdonald (1979) found that red fox pups may be communally reared, and that non-breeding vixens often visit newborn cubs. If the mother is injured, cubs may remain with a non-breeding female. In contrast to our wolves, neither the males nor the female helpers feed the mothers, although they do feed the cubs. Moehlman (1979) suggested that black-backed jackals are more similar to wolves and coyotes in their cooperative care than are foxes. For example, she documented the tendency for some offspring to help provision the young, as well as for males to feed pups and mothers. Interestingly, as in our observations, these tendencies do not appear to vary in an obvious way with food availability, i.e., the behavior is under considerably intrinsic regulation (cf. Fentress, 1976, 1978).

While the evidence for distributed pup care in wolves is still very incomplete, both the captive and field data appear compatible with our findings (e.g., Allen, 1979; Haber, 1977; Medjo and Mech, 1976; Paquet et al., this volume; Peterson, 1977; Zimen, 1976). The data go beyond the issue of pup care, *per se*, in that they emphasize the need for a more critical dissection of various *roles* among individuals within the social group as a function of age, season, sex, reproductive status, etc. For example, simple unitary dominance concepts fail to convey the dynamic and subtle patterns of interaction among animals that may occur, such as the food solicitation of a "subordinate" by a "dominant" female upon the birth of the latter's pups, even though the soliciting animal initiated hostilities only a few weeks before. Related expression of the need to develop more sensitive measures of roles among individuals within social groups of animals can be found in a variety of sources (e.g., Fentress, 1978; Golani and Keller, 1975; Hinde and Stevenson-Hinde, 1976; Kruuk, 1972; Lockwood, 1979; Mason, 1979; Moran and Fentress, 1979; Moran et al, in press; Rowell, 1974; Ryon, 1979).

The most appropriate terminology to use in discussing the shared care of offspring by individuals within a social group has been a matter of some debate and confusion. Brown (1978), for example, argued with some force that the division between communal and cooperative care in the literature may not reflect any important biological distinction. Because we do not wish to add to the problem, we have adopted the unambiguous term of *distributed* feeding. Finer distinctions that hold some promise are now being proposed by others, such as between (a) direct and indirect care (e.g., feeding versus the construction of shelters), and (b) depreciable and nondepreciable care (in reference to the selectivity of investment) (Kleiman and Malcolm, 1979). Vehrencamp (1979) has also proposed the wider use of the term *eusocial behavior*, as developed in entomology, to emphasize the three significant features of (a) cooperative care of the brood, (b) overlap of generations in a group, and (c) reproductive division of labor. We thus can anticipate refinements to our thinking in the years to come.

How broadly one wishes to cast the comparative net is a matter of taste, and certainly the biological, as well as logical, distinction between homology and analogy must be borne in mind. Sometimes comparisons between disparate levels and populations of analysis can be useful merely in that they help us to think more clearly about our own data base (cf. Fentress, 1980). To cite but a single example, Brown (1978) reviewed data on plural breeding bird species, such as in the genus *Aphelocoma*, where individuals switch roles between breeding and helping status as a function of the success of their own reproductive efforts. This, he argues, makes a case for looking for possible universals in "variance-enhancement," with situationally produced "role-switching" in helping species. Similar prospectives may apply to data such as ours on the flexibility of social roles among our wolves, although obviously one should anticipate important differences in the details and mechanisms of organization.

There is a need for the continued accumulation of a more comprehensive data base that focuses not only upon individual species, but also upon individual animals under particular environmental constraints. It is important that this be undertaken with a dual awareness for (a) the divergence of data that may be found, and (b) the possibility of obtaining insights from studies that are, on the surface, quite far removed (e.g., Lopez, 1978; Stephenson and Ahgook, 1975).

The distributed feeding and associated behavior in wolves is obviously an important and subtle class of phenomena that we are just beginning to appreciate. We thus conclude where we began—with the guarded expectation that careful interpretation of the close observations permitted through the long-term documentation of captive animals will provide insights of more general interest.

SUMMARY

Although communal care of young is uncommon in mammals, it is highly developed in wolves (*Canis lupus*). In a project spanning seven years (1973-1979) we have documented 1,102 feedings (by regurgitation and carrying and dropping food) of certain pack members by others. We present data for five whelping seasons (13 litters), and one instance where two unrelated pups were presented to an unmated pair. Our observations suggest certain regularities by sex, age and social role.

Adult males feed pups, subadults, pregnant females and recent mothers. A 10 year old male with no previous pup experience fed adopted pups and bred the following year. With one exception, we have not seen adult males fed by other animals.

Adult female non-mothers feed pups. Some also feed mothers and subadults. We observed a mother that prevented individual females from caring for pups. A four year old female with no previous pup experience fed adopted pups and bred the following year. Adult non-mothers are not fed.

Mothers do not feed other mothers although we observed two females that, after unsuccessful pregnancies, fed other mothers. Pups of all mothers are fed indiscriminately. Certain mothers also feed particular subadults, not necessarily their own. Mothers are fed by both adult males and non-mother females.

Subadults normally feed pups, do not feed mothers, and are often themselves fed by adults.

These observations indicate selective feeding of adults, but no selective feeding of pups. Our observations emphasize the importance of individual differences as well as changes in social structure that are reflected in feeding patterns. The robustness of feedings under these conditions where food is freely available suggests that similar phenomena are likely to be widespread among wolf social units in nature.

Acknowledgements

The various stages of this research were supported by funds from the University of Oregon, U.S. Public Health Service Grant MH-16955, Natural Sciences and Engineering Research Council of Canada A-9787, plus funds from Graduate Studies and Arts and Science, Dalhousie University. In addition to these sources of funding we are grateful for food donated by the Willammette Poultry Company in Creswell, Oregon, plus medical services provided by the Bush Animal Hospital in Eugene, Oregon and Dr. G. Finley of the Nova Scotia Agricultural College in Truro, N.S. Our research in Nova Scotia would not have been possible without the continued cooperation of Mr. E. Pace and the staff of the Provincial Wildlife Park in Shubenacadie and the personnel of the Department of Lands and Forests, Shubenacadie Depot. Many people contributed their time and skills in assisting us through data collection, advice, construction, and maintenance at various stages of the work. They are: W. Barr, J. Bauguess, N. Bennett, S. Devlin, R. Field, R. Finney, D.J. Fisher, I. Golani, B. Grantmyre, F. Harrington, Z. Havkin, J. Hise, W. Loder, B. Lopez, M. Max, P. Meyerding, G. Moran, P. Nau, E. Pace, H. Parr, B. Riggs, and L. White. J. Lord and F. Stephanie assisted with the preparation of figures, and M. MacConnachie helped set up the tables and typed our various drafts of the manuscript.

Reinforcement of Cooperative Behavior in Captive Wolves

Charles A. Lyons, Patrick M. Ghezzi and Carl D. Cheney

INTRODUCTION

A thorough understanding of the behavior of animals requires measurements derived from both observational field and experimental laboratory studies. Such approaches, historically considered in competition, are more profitably regarded as complementary. Observations of behavior within the natural environment provide valuable data concerning the form of the behavior, its natural constraints, and the circumstances under which it most commonly occurs. This is essential information for establishing conclusions about behavior with a high degree of external validity. Conversely, controlled experimentation provides a more molecular analysis, allowing precise study of the fundamental processes and particular environmental stimuli influencing the behavior. Through experimental manipulation, conclusions with a high degree of internal validity may be reached. Together, these two data sources contribute to an understanding of behavioral phenomena in relationship to the overall environment and ongoing behavior of an organism.

The study of wolf (*Canis lupus*) behavior to date has been largely observational. The ethological research that has provided valuable descriptions of wolf behavior in the field, by and large, has not yet empirically identified many functional relationships between aspects of behavior and parameters of the environment, although several components of wolf behavior have been studied in detail (e.g., Klinghammer, 1979; Mech, 1974c).

One interest we have pursued with regard to wolf behavior has been cooperation, and we have arranged a simple laboratory analog which involves a pair of captive wolves (*C. l. pambasilius*) as subjects. This paper discusses the general issue of experimentally analyzing cooperation in wolves, our initial attempts at such analysis in brief and, in addition, describes the type of instrumentation we feel is necessary for controlled experimentation with such "nontypical" experimental subjects.

COOPERATION

Like many species of hyenidae, felidae and canidae, wolves engage in cooperative hunting. As Sullivan (1979) has noted, efficient hunting would seem to require a high degree of cooperative behavior among several members of an organized pack. Independent observers have anecdotally reported witnessing various tactics including the use of: (1) one or more wolves to drive prey toward other wolves waiting in "ambush"; (2) a distracting "decoy" which allows other members of the pack to approach unnoticed; or (3) simultaneous "timed" attacks at a mother and her calf in order to separate the two (Fox, 1971, 1975; Mech, 1970; Zimen, 1981). Assuming that deliberate cooperative behavior among wolves does indeed occur, it should be possible, in theory, to isolate the critical variables controlling such behavior. To date, neither an empirical nor adequate theoretical analysis has been undertaken.

It is our contention that cooperative hunting among wolves, when and if it occurs, is largely a function of advantageous environmental circumstances; that is, circumstances which provide for cooperative behavior in overcoming the defenses of a large (or otherwise inaccessible) prey item (cf. Kruuk, 1975). Our research is designed to evaluate this contention. We first establish an environment which meets the minimum requirements necessary to produce cooperative behavior, and then attempt to quantify the extent to which two animals cooperate.

The most elementary cooperation procedure requires only two participants. The typical arrangement for assessing cooperation is essentially a forced partnership whose necessary conditions are: (1) that there be the opportunity for an equitable distribution of responses and/or reinforcers as a result of participation in the partnership, and (2) that reinforcement of either participant be at least, in part, dependent upon the behavior of both participants (Hake and Vukelich, 1972).

These two procedural requirements allow for considerable methodological variation. For instance, reinforcement to both partners may be largely dependent upon the behavior of only one participant, rather than equally dependent upon both; likewise, reinforcement may occur after each cooperative episode, or only intermittently (Rosenberg, 1959, 1960; Rosenberg and Hall, 1958). Furthermore, considerable deviations from reciprocity (i.e., equitable distribution of responses and/or reinforcers) may still be classified as cooperation (see Hake and Vukelich, 1972), as in the classic work of Daniel (1942, 1943). In those studies, one of a pair of rats was required to sit on a platform for 30 seconds which would, in turn, allow the other rat to cross a deactivated shock grid for access to food. A dependency was established since neither rat could eat without the other at least occasionally sitting on the platform. But, given that this was a free operant arrangement with no reciprocal requirements, considerable deviations from reciprocity would be possible. Yet, the *opportunity* for reciprocity was present throughout the experiment in that both could get adequately fed and both could avoid being shocked (which is what eventually happened).

Functionally, cooperation can be considered as a chain of responses on the part of at least two animals, but initiated by one participant (see Figure 5.1).

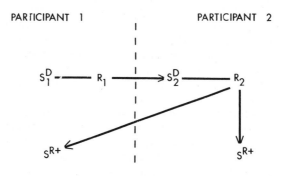

Figure 5.1: A diagnostic episode of cooperation in terms of discriminative stimuli (SD), responses (R) and reinforcement (S^{R+}).

This initial response (R_1) may be a reaction to physical stimuli, e.g., the scent, sight or sound of prey (SD_1). Whatever the stimulus, this initial response also serves as a discriminative stimulus (SD_2) to another animal whose response (R_2), when concurrent with R_1, produces reinforcement (S^{R+}) for the pair in the form of prey capture. It is not necessary that the participants "know" they are cooperating; each may be responding, without regard for the partner, to stimuli in a manner that has produced reinforcement previously. The net result, however, is still cooperative in that both animals' responses were necessary, and both participants received "pay-offs" for their actions.

In order to provide an artificial analog of cooperation in the wild, we employ a discrete trials procedure with our wolves. The opportunity to engage in cooperative episodes is controlled by the experimenter illuminating one wolf's panel operandum once every 45 seconds for a 10-second period. If a single panel depression is made during the 10-second period, the panel light remains on and the light above the other subject's lever is illuminated. The subject at the lever must then complete a designated response requirement within the remaining period of illumination. Upon the completion of this sequence, food is delivered to both wolves. Food is never delivered if the opportunity to cooperate is not initiated, or if subsequent to initiation, the other subject does not complete the response requirement, or both.

This simple arrangement satisfies the minimum requirements necessary to produce cooperative behavior. That is, both animals must participate in order for either to obtain food; they do not earn food independently and, if either fails to satisfy its requirement, food is not delivered even though the other may have completed its requirement. Further, if both participants satisfy their respective response requirements, the frequency, magnitude and time to reinforcement remains identical for each individual. This arrangement therefore satisfies the requirement of reciprocity.

Cooperation in the wild is characteristically initiated by one animal, as determined by the social dynamics of the pack. In our procedure, the opportunity to cooperate is similarly initiated by one subject. This type of leader-follower arrangement need not remain static but may be varied so that, at times, one subject may initiate and the other follow (cf. Skinner, 1953:306). Further, one procedure we have employed requires only one response by the leader to initiate and no less than five by the follower to subsequently produce a reinforcer for both. These requirements may also be manipulated.

Another aspect of the procedure concerns the form of the response required on each operandum. This feature is not particularly critical to the simple analog discussed here, although it may be that most examples of cooperation outside the laboratory involve topographically dissimilar responses.

It is essential, in our analysis of cooperation, that any increase in cooperative responding be due solely to the procedural relations between responses and reinforcers rather than to some other individual or social arrangement. That is, demonstration of control by any cooperation procedure requires that control be established by the reinforcer resulting from the procedure and by the specific procedural relations between the behaviors and their reinforcing consequences (Hake and Vukelich, 1972). We demonstrate control by noting individual response rates during opportunities to cooperate relative to rates during the 45-second inter-trial interval. This is similar to Hake and Vukelich's suggestion of manipulating the presence and absence of the participants and noting the resultant changes in responding. We also note the number of reinforcers earned per opportunity to cooperate. This measure gives an additional indication of whether each participant is responding at the appropriate time so as to maximize its probability of reinforcement.

We have been successful in producing behavior in captive wolves which meets the minimum criterion of cooperation; that is, both animals simultaneously participate in order for either to obtain reinforcement. Further, since neither participant can earn food independently, all available reinforcers are equitably distributed. In its most elementary form, we consider this process to be an analog of cooperation as it occurs in the wild.

INSTRUMENTATION

One of the many obstacles encountered in designing a laboratory-oriented behavioral analysis of any organism larger than the rat or pigeon is the design and construction of experimental equipment (Markowitz and Stevens, 1978). Items such as durable operanda and reliable food delivery mechanisms are important for an objective definition of behavior and stimulus events. Yet, given the recency of this approach to the behavior of non-typical experimental subjects (e.g., wolves), such items have either never before been constructed or, if they have, are generally unavailable from commercial sources.

The experimental enclosure for our research (Figure 5.2) contains two different operanda, each of which is connected adjacently to separate but identical food delivery devices.

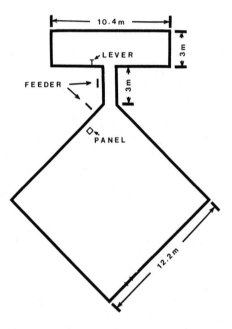

Figure 5.2: Diagram of the experimental enclosure.

One operandum is a steel T-shaped lever mounted on a wooden panel and extending horizontally into the enclosure (Figure 5.3). Tension on the lever is maintained by a spring attached to the wooden panel at one end, and to a lever behind the panel at the other. Depression of the lever increases tension on the spring such that the lever returns to its original position following a response. Lever depression operates a microswitch located near the base of the panel, producing electrical impulses which then may be counted or used to control experimental events through standard 28-V electromechanical equipment. In addition, a 15-watt lightbulb mounted on the wooden panel directly above the lever functions as a discriminative stimulus (SD_2).

The entire device is attached to standard cyclone fencing by means of two steel braces fastened to the top and bottom of the front of the wooden panel, and secured by wing nuts. The distance from the floor of the enclosure to the lever is therefore adjustable to accomodate the size of the subject: for our wolves, 22 cm seems ideal for depression by either front paw. Also, Jones plugs may be used so that wires can be easily connected and disconnected in order to remove the device for cleaning, adjustment or relocation.

The second operandum is a permanent, ground-level, flat, translucent plexiglass panel (27 x 27 cm) hinged to a buried wooden box (Figure 5.3). Springs mounted between the plexiglass and frame provide resistance to downward motion and return the plexiglass panel to its original position after depression. A microswitch which is mounted inside the frame of the box and below the panel operates when the panel is depressed. A 15-watt lightbulb is located at the center of the box beneath the plexiglass panel, and serves as a discriminative stimulus (SD_1). Wires from the microswitch and lightbulb lie underground in conduit, protected both from the animals and from the weather.

Two identical food delivery mechanisms, one located near each operandum outside the enclosure, consist of endless rubber belts mounted on wooden tables with a take-up wheel at one end. Motors are fashioned from automobile windshield-wiper motors (12-V) containing a shaft which turns in one direction when operated. Each food delivery mechanism is connected to electromechanical controlling equipment allowing control over the length of time each motor operates. Small "bite-sized" food cubes are spaced along the rubber belt such that a one-second operation of the motor advances the belt sufficiently to dispense a food cube into a tube, allowing the food to fall within the enclosure beside the operandum.

Using standard 28-V controlling equipment and a 12-V auto battery switched through a relay to the food delivery mechanisms, each light, operandum and food belt may be programmed to operate independently or concurrently. This arrangement allows for system flexibility, permitting, for example, the shaping of each subject on a separate schedule of food reinforcement or requiring responses from both animals within a single compound schedule. Further, the panel operandum described above is ideal for an autoshaping procedure (Brown and Jenkins, 1968).

The entire system is inexpensive and easily constructed, requires a minimum of service, and is relatively impervious to inclement weather or destruction by experimental subjects. The system may be used with a variety of species under an assortment of experimental conditions.

Figure 5.3A: One of the pair of wolves responding at the panel. Food is delivered into the exercise area near the panel.

Figure 5.3B: One of the pair of wolves responding at the lever. Food is delivered into the home pen area near the lever.

DISCUSSION

There has been a noticeable increase, recently, in the use of laboratory analog approaches to investigate traditional ethological issues (e.g., hunting strategies of owls and hawks [Cheney, 1979], social interactions within a family of diana monkeys [Markowitz and Stevens, 1978], and prey detection by blue jays [Pietrewicz and Kamil, 1981]). Much of this research has been initiated without the benefit of previous studies involving non-typical laboratory subjects. Our goal has been to alleviate this problem by describing the procedural framework within which cooperation in wolves and other species may be fruitfully studied. It is our contention that this approach is necessary for an adequate assessment of these types of issues.

Several objections may be raised regarding the present analysis of cooperation. To some, "real-life" cooperation would seem to imply much more than what our wolves exhibit. Actually, it is not intended that the situation be directly comparable to naturally occurring encounters between conspecifics. The task of reducing cooperation to the most elemental components necessitates a degree of oversimplification which may seem artificial and contrived. A systematic increase in the complexity of the situation to make it more reflective of the natural environment would remove those objections, but would make the task of isolating and analyzing the fundamental processes involved in cooperation extremely difficult. Consequently, the addition of such complexity must await the more fundamental task of determining basic functional relations.

A related objection may be raised regarding methodology. The issue is whether the behavior of each wolf is controlled individually by mechanical stimuli (e.g., lights), or by the behavior of its partner and, therefore, "truly" cooperative. Recall that in our procedure, the opportunity to cooperate begins by the periodic illumination of one subject's apparatus. If this subject responds during this brief period, the light above the other subject's apparatus is illuminated. This second subject must then complete its response requirement in the time remaining for either to be reinforced. The mechanical stimuli in this situation serve to set the occasion for each other's behavior (much as a prey item in the wild), which, if emitted before time expires, results in reinforcement for both animals. This appears comparable to the stimuli which serve as discriminative events for "natural" cooperation. The common consequences experienced by each individual in the presence of these stimuli are of primary importance in the laboratory analog, and this is functionally identical to the situation which holds in the wild. The use of mechanical, rather than biological, controlling stimuli does not, therefore, represent a substantial departure from the natural situation.

The work of Sidowski, Wyckoff and Tabory (1956), and Sidowski (1957), may help clarify the present argument. In those studies, partitions and individual operandi were used to prevent human subjects from learning that reinforcement for either was dependent upon the behavior of the other. The rationale was that the main factors controlling social behavior are the same factors which govern the behavior of individuals—namely, reinforcement. The question of interest was whether an increase in cooperative responding would be obtained in the absence of face-to-face contact or "understanding" between participants concerning the relationship. The results showed that it made little difference whether or not the participants were informed regarding the presence of another subject or his/her role in determining reinforcers. Once again, the critical variable responsible for producing cooperative behavior was the individual availability of a response which resulted in reinforcement.

The present methodology represents an initial attempt to experimentally analyze the fundamental conditions which engender patterns of behavior that have been labelled cooperation. As with any initial attempt, several aspects of the procedure may be improved to reduce response variability, or to produce behavior more topographically similar to traditional notions of cooperation. It may prove more profitable, for example, to employ auditory or olfactory stimuli as additional discriminative events for cooperation to more closely mimic conditions in the wild. It is our contention, however, that the basic methodology of using captive subjects in controlled laboratory settings enables a more complete study of the necessary and sufficient conditions that produce cooperative behavior, and allows a determination of the range of conditions and situations under which cooperation occurs. The conscientious use of such experimental methodology, in combination with field description and research, will enhance our understanding of ethological phenomena such as cooperation to a greater extent than either approach employed exclusively.

Acknowledgements

We gratefully acknowledge the assistance of
Lisa Godzac, Ed Blake and Matt Nichols.

Probability Learning in Captive Wolves

Carl D. Cheney

INTRODUCTION

When a predator hunts, the question of what determines where it forages arises. We have been interested in the experimental analysis of this question, utilizing a probability learning, or choice behavior research design from experimental psychology (e.g. Mackintosh, 1974:190–195). Probability learning involves a situation where two or more alternatives are simultaneously available but only one is rewarded. The animal is allowed to choose one alternative. If the first choice is incorrect, then correction choices are allowed until the trial terminates with a reward. The situation we use involves a modestly food-deprived subject being confronted with two or three possible "hunting" areas or response operanda (Skinner, 1962). Each response possibility has a different but consistent probability of paying off. For example, one of three spatially distinct locations may provide food on six out of 10 trials, the second location may pay off on three of 10 trials, and the third location on only one out of 10. The animal is free to choose how it will distribute its choices over a series of trials.

Of interest is the development of a consistent pattern of first choices over an extended series of trials. There are a variety of options possible. (1) A simple position preference may develop (i.e. the north location is always investigated first). (2) The first choice may be the location that was rewarded most recently (reward following). (3) The matching of first choices to reinforcement probability may occur. (4) The animal may choose the highest probability area (maximizing). (5) Random selection may occur as no consistent pattern is acquired.

Matching is said to occur when the subject's selections come to closely approximate the reward probability of each choice. That is, an alternative rewarded 60 percent of the time is chosen first very nearly 60 percent of the time. This would be a likely strategy if an animal win-shifts (i.e. selects an alternative *other* than the one most recently successful) while continuing to maximize profitability. Apparently such matching is uncommon, however; only fish are reported to show relatively consistent matching (e.g., Mackintosh et al., 1971).

Maximizing is the consistent choice of the highest probability food source. Many species develop a maximizing strategy (win-stay) rather than the tendency to win-shift (Mackintosh, 1969). They go where prey have been found most often in the past. For example, given a two-choice situation with probabilities of 70 percent and 30 percent, a maximizer will choose the 70 percent alternative on more than 90 percent of the trials (Bitterman et al., 1958; Johnson, 1970).

The probability-learning, discrete-trials procedure allows one to observe the presence of a win-shift or a win-stay strategy, and the development of matching, reward following, or whatever else might occur. We have found that some coyotes *(Canis latrans)*, red foxes *(Vulpes vulpes)* and raptors initially have a clear win-shift strategy that develops into quasi-maximizing at differential rates between species (coyotes being the fastest) (Loether, 1978; Mueller, 1977; Snyder, 1974). However, all our captive wild subjects have shown a marked tendency to switch among resource choices, even with 100 percent reward at some alternatives. This win-shift tendency is therefore present even when food appears inexhaustible at one resource location.

Switching, of course, may allow the free-ranging predator to constantly monitor prey densities throughout its range. If a change in prey density is detected, the more profitable prey resource can then be exploited (Smith and Dawkins, 1971). It may also allow for dietary variety and certainly provides a change in conditioned reinforcers (i.e. scents, sights, etc.). Furthermore, switching hunting routes provides an opportunity for surprising prey items that are unable to predict predator habits. Consistently hunting one area not only may deplete the resource but will allow prey species an opportunity to avoid predators by relocating.

The purpose of the present study was to observe the strategies used by three communal living, food-deprived wolves when simultaneously exposed to a three-choice, correction-possible probability learning situation. We assume that this design and procedure is analogous (but simplified) to the conditions facing a free-ranging predator. The animal is free to respond or not, and our utilization of only three alternatives is not overly restrictive. As described elsewhere, the predatory episode can be considered a chain of discrete stimulus-response-reinforcer elements (Cheney, 1979; Cheney and Snyder, 1974). In order for the controlling elements of a predatory episode to be made explicit, specific and replicable manipulations are required. Although observations in the field may eventually yield functional relationships, given the many potential influences, useful and valid statements may be difficult to acquire despite exhaustive efforts. In captivity, however, repeated and objective observations made under similar conditions may allow us to derive predictive statements concerning the variables controlling foraging (Cheney, 1978).

METHODS

Subjects

The three wolves *(C. lupus pambasileus)* involved in this study included a female named Laska, age six years, taken as a pup in Alaska and reared at the Washington Park Zoo, Portland, Oregon, a second female (Roz), age four years, born at the Zoo, and a three-year-old male (Mac), also zoo-born. The oldest wolf, Laska, was the most assertive and the one from whom all the data for this report came. The wolves were housed together in a 4 x 10 m kennel and allowed into the 12 m² test arena during the day when trials were conducted. Access to water and a den in the kennel were always available, even during sessions.

Apparatus

Three spatially distinct food sources were used to provide choice alternatives (Figure 6.1).

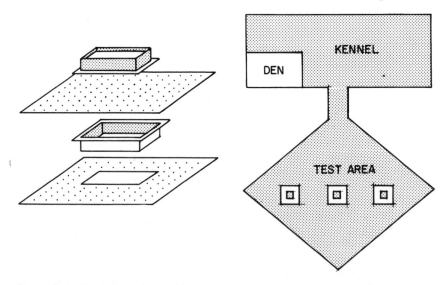

Figure 6.1: Exploded view (left) of the apparatus used to provide a food source. Three such units were in use simultaneously. Right view indicates the living kennel area and the test arena. Dimensions are given in the text. The relative location of each food apparatus is indicated.

These were constructed of perforated metal plates and stainless steel containers. The bottom plate held a container of food. This food was made inaccessible by the upper plate. The upper container could conceal a bit of food when placed upside-down on the top plate. The bottom container provided food odors at all locations so that olfactory cues to the one location containing accessible food under the top container were removed. Food reinforcers were tennis-ball-size pieces of frozen carnivore chow of local creation which also served, in larger portions, as the maintenance chow.

Procedures

The three wolves were deprived of food for 24 hours before each session, and fed only a subsistence amount following each session and on days when sessions were not conducted. Trials were repeated each session until a period of 10 minutes elapsed with no responding. Sessions were run nearly every consecutive day with between five and 20 trials in each session. Session length averaged 30 minutes. Over 400 trials were conducted in collecting the data reported here.

Three sets of probability learning sessions were conducted. Table 6.1 shows a sample distribution of food per location and trial. The first set of 60 trials distributed reinforcers at the north location 36 times (60% alternative), the center location 18 times (30%), and the south location six times (10%). Only one location held accessible food on any single trial. The food location probabilities were changed for both latter sets of 60 trials so that each geographically different location had its turn as the high, medium and low probability choice.

Table 6.1: Example Distribution of Food for 10 Trials*

Available Chamber Location	. . . . .Trials.									
	1	2	3	4	5	6	7	8	9	10
N	X			X	X		X	X	X	
C			X			X				X
S		X								

*The only restriction was that a single location could not hold food more than three times in succession. Distribution is N = 60%, C = 30% and S = 10%.

On every trial, the experimenter approached the feeders and exited the arena in a semi-random pattern (north, center, south), and also lifted each top container. While the experimenter was in the arena, the wolves were in the kennel. The wolves entered the test arena only when the experimenter completed apparatus manipulations and left. The experimenter or an observer recorded the sequence of choices as to position (first north, second south, etc.) as well as the single food location on every trial. Hits were reported as that position selected which also contained the food in the trial.

When a correct choice was made, the wolf was allowed to consume the food before the experimenter entered the arena to set up the next trial. Before switching to the next set of 60 trials, a series of nine trials was conducted to force three selections to each position. All containers held food and the subject was allowed one choice. That choice was then reloaded and a second trial conducted, and so on. Once three choices were made to a single position, that position apparatus was removed from the arena. These trials ensured equally reinforced experience at every position prior to initiating the next differential probability series.

RESULTS

We assumed at the start of this study that there would be rapid modeling as one wolf began responding and obtaining food. However, over the number of trials conducted, this proved to be wrong. All choices were made by the older female. Not once did either of the other two lift a top container. They both frequently came into the arena with Laska; they were often nearby when Laska made choices; they occasionally investigated the apparatus, but never made a correct response. The male did, on two occasions, grab the food away as Laska removed the container. Also, on three occasions, Mac lifted the entire apparatus out of the ground but was unable to obtain the food. The younger female, Roz, never attempted to lift a container or steal food. We did, on limited occasions, allow only the younger two wolves into the arena and still they did not respond. We conclude from this experience that modeling is not an immediately likely activity under these conditions, possibly due to social status within the pack and the timidity of the subordinate members.

Within 10 shaping trials where food was only partially covered by the container, Laska completely mastered the technique of removing the container. She frequently went to and investigated each position before making a choice. The fact that her first two choices were often wrong, actually 90 percent in one series of trials, and that her second choices were also wrong about one-fourth of the time indicates that although Laska appeared to try to use odor as a cue, it was difficult if not impossible for her to locate the loaded container by smell alone.

Figure 6.2 contains the series of choices in blocks of 10 trials for the initial series of 60 trials. If choices had matched reinforcer probability, the curves would have each approximated the first series of 10 trials in the upper left; that is, 30 percent to the north, 60 percent to the center, and 10 percent to the south location. As can be seen, the distribution of choices varies across blocks. However, by the final 10 trials after 50 trials of experience, she was almost matching.

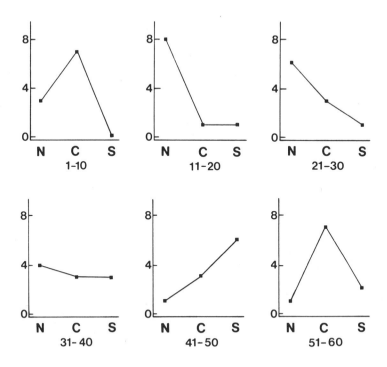

Figure 6.2: The distribution of first choices in the first set of 60 trials in 10 trial blocks. The top left graph indicated three responses to the north (N) feeder, seven responses to the center (C), and zero responses to the south (S). Probability of food: N = 30%, C = 60%, S = 10%.

Figure 6.3 shows the second series of 60 trials, again in 10 trial blocks. She was very close to matching throughout this set, especially in the second 30 trials. These results suggest she had learned the payoff probability and had developed a matching strategy. She had a very high run of hits during the last half of the set (23 out of 30) and only two double misses. These were two choices in a row on the same trial to the wrong location.

As shown in Figure 6.4, Laska was very consistent in the final series and approximated matching throughout. Her hit rate was a bit over 50 percent in the final 30 trials. Her selection of the highest probability location first on 16 out of the final 30 trials was close to matching, as 18 would have been perfect. Table 6.2 summarizes the results of the 180 probability learning trials. Clearly, the overall strategy was not maximizing.

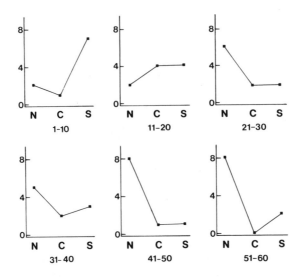

Figure 6.3: The distribution of first choices in the
second set of 60 trials. Probability of food: N = 60%,
C = 10%, S = 30%.

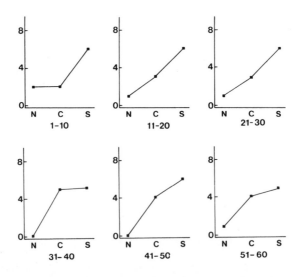

Figure 6.4: The distribution of first choices in the
third set of 60 trials. Probability of food: N = 10%,
C = 30%, S = 60%.

Table 6.2: Distribution of Successful Choices (Hits)
During the Three Sets of 60 Trials

| | . . . . Set . . . . | | | | Percent of Maximum Possible |
	1	2	3	Total	
First choice hits					
In last 10 trials	5	8	2	15	50
In last 15 trials	6	12	5	23	51
Total in 60 trials	23	33	31	87	48
Second choice hits	23	14	15	52	56
Win-stay sequences*	25	26	20	71	39

*Defined as sampling the last rewarded location on the next trail. Random choice of location regardless of previous success would result in a win-stay ratio of 1/3. The observed ratio does not differ from that expected by random choice of location ($X^2 = 3.03$, $\underline{p} > 0.10$).

We then ran a series of 120 trials in which all chambers were loaded on every trial. This series was conducted to ascertain if a reward following a random, or a switching strategy, would be obtained in the face of continuous success. As seen in Table 6.3 (Series A), Laska exhibited a strong place (win-stay) preference. Of 120 single choices, 95 were to the south location. She did not choose this chamber exclusively, but certainly significantly. This contrasts to her performance in the previous condition where she appeared to switch at random after a reward (Table 6.2).

Table 6.3: Distribution of First Choices
When One, Two or All Three Locations Always Held Food
While the Others Were Always Empty

Series	. .A (N = 120). .			. .B (N = 45). .			. C (N = 45) .		
Location	N	C	S	N	C	S	N	C	S
Food	Yes	Yes	Yes	Yes	Yes	No	Yes	No	No
Trial Block									
1	1	8	21	0	4	11	4	0	11
2	2	4	24	0	11	4	10	0	5
3	6	1	23	0	11	4	10	0	5
4	2	1	27	—	—	—	—	—	—
Total	11	14	95	0	26	19	24	0	21

Within six days we conducted 45 trials with the south location always empty and both north and center always loaded. Table 6.3 (Series B) shows the preference shifting to the center location by the second set of 15 trials. Throughout this series, however, she continued to choose south about one out of every three trials.

A third series of 45 trials conducted immediately after the second, with only north loaded, showed, first, a return to south when center failed, and a final exclusive choice of north on the last five trials (Table 6.3, Series C).

Finally, a complete extinction series was conducted with criterion arbitrarily defined as 15 minute access to all chambers with no choice being made. She met this criterion in 12 trials. On Day 1 of extinction she made five choices, with no location receiving more than two choices. On Day 2, she distributed six choices evenly over the three chambers. On Day 3 she did not make a choice in 15 minutes and the trial and experiment were terminated.

DISCUSSION

As Mackintosh (1974) put it, "the interest shown in discrete-trial probability learning experiments arises from the light they throw on models of choice behavior." It appears that we are concerned with exactly this type of choice when we inquire as to where a predator will hunt on its next foraging episode. Therefore, it seems that this design is appropriate in an analysis of foraging if we are eventually to locate the controlling variables for such behavior in the environment.

Although it has been found with some animals that choices are made which maximize reinforcer payoff, not many wild carnivores have been extensively studied in a controlled situation (Loether, 1978). When maximizing has not been found, the performance has been attributed to a failure on the subject's part to attend to the relevant stimuli. Ginsberg (1979) has stated that wolves are extremely attentive; however, Laska did not maximize, suggesting she may have failed to attend to salient and critical stimuli. On the other hand, she may simply develop different strategies for different situations.

Our coyotes and red foxes, in a similar three-choice discrete trial situation, came to maximize quite clearly. On the other hand, Laska switched choices too often and consequently undermaximized. She tended to win-shift except during the 120 continuous reinforcement trials, wherein she maximized. Possibly given more probability trials, a definite maximizing strategy would emerge, or it might be that the conditioned reinforcing effect of switching would continue to maintain win-shift behavior except in a continuous reinforcement situation. And even then, she did not attach to one position exclusively.

The social dynamics emerging during trials were that of a dominant animal assuming the active role of participant with the two subordinate animals present but uninvolved. On only two or three occasions did either subordinate, even with Laska physically excluded, approach within one-half meter of the chambers. They appeared frightened of the apparatus and clearly dominated by Laska. This may indicate a problem in working with such a socially organized carnivore. No such failure to participate has been observed before with our other canids.

We believe this procedure is an adequate paradigm to study foraging, and with it we hope to eventually specify some of the relevant variables involved.

Acknowledgements

The invaluable assistance of Charles Davis, George McCulloch and David Thurber in conducting this research is gratefully acknowledged.

A Wolf Pack Sociogram

Erik Zimen

INTRODUCTION

This chapter presents results from ten years of continuous observation of social behavior within a captive wolf *(Canis lupus)* pack. Descriptions of social behavior of individual pack members, reflecting sex, age and rank position, are provided. Understanding these associations yields not only a better comprehension of the wolves' social organization, but may also help unravel one of the central questions of recent wolf research—the subject of population regulation.

It is generally agreed that a system of self-regulatory population control exists in wolves (Mech, 1970). Despite a high reproductive potential, wolf numbers seem to maintain a dynamic equilibrium with prey availability, at least in areas not greatly disturbed by man. Apparently the following factors contribute to regulation of wolf numbers: (1) territoriality of wolf packs (Mech, 1972); (2) stability of territory size (Mech, 1972); (3) optimization of pack size to prey availability within territories, achieved by: (a) breeding inhibition of adult females (Mech, 1970; Zimen, 1976), (b) number of pups per litter (Pimlott, 1969), (c) pup sex ratio (Mech, 1975), (d) pup survival (van Ballenberghe and Mech, 1975), (e) dispersal of pack members (Mech, 1978).

Although important insights have been gained into wolf population ecology, the operational mechanisms of these regulatory variables remain unclear. Field studies have their limitations for illuminating these mechanisms, while studies of captive wolves provide little opportunity to evaluate environmental influences. Fortunately, some of the regulatory factors cited above appear to be partly independent of environmental influences, and affected to a large degree by social behavior within the pack. These are: (i) territorial behavior; (ii) dispersal tendencies of individual pack members; and

282

(iii) the number of breeding females in a pack. This paper deals with the influences of social behavior on these regulating variables. The pack sociogram presented (who does what, when, to whom) allows for the formation of an initial hypothetical model based on behavioral strategies of individual pack members as they relate to reproductive fitness. Unless we endeavor to comprehend the reasons individual wolves behave as they do, attempts to fully understand wolf population regulation will remain superficial.

MATERIALS AND METHODS

A pack of Euro-Asian wolves was continuously observed from 1967 to 1978. They were initially maintained at Kiel University field station in northern Germany. From 1971 on, they were kept in a six hectare enclosure in the Bavarian Forest National Park. During the first four years artificially raised pups were added to the pack. However, from 1972–1978 the pack produced and reared all of its own pups. The study involved 49 wolves (Figure 7.1), of which eight were removed as young pups and eight as adults after becoming pack scapegoats.

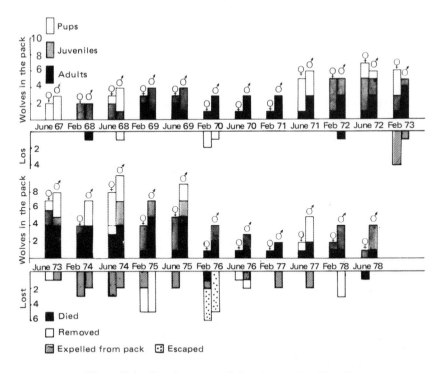

Figure 7.1: Development of the observed wolf pack.

A group of nine wolves escaped in February 1976, of which all were eventually killed, the last in January 1978. Five wolves died as pups and three wolves died of wolf inflicted injuries. Alpha wolves were removed on two separate occasions. The first time was in December 1969 when the tame alpha female became too dangerous for humans, whereupon the alpha male also was removed. The second time was in February 1978 when the alpha male was shot inside the enclosure without my knowledge. With the exception of the death of an alpha female, all other changes in alpha positions were the result of rank conflict.

The wolves were observed mainly during the early morning and early evening activity periods. They were observed for at least an hour daily when possible, and often continuously around the clock during breeding season, pup raising or other periods of intense social activity. A platform in the center of the large enclosure allowed observation of all animals simultaneously. Observations were facilitated because the wolves tended to concentrate their social activities. Whenever two or more wolves interacted they attracted others.

Forty-eight behavior patterns were recorded, e.g., fur sniffing, threat, attack, active submission, social play, sexual initiatives (see Table 7.1). The initiator and recipient of the contact were noted. Data were combined for each month and dyad combination.

Table 7.1: A Short Description of the Observed Behavior Patterns and the Frequency of Their Occurrence per Wolf and Hour of Observation

Behavior Pattern	Short Description	Frequency (obs./hr/wolf)
Neutral social contacts		
Muzzle to muzzle contact	Muzzle sniffing, seldom licking	2.64
Muzzle to fur contact	Fur sniffing	3.31
Fur biting	Biting or pulling of fur	0.25
Fur licking	Often licking of wounds	0.30
Genital sniffing	—	0.21
Genital licking	—	0.12
Anal sniffing	—	0.05
Anal licking	—	0.03
Standing over partner	Standing transversely over lying partner	0.15
Submissive behavior		
Crowding	Friendly jostle around dominant	0.44
Active submission	Face licking, etc.	2.35
Passive submission	Falling and/or lying on back with one slightly raised hind leg in front of dominant	0.59

(continued)

Table 7.1 (continued)

Behavior Pattern	Short Description	Frequency (obs./hr/wolf)
Agonistic behavior		
Aggressive behavior, generally without body contact		
Offensive threat	Snarling with raised lips; many combinations with other inhibited aggressive behaviors possible	1.18
Ambush	Sudden threat attack	0.03
Following	Walking or running behind opponent, often in imposing posture, without trying to catch up	0.22
Mobbing	One, but more often two or more, wolves encircling defensive opponent	0.58
Attack	Sudden real attack from some distance	0.03
Chasing	Full speed chase in order to catch opponent	0.11
Aggressive behavior generally with body contact, but biting not involved		
Imposing	Stiff-legged walk toward, parallel or perpendicular to opponent, with raised hackles	0.61
Impose-shoving	Lateral body shoving of opponent, often with neck display	0.08
Front feet on back	Standing with stiff front feet on back of opponent	0.06
Pinning down with head	Lying opponent is pinned to the ground by dominant, keeping head or wide-open muzzle over head, neck or throat	0.12
Pinning down with body	Opponent is pressed to ground by dominant lying on top of it	0.06
Jump on	Sudden jump on opponent, who normally rolls onto back	0.06
Standing over opponent	Standing stiff-legged over, or with one or two feet on lying opponent	0.03
Shoving	Tactic to ward off bites when closing in on opponent	0.09
Hip slam	Full lateral body slam when running past opponent	0.01

(continued)

Table 7.1 (continued)

Behavior Pattern	Short Description	Frequency (obs./hr/wolf)
Aggressive behavior involving inhibited biting		
Biting over muzzle	Highly inhibited aggression toward pups or low ranking pack members, sometimes a friendly social contact	0.29
Lunge and snap	Jumping forward, directing snaps; sometimes more severe bites, mostly toward hind quarter of opponent	0.50
Inhibited body biting	More or less inhibited biting, mainly toward front parts of opponent; loud vocalizations	0.19
Aggressive behavior with uninhibited biting		
Severe biting	Getting a strong hold of any part of opponent's body; biting with full strength	0.11
Head shake	While biting, vigorous lateral head shaking	0.02
Defensive behavior		
Defensive threat	Full snarl, ears back	1.05
Defensive lunge and snap	Short jumps forward, directing snaps mostly toward face of attacker, mobber	0.48
Snap clatter	When snapping in the air, a noisy beat made by the teeth when jaws hit together	0.02
Keeping distance	Walking or running away from follower without trying to enlarge distance	0.27
Flight	Full speed run away from attacker, chaser	0.13
Defensive circling	When mobbed on the spot, circling, trying to avoid being attacked from behind	0.05
Defensive bite	Full power bite when severely attacked	0.06
Play behavior		
Initiating play	Play bow, playful approach, playful run away, etc.	0.67
Play with body contact	Wrestle, play biting, etc.	7.47
Play running	Playful chase	1.15
Sexual behavior		
Presentation	Females walking or standing in front of male, holding tail sideways	0.03
Mounting	Almost exclusively males mounting females from behind	0.08
Pelvic thrusts	While mounting	0.04

The social rank order in the pack was noted each month on the basis of expressive behavior shown during wolf encounters; that is, the rank order was assessed independently of all quantitative aspects of wolf behavior. In this way it was possible to correlate the behavior of each wolf to its sex, age and rank position (Figure 7.2).

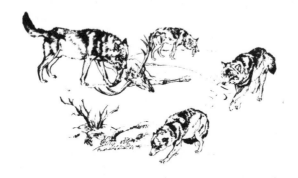

Figure 7.2: The expressive behavior of wolves is highly developed. In front are three males expressing their rank relationships (all drawings by Prill Barrett).

RESULTS

Social Rank Order

The Concept of Dominance and Social Rank Order: There has been much confusion concerning the concept of dominance and social rank order (Lockwood, 1975). Therefore, dominance, in this paper, is understood as follows:

A dominance relationship between two animals is based on their mutual experiences which have allowed each animal to assess the relative strengths and weaknesses of the other. The dominance relationship is not necessarily based on actual strengths, but arises from an assessment of each animal by the other, and so need not be the outcome of a direct aggressive confrontation. The dominance relationship between two animals is expressed by the degree of social freedom each animal allows itself during an encounter. The more that initiatives of the subdominant animal are suppressed by the dominant, the stronger the dominance. A dominance relationship between two animals is also often dependent on individual relationships with other group members. Thus, the rank order within a group is more than just the sum of all pair relationships.

Dominance relationships ultimately serve to minimize dangerous direct encounters, particularly in conflicts over limited resources. Even so, the establishment and maintenance of dominance, and the degree of freedom permitted the subdominant, are in themselves sources of conflict. The tendency for animals living in non-anonymous groups to form dominance relationships can therefore be understood as a tactic to avoid probable future conflicts.

Structure of a Wolf Pack Rank Order: The social rank order in wolf packs has been studied in detail (Schenkel, 1947; Rabb et al., 1967; Zimen, 1975, 1976, 1978; Lockwood, 1979). Although different methods have been used to evaluate dominance, it is generally agreed that rank order structure has the following features:

> There are separate rank orders for males and females. This division is especially evident in the higher ranking wolves, less so in the lower ranking wolves.

> Each sexual rank order is mainly structured according to age, that is, older animals are normally dominant over younger animals.

> Rank differences are most prominent among the higher ranking wolves, less distinct among the lower ranks and younger wolves, and non-existent among pups.

> Severe suppression by the dominant wolves tends to equalize rank differences among subordinates.

> With the exception of the pups, wolves within a particular age group display rank orders similar to that of the whole pack.

> There are no cross-sex dominance relationships between males and females, as long as these animals occupy the same levels within their respective rank orders. Where these positions are different, however (e.g. dominant group versus low ranking group), or where the age gap is significant, there are clear cross-sex dominance relationships (Figure 7.3).

The dynamics of rank relationship within the wolf pack will be discussed later.

Social Behavior Within the Pack

Many years of continuous observation of social interaction among wolves of known sex, age and rank has provided sufficient information for construction of a wolf pack sociogram which is independent of individual animals and their specific characteristics. The alpha male position, for example, was held by six different wolves during ten years of observation. The behavioral data for these six males was summed for each month they were observed in the pack. To construct the pack sociogram the monthly data for each of these males, for as long as they occupied the alpha male position, were lumped together. Data for lower ranking males and females, juvenile males and females and the pups were treated similarly. Thus, for each monthly period all wolves in the pack were allocated to one of the following age, sex and rank positions: (A) adult male #1 (alpha male), #2 (beta male), #3, #4; (B) adult female #1 (alpha female), #2, #3, #4; (C) juvenile male #1, #2; (D) juvenile female #1, #2; (E) pup.

Figure 7.3: Model of the wolf pack rank order. At top, the alpha male and alpha female, then the beta male (the corresponding position in the female rank order is often missing), eventually followed by a group of adult sub-dominant males and females. Within the group of juveniles, there is often an inferior rank order while the pups normally show no rank-related behavior.

To simplify presentation of the results, some social positions have been combined. This is permissible as detailed analysis has shown there exists a high degree of conformity within certain groups; specifically, the low ranking adult males; all adult females except the alpha female; male juveniles; female juveniles; and finally, all pups independent of sex (Zimen, 1978). Therefore, the following positions in the pack will be considered for each behavior pattern matrix: (1) alpha male, (2) beta male, (3) alpha female, (4) the group of low ranking adult males (all adult males except the alpha and beta males), (5) the group of low ranking females (all adult females except the alpha female), (6) juvenile males, (7) juvenile females, and (8) pups.

To further avoid overwhelming the reader with details, only the most prominent of the 47 registered behavior matrices compiled will be presented. Consequently, numerous details must be excluded.

However, the selected matrices fully suffice to demonstrate how individual wolves behaved in relation to their sex, age and rank positions.

Neutral Social Contacts: All forms of non-agonistic, submissive, sexual or playful interactions are referred to as neutral social contacts. A typical form would be as follows: one wolf approaches another, quickly pushes its nose into the other's fur and walks on (Figure 7.4). This is probably an olfactory control. Another example would be two wolves, standing parallel, turning their heads and touching noses briefly, probably for less than a second. Again, this is probably an olfactory control as well as a tactile contact. The rarely observed genital or anal controls are also considered neutral contacts, as is fur licking of a partner (this sometimes lasts up to 30 minutes and is often directed towards an open wound). Except for social play, neutral contacts were the most frequent behavior patterns observed (Table 7.1). On average, a wolf showed muzzle-to-muzzle contact 2.64 times per hour and muzzle-to-fur contact 3.31 times per hour. If less frequent forms of neutral behavior are included, the average pack member showed neutral contacts about six times per observation hour, including resting and travelling time.

Figure 7.4: Fur sniffing.

The frequency with which different wolves were engaged in neutral contacts, as well as the direction of the behavior, was highly dependent upon their own and their partner's sex, age and rank position. For example muzzle-to-fur contact (fur sniffing) was shown most frequently by the beta male (Matrix 7.1). Juveniles and pups, and the other two dominant adults, the alpha male and alpha female, also showed this behavior often, but low ranking males and females showed it less frequently. The alpha female and the pups were by far the main recipients of muzzle-to-muzzle contacts, the low ranking adults again being less frequently involved. Matrix 7.1 shows graphically that muzzle-to-fur contact was most frequent:

among the three dominant adults, the alpha male, the beta male and the alpha female,

between dominant adults and juveniles,

among younger wolves of the same age class, i.e. juveniles and pups,

by all pack members, especially the alpha female, toward the pups.

Fur sniffing

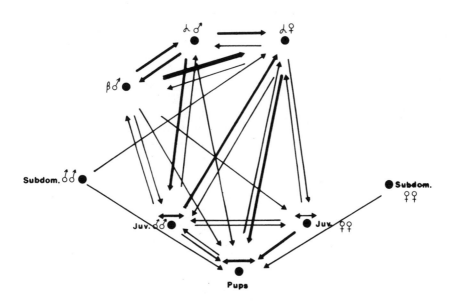

Frequencies <0.25/h not considered

Matrix 7.1: Number of observations per wolf* and hour:

. . . . Observations/hour . . . .

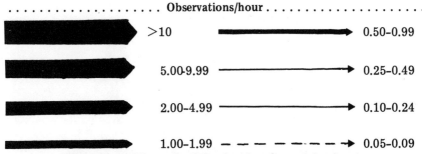

*For the positions of subdominant males, subdominant females, juvenile males, juvenile females and pups, the average frequency per wolf was calculated.

As will be discussed later, wolves less frequently involved in neutral social contacts (e.g. low ranking adult males and females), often showed a tendency to voluntarily remain apart from the pack, or alternatively were forced to leave by individuals or concerted pack action. As neutral contacts were more or less equally distributed among other pack members, it seems that these behavior patterns, aside from being informative, play an important social role within the pack, ensuring and reinforcing the bonds necessary to keep the pack together.

Submissive Behavior: The two most prominent forms of submissive behavior observed in wolves, active and passive submission, have been described in considerable detail (Figures 7.5 and 7.6) by Schenkel (1947, 1967). Intense active submission plays a frequent and important part in many social interactions. For example, active submission is commonly observed in the general friendly get-togethers which were first described by Murie (1944). These "ceremonies," often observed at the beginning of an activity period, are usually initiated by younger wolves, showing intensive active submission towards older animals. Commonly, other pack members join in, turning the "ceremony" into one of communal submission and friendly contact. This activity often develops into a general chorus howl. In the pack I observed, active submission was directed toward (Matrix 7.2):

> the alpha male (40% of all observations). All pack members, including the beta male and even the alpha female, showed the alpha male active submission. In addition, he was the obvious hub of friendly get-togethers;
>
> the alpha female, primarily by the female pack members;
>
> the beta male, primarily by the younger male pack members;
>
> the juveniles, by the younger pups.

Figure 7.5: Active submission.

Figure 7.6: Passive submission.

Active submission

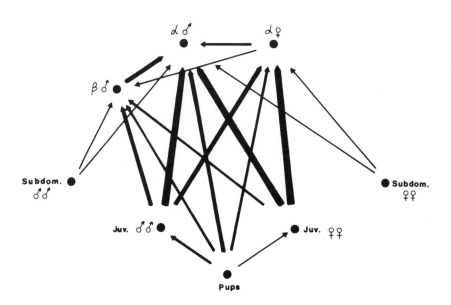

Matrix 7.2: See Matrix 7.1.

Thus, the direction of all active submission was clearly from low ranking to high ranking animals. The only exception was the frequent submission of the alpha female to the alpha male, and sometimes to the beta male as well. Her submissive behavior, however, was mainly related to sexual stimulation before the breeding season, and food begging during pup raising, and cannot therefore be considered subdominant. Rather, it appears to be associated with a juvenile mimicry tactic typical for dominant females in groups exposed to kin selection (Geist, 1978). Furthermore, individuals showed submissive behavior only to wolves with whom they had a distinct dominance relationship. Thus, such behavior was seldom observed between young wolves within an age group, where dominance relationships were weak. This behavior pattern also decreased whenever a rank relationship between two wolves became unstable. This happened, for example, when a low ranking wolf displayed intense expansion tendencies towards a dominant animal, or when a dominant individual severely suppressed a low ranking wolf. Lessening of submissive behavior between partners was invariably a sign of dominance instability, and frequently preceded fighting.

Of interest is the predominance of active submission directed toward the alpha male, not only by older pack members but by pups as well. Detailed observations of the ontogenetic development of submissive preferences show that three to four month old pups are able to recognize dominance relationships among older wolves (Zimen, 1974). Beginning at that age submission is increasingly directed toward the outstanding dominants in the pack. For the pups, these are the dominant adults and the dominant juveniles, while the low ranking adults soon lose attraction. Although still dominant over the pups, these low ranking wolves are often suppressed within the pack. This is recognized and acted upon by the pups (a clear indication that pair relationships are highly dependent on each animal's social relationship with all other pack members).

The direction and intensity of active submission reveals its function. Generally, it is a form of appeasement behavior, demonstrating a subdominant animal's inferiority and non-competitive intentions. It helps, therefore, to stabilize dominance relationships and ultimately reduce aggression among pack members.

Unlike active submission, passive submission was directed primarily toward the beta male. Passive submission regularly took the form of rolling on the back, and as will be seen later, occurred in response to the general, frequent, moderate intensity, aggressive behavior displayed by all beta males. Thus, passive submission is also a form of appeasement, serving to terminate a moderate form of aggressive attack already in progress and threatening to increase in intensity. Like Schenkel (1967), I never observed passive submission used as a tactic to terminate severe attacks, the only reactions to which were flight or intensive defense.

Agonistic Behavior: 21 percent of all observed social interactions within the pack were of aggressive nature. However, of the 45 classified social behavior patterns, 27 (60%) were agonistic (Table 7.1). Aggressive actions are certainly more spectacular than most neutral or friendly forms of behavior, and one sequence may therefore be described in terms of several basic patterns. However, the high representation of aggressive behavior patterns in the wolf ethogram also illustrates the differentiated nature of wolf agonistic behavior.

Agonistic behavior can be divided into four stages of generally increasing intensity. Recognized in this study were agonistic behavior: (1) without body contact; (2) where opponents may have body contact but do not bite; (3) with inhibited biting; (4) with uninhibited biting. Typical agonistic behavior patterns are described below.

Offensive Threat – Threat behavior is typical for the first stage of agonistic behavior where opponents, in most cases, are not in body contact. It normally occurs between two animals near one another, but occasionally the threatened animal may be more than 100 m away (Figure 7.7). Of 688 analyzed threats, 354 (51%) were of defensive character, i.e. a reaction to distinct aggression. The others might be termed offensive, although only 75 (23%) occurred spontaneously. The remainder were also reactive, not to aggression but to other forms of molestation. For example, a wolf feeding on a small piece of meat may have reacted with threat behavior to the approach of another pack member, or a plagued adult might have reacted with threat toward a group of high spirited, enthusiastically submissive pups. While 83 percent of all defensive threats were followed by more intense forms of defense such as snapping, or flight behavior, only 33 percent of all offensive threats were followed by more intense forms of aggression, and these primarily consisted of muzzle inhibited biting of molesting pups. Threat demonstration terminated the immediate confrontation in most other cases. Threat behavior was observed relatively frequently, occurring at an average rate of 1.18 threats per hour per wolf. This behavior accounted for approximately 30 percent of all observed agonistic encounters (Table 7.1).

Figure 7.7: Offensive threat.

In its offensive form, threat was shown most frequently (Matrix 7.3):

> by high ranking adults toward pack members of the same sex in close-ranking positions.
>
> by high ranking adults toward juveniles, again predominantly to those of the same sex,
>
> between pups.

Offensive threat

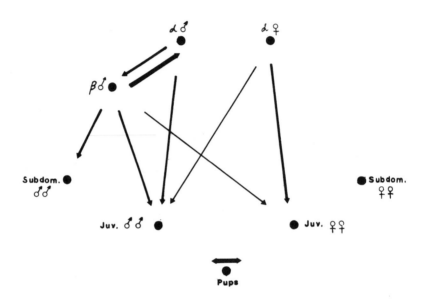

Frequencies <0.25/h not considered

Matrix 7.3: See Matrix 7.1.

Offensive threats were mainly shown by core members of the pack. This behavior could be regarded as an expression of tension that occurs among animals living in close proximity, and probably accounts for many of the observed incidents. Threat behavior between pups was exclusively of this nature. Very young pups showed an especially high frequency of low to moderate intensity aggressive behavior. This probably is the result of a missing dominance system and does not reflect an attempt by the pups to establish a hierarchy (Zimen, 1975, 1978). However, among older animals threat behavior constituted more than merely the expression of tension. The fact that threats were especially frequent between close ranking pack members suggests this form of low intensity aggression was a form of dominance maintenance. This also applies to the frequent threats of dominant animals toward juveniles, their potential future competitors. Increased threatening of juveniles as they matured and integrated into the adults' existing rank order indicates a gradual decrease in tolerance by older animals.

Wolves in the beta male position showed approximately three times as many offensive threats as pack members in all other positions (Matrix 7.3). A possible explanation for this phenomenon is discussed in the section on "Behavioral Strategies in the Wolf Pack." Beta males were also the only wolves which frequently threatened higher ranking wolves. The vast majority of these threats, however, were linked with sexual behavior during the breeding season. Outside the breeding season the beta male directed threat behavior primarily at lower ranking wolves.

Imposing – Imposition behavior is a typical intense agonistic pattern, involving body contact, usually without biting. Commonly, a wolf approaches another wolf with legs stretched and stiff, tail straight back and rigid, head held high, and the hair on its back erect. A fixed stare at the opponent is adopted, sometimes accompanied by growling and visual threat. The animal imposed upon usually withdraws. Two wolves imposing simultaneously may walk and stand parallel or perpendicular, alternately turning their heads away when stared at (Figures 7.8 and 7.9). The direction of imposing behavior makes it clear that it is restricted exclusively to adjacent ranking pack members (Matrix 7.4), either:

from dominant wolves toward the next lower ranking wolves of the same sex, or,

between the male juveniles, who so far have not established stable dominance relationships.

Figure 7.8: Imposing behavior.

Figure 7.9: Imposing with lateral body push.

Imposing

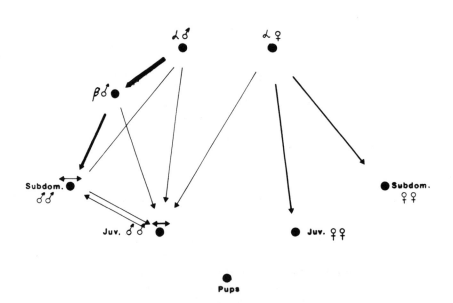

Frequencies <0.10/h not considered

Matrix 7.4: See Matrix 7.1.

Thus, imposing is a behavior pattern typically associated with dominance formation and maintenance. It is rarely observed in direct conflicts over preferred sleeping places, or resources such as food. Such encounters are usually solved by the dominant's exercise of previously established superiority. Again, this clearly demonstrates the function of dominance relationships as a tactic anticipating the result of conflicts inevitable among group living animals.

Inhibited biting — Threats or imposing behavior between equal ranking opponents may sometimes lead to agonistic encounters of higher intensity, involving biting. In the vast majority of cases, however, biting was clearly inhibited. The action was typically accompanied by loud growling and threats. Often the opponents reared up, front legs straddling each other's shoulders, and struggled standing upright (Figure 7.10).

Figure 7.10: Inhibited biting.

As would be expected, the many different forms of inhibited biting encounters were restricted to those wolves not yet having established dominance relationships, e.g. the juveniles and young subdominants (Matrix 7.5). The few inhibited biting sequences observed in cross-sex juveniles indicated that these young animals had not yet fully recognized sexual differences. As they matured, such typical dominance-establishment behaviors became increasingly sex-restricted. As dominance relationships were slowly established, inhibited biting disappeared. As will be seen below, renewed dominance conflicts between adults whose dominance was previously established were usually more serious, and biting was rarely inhibited.

Encircle, lunge and snap — The agonistic behavior patterns thus far discussed were closely related to dominance formation and maintenance. The following are of a different nature, serving to suppress individual pack members and often leading to their exclusion from the pack. The intensity of the behavior again reveals it as clearly sex and rank related, and ultimately exposes the degree of different interests of individual pack members.

A rather mild form of mobbing a pack scapegoat is to encircle the targeted individual. This behavior may easily become exacerbated when some wolves close in, and lunge and snap at momentarily un-protected body areas (usually the hind quarters) of the victim (Figure 7.11).

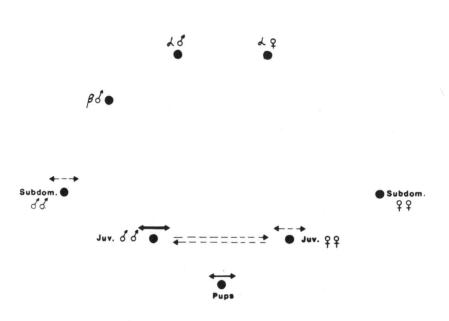

Matrix 7.5: See Matrix 7.1.

Figure 7.11: Encircle and push forward and snap.

The matrices for Encircle (Matrix 7.6) and Lunge and Snap (Matrix 7.7) reveal the difference in intensity of these two mobbing behaviors. Encircle was shown:

by all but the alpha male toward any pack scapegoat, often independent of its sex,

toward low ranking adult males and females who in the great majority of cases became scapegoats.

Encircle

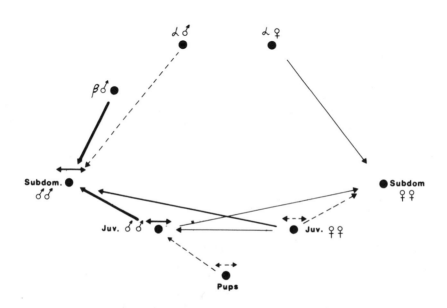

Frequencies <0.05 /h not considered

Matrix 7.6: See Matrix 7.1.

The more severe lunging and snapping was much more highly sex-related, and was shown:

toward a low ranking adult or sometimes juvenile scapegoat by high ranking adults, also by younger wolves of the same sex as the victim.

The difference in these two forms of attack behavior reveal some interesting details of wolf pack sociology: (1) The most dominant alpha male seldom joined in maltreating scapegoats; (2) Group attacks were often initiated by the beta male; (3) Young pack members were always ready to join whenever there was a wolf to attack; (4) When the alpha female occasionally joined or initiated group attacks on low ranking females, her biting was rarely inhibited. She exhibited severe agression far more frequently than any of the other pack members (see below).

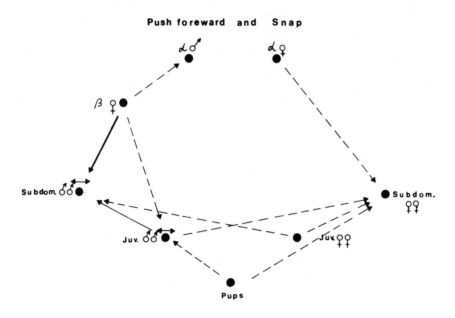

Matrix 7.7: See Matrix 7.1.

Severe Biting – Severe biting was observed in fights between two animals (Figure 7.12) or during group attacks, when the victim was distracted by encircling wolves and the attacker could lunge and bite at unprotected body areas. The attackers intense tugging and head shaking made it readily apparent that their primary intent was to inflict the severest possible wounds. This form of all-out aggressive behavior was practiced almost exclusively by the alpha female in her attempts to suppress low ranking adult females (Matrix 7.8). In some very rarely observed instances severe biting was also shown in the final stage of dominance fights for one of the alpha positions, or in dominance fights between same-sex juveniles.

Following, Keeping Distance, Hunting and Flight Behaviors – A dominant wolf suppressing another pack member may follow that animal at a walk or run, not attempting to catch up, but forcing the suppressed wolf to remain at a distance. The suppressed wolf does not try to enlarge the distance, but maintains it. Other wolves, mainly the juvenile "gang," may join in the following. This behavior was most often observed between the beta male and any low ranking male he had chosen or managed to suppress (Matrices 7.9 and 7.10). Low ranking females showed a tendency to remain at a distance from the alpha female, and from her juvenile helpers. The alpha female, however, often engaged in a more severe form of aggression when pursuing a low ranking female (Figure 7.13). Her intent was to apprehend the victim (Chase, Matrix 7.11). In this endeavor she was sometimes joined by juveniles, especially the females. If they succeeded, the hapless wolf was severely bitten. When attacked, low ranking females ran desperately, trying to lengthen the intervening distance (Flight, Matrix 7.12), and continued running even when the persecutor(s) stopped, disappearing out of sight of the alpha female and her helpers. Comparisons of Matrices 7.11 and 7.12 show that hunts were initiated by the alpha female, and that the juveniles alone did not as a rule put a low ranking female to flight. These observations again illustrate the infrequent but severe aggression practiced by the alpha female, while aggressive encounters between the males were usually less intense. All males occupying the alpha position showed an unusual degree of tolerance.

Figure 7.12: Intensive biting.

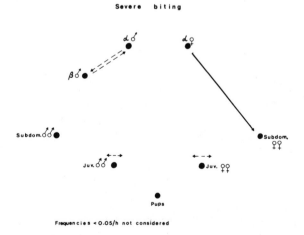

Matrix 7.8: See Matrix 7.1.

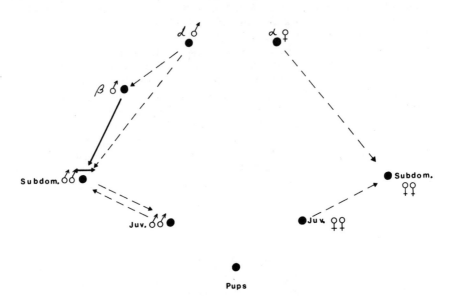

Following

Frequencies <0.05/h not considered

Matrix 7.9: See Matrix 7.1.

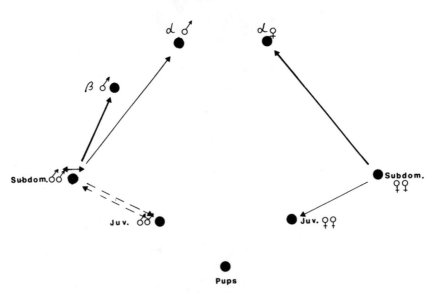

Keeping distance

Frequencies <0.05/h not considered

Matrix 7.10: See Matrix 7.1.

Hunt

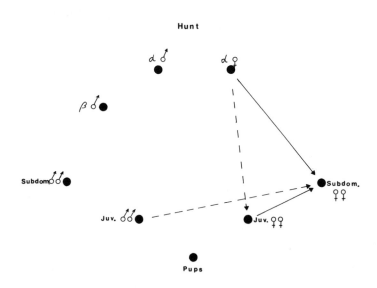

Frequencies <0.05/h not considered

Matrix 7.11: See Matrix 7.1.

Flight

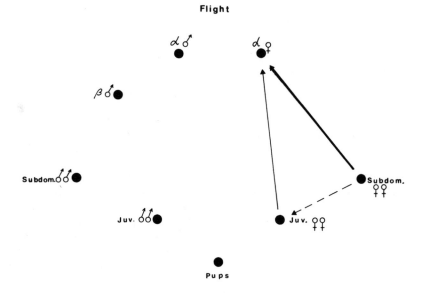

Frequencies <0.05 not recorded

Matrix 7.12: See Matrix 7.1.

Figure 7.13: Flight and hunt.

Play Behavior: Play was the most frequent form of social behavior observed. Almost all play behavior was social (Figure 7.14). Play initiatives, play with body contact, and play-running were recorded separately. Of these categories, play with body contact (biting and wrestling) was by far the most frequently observed form, and will therefore be considered here. The most frequent play encounters (Matrix 7.13) were:

between pups;

between juveniles, whereas a slight tendency to play with litter mates of the same sex was observed as age increased;

between low ranking adults exclusively of the same sex, the males playing very much more than females;

between all pack members and pups, the low ranking adult females being especially active;

between all pack members and juveniles, low ranking adult males being especially active;

between the beta male and male juveniles;

between the alpha female, pups and male juveniles.

The function of play has been discussed by Meyer-Holzapfel (1956), Loizos (1966) and Bekoff (1974) among others. For young animals it provides experience in body movement coordination, and furthers adjustments of social behavior patterns. This, however, does not explain why adult animals play. Some details of behavior observed during my study may be useful for clarifying the role of adult play: (1) Adult wolves with established dominance relationships were rarely observed playing with one another. (2) Young, low ranking adults who were targets of aggression were amazingly active when it came to playing, either with each other or with younger pack members. Hereby, a clear age and sex differentiation was observed:

low ranking males played with one another but not with low ranking females. They played with all juveniles, slightly preferring males, but seldom with pups.

low ranking females seldom played with one another, nor with male juveniles, but only with female juveniles and with pups, i.e. with wolves the same size or smaller than themselves, but not with larger males.

Figure 7.14: Play invitation.

Social play with body contact

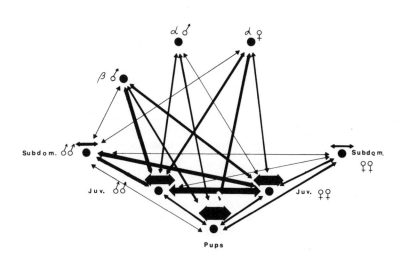

Frequencies <0.10/h not considered

Matrix 7.13: See Matrix 7.1.

These different play preferences indicate that the function of play behavior is slowly transformed from an exercise in coordination and social adjustment in pups to a dominance related behavior in juveniles and adults. In play, dominance can be demonstrated and strengths evaluated. Additionally, play can be used as a tactic to redirect aggressive tendencies into playful activities. Such a tactic was used extensively by low ranking adults to dampen or delay an increase of intolerance in the dominants. If a dominant animal's aggression increased in spite of a subdominant's pup mimicry efforts, the low ranking animals reacted with an even stronger tendency to play. A transfer to intolerance, and group attacks on younger low ranking adults were regularly preceded by an increase of play behavior by the victim. This also explains the high irregularity of adult play behavior, with periods of intense play and long periods when play was infrequent or not observed at all. If the play appeasement tactic was abortive, play behavior of the attacked wolf stopped suddenly and other forms of defensive behavior took over. This explains why adult females and juveniles played less frequently than male adults and juveniles. During long periods of the year they were more heavily suppressed by the alpha female and her juvenile helpers, than were the subdominant males. When being attacked, play behavior was for them a useless appeasement tactic.

These observations show that we must revise our concept of play behavior as having no specific serious context ("Ernstbezug," Meyer-Holzapfel, 1956), and occurring only in relaxed situations ("Entspanntes Feld"). Play may indeed be a most serious occupation. In addition, play behavior in adult wolves lacks another criterion believed to be specific to play. It is rather stereotyped, and not at all irregular in its sequences – while irregularity of sequence is typical in play of young animals (Zimen, 1978).

Sexual Behavior: A summary of sexual activities and their outcome over 12 successive breeding seasons is provided in Table 7.2. The most striking phenomenon is the limited reproduction of the pack. Of 22 potential litters, only five occurred. All were born to the alpha female. It is interesting to note that the alpha female in some years copulated with more than one male, especially at the beginning and end of the copulatory period, when she was either not yet or no longer able to conceive.

Although each year and breeding season had its own peculiarities due to changing numbers, ages, sex composition etc. of the pack, a typical "wolf pack year" could be extracted from the data. During the summer pup raising period the general mood in the pack was friendly and tolerant. In addition to the pups' parents (usually the alpha pair), many other pack members, especially the juveniles and low ranking adult females, helped feed or care for the pups. Howling, and general friendly gatherings of all pack members, however, were rare, as were all group activities involving play sessions or raised-leg urinating ceremonies.

Table 7.2: Sexual Behavior and Results in the Observed Wolf Pack

Year	Females 22 mo.	Females 34 mo.	Males 12 mo.	Males 34 mo.	Copulations Between	No. of Litters	Mother	Father	Remarks
1968	—	—	—	—	None	0	—	—	No sexual activities
1969	2	—	1	—	None	0	—	—	Both juvenile females and one pup had vaginal bleeding, but male showed no interest
1970	1	—	3	—	None	0	—	—	Female had vaginal bleeding but males (all brothers) showed no interest
1971	—	1	—	3	None	0	—	—	2 subdominant males showed interest but were rejected, alpha male showed no interest (all littermates)
1972	—	1	1	3	None	0	—	—	2 subdominant males showed interest but were rejected, alpha male showed no interest (all littermates)
1973	4	—	3	3	Alpha female + beta male Alpha female + juvenile male	1	Alpha female	Beta male	Old alpha male showed no interest

(continued)

Table 7.2 (continued)

Year	Pack Members				Copulations Between	No. of Litters	Mother	Father	Remarks
	Females 22 mo.	Females 34 mo.	Males 12 mo.	Males 34 mo.					
1974	2	4	1	5	Alpha female + alpha male Alpha female + juvenile male	1	Alpha female	Alpha male	At mid-estrus, copulations only with alpha male who showed intensive guarding of alpha female
1975	—	3	3	2	Alpha female + alpha male Alpha female + beta male Alpha female + 2 juvenile males	1	Alpha female	Alpha male	At mid-estrus, copulations only with alpha male who showed intensive guarding of alpha female
1976	—	2	2	4	Alpha female + alpha male Alpha female + beta male Alpha female + juvenile male Beta female + subdom. male Beta female + juvenile male	0	—	—	Social turmoil, rank order broke down in January. Frequent and intensive aggression. Many copulations but no pups born
1977	—	1	—	4	Alpha female + alpha male	1	Alpha female	Alpha male	
1978	—	1	—	1	Alpha female + alpha male	1	Alpha female	Alpha male	
1979	1	1	3	1	None	0	—	—	All dominant pack members are littermates

Pup feeding and care gradually diminished during autumn, while other social activities such as communal howling and raised-leg urination by dominant pack members increased. Social tension within the pack also increased markedly. In late autumn, sudden eruptions of group aggression toward low ranking females occurred, mainly initiated by the alpha female. These low ranking females usually left the pack within a few days to live in areas of the enclosure seldom utilized by the pack. Dominance relationships among the other animals also began to destabilize, and a period of general instability followed, until new relationships had evolved and settled.

Sexual activities initiated by the alpha female usually began in December or January. She showed intense submission and obtrusive behavior toward the dominant males, who at first barely reacted. Slowly, however, their interest increased and they investigated the urine she frequently squirted at bushes, trees or rocks along trails in the enclosure. The dominant males also urinated on these places, often followed by another urination by the female. Gradually almost all males, including the pups, followed the female on her travelling routes, making her the dominant center of pack activities. Initial mating attempts, often by younger males, were rejected, or prevented aggressively by one or more of the dominant males.

During this period of early sexual contact I always had the impression that the males were unaware of what was occurring. They all made a helpless impression. By the time the female started to present, however, one of the dominant males had usually taken the lead. He prevented others from approaching close to the female. This wolf was not necessarily the alpha male. In fact, in one year (1973, see Table 7.2), the alpha male showed no interest whatever in the female. Generally, the alpha male took the initiative in the 10–17 days of intense sexual contact, copulating with the female approximately three to four times in 24 hours, each copulation lasting between five to 20 minutes. Other males were able to copulate with her only during the first and last few days of the receptive period. If young males were involved (once even a 10 month old pup was observed copulating with its mother), the copulation seldom lasted longer than a minute. Although these peripheral copulations never resulted in a pregnancy, they demonstrated that it was not the alpha female who played the important role in choosing a partner. She did display definite preferences, usually for the alpha male, but if another strongly motivated male succeeded in occupying the position closest to her she would, after initial rejections, accept him. She was easy to seduce, especially at the end of the receptive period. Who was going to father the pups was resolved mainly among the males. Sexual motivation and dominance were the two outstanding factors in this process.

The intensity of sexual motivation seemed to be dependent on various factors. Firstly, not all males reached full sexual motivation simultaneously. Sexual contacts also seemed to have an intensifying effect. Secondly, individual relationships with the female seemed to be significant. Litter brothers showed less interest in copulating with their sister, and the female showed stronger rejections if one of her brothers attempted to copulate with her, indicating a slight litter-mate incest barrier (Zimen, 1976, 1978).

Almost all males concentrated their interest on the alpha female, even if they never had an opportunity to approach her. Low ranking females, or females expelled from the pack, were rarely followed and copulation with them was seldom attempted. However, all normally showed vaginal bleeding, and the behavior of my domesticated dog indicated they were sexually attractive.

If sexual, or sometimes merely social contacts occurred between a low ranking female, the alpha female increased her attacks and prevented the normal development to a copulation, and ultimately to pregnancy and birth of an additional litter in the pack. Only once did more than one female copulate, and that was during the winter of 1976, when there was great social turbulence due to the escape of nine juveniles and subadult wolves from the enclosure, and a subsequent breakdown of the hierarchy. No pups were born however. Probably the intense social stress resulted in early abortions.

In the wolf pack I observed there was no active killing, or mistreatment or neglect of pups resulting in their deaths. However, such occurrences have been observed in captive wolf packs where two females produced litters (Altmann, 1974; Fentress and Ryon; this volume; Klinghammer, pers. comm.), as well as in other canids (Macdonald, 1977).

After the copulatory period the high arousal level of all pack members declined, especially that of the alpha female. The result of this was not a decrease in frequency of aggressive behavior that might be predicted, but rather a temporary increase (Zimen, 1978). A possible explanation for this increase in aggressive behavior, of mainly low and medium intensity, could be that weakening pressure of the dominants led to a new differentiation in relationships between younger and lower ranking wolves. Excluded wolves were also being allowed back into the pack at this time, and had to be integrated. A period of generally unstable relationships resulted, until a new equilibrium was established, and social tensions eased in time for a new pup raising period.

Territorial Behavior: Wolf pack territoriality seems highly dependent upon the spatial behavior of the main prey species. In areas of resident prey most wolf packs are territorial (Mech, 1970), whereas in areas of migrating ungulates wolf packs are spatially separated only during the summer pup raising period but partly highly concentrated around prey winter yards (Stephenson, 1979). Naturally, observation of a captive wolf pack can reveal little about territorial behavior. However, several circumstantial observations and one experiment are of interest.

During the winter 1972/73, I took various domesticated dogs of all sizes, one at a time, to the outside of the large wolf enclosure, to observe the reactions of the wolves (Zimen, 1976). Generally, the dominant adult wolves reacted vigorously, displaying signs of attack. Most of the dominant juveniles also reacted aggressively, while the low ranking wolves either showed no reactions, or behaved in a friendly, playful manner.

These observations correspond closely with the behavior of my tame wolves outside the enclosure. The tame alpha wolves were always highly excited in areas they knew well, i.e. on routes they covered regularly. Urinations were frequent, and subsequent scratching at the ground intense. If they observed a dog, they were difficult to control. The low ranking wolves, on the other hand, were generally friendly, attempting to play with any dogs encountered. When a change in rank order occurred, the new alpha wolf showed aggressive behavior within a very short time, while the subordinate's behavior immediately became friendly-submissive. These rapid emotional transformations in the adult wolves were unexpected.

Young wolves or domesticated dogs usually elicited no aggressive reactions in the pack. It was possible to introduce unfamiliar pups up to six months of age into the enclosure without any difficulties. They were usually quickly accepted, and fed and cared for as pack-raised pups.

Adjacent to the large enclosure was a smaller one where wolves which had been expelled from the large pack were kept. Gradually, an organized pack with a typical rank structure was also formed in the smaller enclosure. Although no experiments were conducted, it was my clear impression that these wolves, as long as their numbers did not exceed two to four animals, were less aggressive toward strangers than the dominant wolves in the large pack.

These observations indicate that aggressive behavior toward pack strangers is variable. Besides being dependent upon the availability of a defendable resource, territorial aggression is related to several additional factors:

> the area, whereby well-known terrain increases aggressivity, and therefore excitement over the scent of strangers [in natural situations, territorial boundaries are well-known areas of increased scent marking (Peters and Mech, 1975)],

> age and rank of pack members, as aggressivity increases with age and rank,

> the size of the pack, as wolves in small or newly formed packs are less aggressive,

> sex of the intruder, as well as the sex structure of the pack.

Most striking was the friendliness shown by low ranking or pack excluded wolves in their attempts to contact strangers. These were the wolves who, in winter, mainly howled alone and away from the pack, possibly in attempts to contact other wolves.

Spatial Organization, Social Bonding and Dispersal

In discussing spatial organization and dispersal, observations of an enclosed wolf pack are again limited. This enclosure, however, was large enough for the wolves to show distinct differences in their spatial behavior. The average distance between any two pack members should indicate the bond linking them. Wolves maintaining close contact are probably more strongly bonded than those which are seldom together.

The enclosure was large enough for single wolves to leave the pack and the area generally used, either temporarily or for extended periods. Wolves who showed distinct divergence from pack members in spatial behavior and in activity periods, and rarely had neutral or friendly social contacts with the large body of pack members, will be referred to as emigrants. A brief review will be given on the most outstanding results of over 100,000 distances registered between wolves in the enclosure over the years (see also Zimen, 1976, 1978).

In winter months, pack members were closer together than in summer, and chorus howling and general friendly get-togethers were much more frequent. These observations are matched by those in the wild, where packs seem to stay and travel closer together in winter, while in summer shorter or longer pack splittings are more frequent (Mech, 1970). The focus of the pack was clearly the group of dominant adults who stayed near one another and had frequent social contacts (see section on social contacts). Outstanding in this group was the alpha male who, for most of the year, was the hub of pack activity. Only during breeding season when the alpha female became the primary attraction was the alpha male's role less prominent. He was also in the background, to a certain extent, in the early summer pup-raising period when the pups were the attraction. However, the dominant adults were the center of the pack for the remainder of the year, particularly the alpha male who, as well as attracting others, appeared to play an active part in keeping the pack together.

This pack core group was joined by the juveniles and a varying number of young adults, and eventually by the pups once they were mature enough to travel with the pack in autumn. These additional pack members, including pups over approximately nine months, tended to temporarily leave the pack on their own or in small subgroups, so that they were much more often alone or in small groups than the dominant adults who rarely slept or travelled apart from the pack.

These observations in the enclosure correspond closely with the behavior of my tame, free-running wolves. On walks, the dominant adults seldom escaped from my presence. They were eager to remain close together and to keep me in the pack. The low-ranking wolves, however, behaved more like older pups or juveniles, often seizing a chance to go and explore the countryside alone. Normally, they were back within a day or two (Zimen, 1978).

A detailed analysis of the data for individual peripheral pack members shows that the amount of time they spent with the core group, as well as their average distances from the dominant wolves, was again dependent on age, sex and rank:

> the older the wolves, the more independent their behavior. Marked changes occurred in the 8-10 month old pups when they suddenly showed strong tendencies to leave on their own (Zimen, 1972).

> the higher a wolf's rank within its age group, the stronger was its tendency to stay close to the pack, and especially to the core group.

> females not expelled from the pack as a rule showed less tendency than males to go off alone.

Despite a high degree of independence, a tendency to return to the pack and stay with it distinguished these peripheral pack members from the pack emigrants. The latter left the pack completely — either voluntarily or after being ejected — the difference being somewhat unclear. Older low-ranking adult emigrants showed an increasing tendency to remain away from the pack. They would probably have had some contact had it not been for the aggressive reactions of certain pack members when they approached. Other voluntary emigrants became pack scapegoats after they left, and were subsequently singled out to be attacked, especially by the juvenile "gang."

This also applied to some of the aggressively expelled wolves. Others withdrew so completely within the enclosure that they were seldom seen, and were, for the most part, unmolested. Only if they attempted to reapproach the pack, usually during the spring decrease of social tension, were they likely to be attacked by the group. Some endured such attacks and were eventually reintegrated. Others withdrew and continued living in isolation. This was the rule for wolves who had been expelled more than once, and who had already lived as emigrants for an extended period.

It is interesting that the emigrants did not amalgamate to form a new group within the enclosure, nor did they show any form of communal defense. Solidarity of the suppressed does not seem to exist in wolves.

DISCUSSION

The number of potential behavior patterns available to a wolf is numerous. However, these patterns are not used indiscriminately. Each wolf, in a specific phase of its life, employs only a select combination of behaviors and space usage in relation to other pack members. The possible combinations of behavior and spatial patterns, i.e. an individual's behavioral tactics, can be related to its age, sex and rank, as well as to that of the pack as a whole. For example, typically a six month old female is playful, unaggressive, submissive, strongly bonded to the pack, and shows little, if any, tendency to roam. She is provided food, over which she may have to fight with littermates. This highly adaptive behavior can change drastically if she is suddenly deprived of her parents and other older pack members and placed with a group of younger pups. Instead of begging for food, she starts to provide and regurgitate it for the pups. Instead of being licked, she lies with the pups and grooms them. She becomes protective, more aggressive and less submissive as she gradually adopts the role of an adult wolf.

Such a transformation of behavior was always observed when one of the pack positions became vacant, either by chance or experimentally. Typical was the early increase in aggressive behavior and independence of pups raised in a pack without juveniles. Because of the lack of juveniles, these pups experienced adult aggression earlier. In early winter, the alpha female, instead of increasing her suppression of juveniles and subadult females, directed this behavior toward female pups, especially the dominant ones. Vaginal bleeding and sexual behavior during breeding season could then be observed in 10 month old pups, whereas 22 or even 34 month old females in large packs with older females seldom displayed any sexual behavior. Obviously, all wolves, with the exception of the very young pups, have the potential to behave in any way specific to their sex. Which of the potential behavioral tactics is, in fact, used depends not so much upon the wolf's actual age, but on the specific social context it experiences. In the light of this finding, the observed general tendency of adult subdominant female wolves not to breed can be understood as a form of delayed maturation. Once a female has reached a dominant position and begins to reproduce, it is very difficult for other wolves to prevent her from doing so, even after possible loss of rank. (This corresponds with observations by Packard, pers. comm.)

In evolutionary terms, natural selection must operate differentially on wolves occupying different positions in the pack (see also Bertram, 1976). The behavioral strategies linked with each position function to maximize the reproductive fitness of the wolf occupying that position by increasing the proportion of its own genes and those identical to its own carried by other wolves (Hamilton's kin selection hypothesis, 1964) represented in the next generation. That is to say, to maximize its inclusive fitness, each wolf must have distinct interests, and the behavior it exhibits furthers these interests.

What can be learned from observations of social behavior about the selective forces which operate on the wolf and shape this social behavior?

Behavioral Strategies in the Wolf Pack

The Alpha Female: The interests of the alpha female in a wolf pack are obvious. She must attempt to raise as many pups to adulthood as possible while simultaneously trying to prevent other females from reproducing, thus increasing the proportion of her genes. Pups have high energy requirements. They must grow fast to be able to follow the pack on travel routes by about six months of age. It is therefore advantageous for the alpha female to have many helpers in raising her pups, and only her pups. We now know how she achieves this goal. Before and during the breeding season, she is very aggressive to all other mature females. For theoretical reasons, one would expect her to be especially aggressive if the subdominant females are not her own daughters or littermates. Unfortunately, my observations cannot shed light on this question as the numerous disturbances occurring in the pack prevented a daughter of the alpha female from ever reaching maturity during her mother's reign.

After the breeding season, aggressiveness in the alpha female drops drastically, allowing suppressed or excluded females to rejoin the pack. It also helps to create a friendly atmosphere in the pack for pup-raising.

For breeding, the alpha female favors the alpha male, probably the strongest and most experienced male in the pack. This is to her advantage in securing food, as well as territorial rights for herself and her offspring. She may also copulate with other males, however, thus bonding them to herself and her pups. The long estrus in female wolves seems adaptive to polyandric bondings. (In our pack, copulations occurred for a minimum of 10 and a maximum of 17 consecutive days; nonbreeding females showed vaginal bleeding up to five weeks.) If her primary partner weakens and his alpha position is unstable, the alpha female may join in the attacks on him, thus speeding up the process of finding a new alpha male and shortening the period of pack instability (see section on Pack Stability and Inbreeding).

The Alpha Male: Pup growth must be in the interest of all pack members, as one day the pups will fill vacancies in the communal hunt. A pack that fails to raise pups for some years will eventually disappear. With the exception of the alpha female, however, no wolf will be more interested in pup survival than the alpha male, their most probable father. He is also likely to be the father of the juveniles, and may also be the brother or uncle of other pack members. As in the case of the alpha female, it must therefore be in his interest to see the pack united. For these two, as well as for the pups, the optimal pack size will therefore be larger than that for other pack members as the alpha animals and the pups are the first to feed (Zimen, 1976). The position of the alpha male is rather secure (as is the position of the alpha female), as only severe aggression can force him to relinquish it.

Unlike the alpha female, however, the alpha male has no interest in suppressing other wolves as he can secure his rights to father the pups quite easily. Therefore, with the exception of the pups, no wolf in the pack can have a stronger interest in seeing a friendly, co-operative atmosphere among pack members. On the other hand, he will be especially keen to secure the hunting grounds of the pack and keep all pack strangers away. In actual fact, all alpha males I have observed have been surprisingly friendly and tolerant toward pack members as well as extremely aggressive against intruders.

Besides being highly attractive for other pack members, the alpha male also took active part in keeping the pack together, often preventing single wolves or subgroups from splitting off. He was the center of many group activities, such as chorus howling or greeting ceremonies, as well as being the dominant individual in most deci-sion-making processes. The alpha male was the "tolerant boss" of the pack.

The Beta Male: The beta male is the most outstanding, probable successor of the alpha male. He is possibly also related to the alpha male or to the alpha female, being a brother or son of one or both of them. There is a fair chance, then, that he is also closely related to the pups, being either uncle or older brother. This should give him an interest in the successful raising of the pups. In fact, the beta males in our pack were all strongly bonded to the two dominant wolves, as well as being quite involved with the pups, although the pups were usually not his own. Nevertheless, the beta male might have "be-lieved" they were. The fact that he copulated with the alpha female in some years, although he fathered her pups only once, seems to have increased his bond to her and also the pups, born two months later. This is certainly in the alpha female's interest. In fact, her promiscuous behavior could possibly be understood as a strategy to bind more than one male to herself and her pups, which could also be interpreted as a behavioral deception analogous to "prostitution" behavior in hummingbirds (Wolf, 1975). From the work of Maynard-Smith (1978), however, we know that a "lie" cannot become an evolutionary stable strategy (ESS). The fact that the female can continue cheating, therefore, can only be explained by the possible close relationship between the deceived and the offspring of the deceived.

Our knowledge of sexual behavior in free-living wolves is limited, but they are probably not, as a rule, promiscuous. However, the alpha female is certainly not the "true loving wife" described in some modern popular accounts of wolf behavior — an interesting fact in light of evolutionary theory.

Another interesting aspect of the beta male behavior is his frequent low-intensity aggression, mainly directed down, but occa-sionally also up the rank order. This may be because the beta male position is the only other permanent adult male position in the pack beside that of the omnipotent alpha male. For juveniles and young adults, therefore, the beta male position must be a highly desirable one. To attain it is possibly their only chance of being able to stay in the pack and eventually become alpha males. From the point of view of the beta male, therefore, all low-ranking males constitute a potential threat so that any latent expansion tendencies are promptly blocked by him, often long before they become manifest.

The occasional slight aggression by the beta male toward the alpha male, on the other hand, can be explained as optation for that position, thereby continually testing the alpha male's strength. Thus, the beta male position becomes the "aggressive" turntable of the pack.

The Low-Ranking Males: Normally, low-ranking adults are either wolves having just reached maturity or old animals who have lost their dominant position. Whether or not they will be expelled from the pack seems to depend on food availability, while a voluntary exile seems partly independent of food (Zimen, 1976). If the pack is successful in hunting, the low-ranking male's survival chances are good as long as he stays with the pack and can profit from communal hunts. But if hunting results are poor, he might do better to leave and attempt to live on his own. Indeed, packs with declining hunting successes tend to decrease. In extreme circumstances, only the alpha pair survives (Mech, 1978). But even given a fair or good food supply, staying with the pack has disadvantages for low-ranking males. Their chance to reproduce is limited; it may take years before one of the dominant positions becomes vacant. If they leave, their chances of meeting a strange wolf of the opposite sex, or finding a vacant territory and breeding, are small. The best strategy, therefore, must be to stay with the pack as long as possible, helping to raise the pups which are generally younger kin. At the same time, they should be alert for any opportunity to join with other wolves in forming a separate pack.

It is evident that subdominant males are highly independent, often going off by themselves or in small groups. They are unaggressive toward strangers and may even seek their company. Within the pack, they show a form of pup-mimicry behavior, being very submissive and playful, attempting to keep the atmosphere friendly and cooperative.

The Low-Ranking Female: For low-ranking females, the situation is similar to that of low-ranking males. However, their chance of surviving alone is probably lower due to their smaller size and negligible prospect of joining a strange pack if it already contains a female. Their prospects for staying in the pack are also somewhat limited. Because of the high aggressiveness of the alpha female, pup-mimicry behavior is of little value. Their only option seems to be to behave as unobtrusively as possible outside pup-raising periods. They may even stay away from the pack, trailing it and living off scraps left at kills. In summer, however, they may be able to appease the alpha female by willingly helping to raise her pups.

The cryptic behavior of low-ranking females contrasted noticeably with the conspicuously playful behavior of low-ranking males.
Also, the females showed less tendency to leave voluntarily. If they
were integrated into the pack, they always stayed very close.

The Juveniles: Exclusion from the pack carries an even greater
risk of juveniles than for other wolves. They are too inexperienced
to have much chance alone. There is still much to learn in the pack,
e.g. hunting and pup-raising. Additionally, the probability that the
pups are their full siblings is higher than that for older subdominants.
Juveniles are usually very active helpers in pup-raising.

As juveniles need not ordinarily fear severe aggression from
adults nor pack exclusion, they can seize every opportunity for im-
proving their own position either among littermates or with older
subdominants. Conversely, they must beware of taking too great a
risk, as engaging in a dominance fight increases the chance of incur-
ring a debilitating injury. In our pack, the juveniles never missed an
opportunity to join in attacking individuals. They acted only in a
group, never alone, and primarily remained on the fringe of an
assault.

The Pups: For the pups, remaining in the pack is a matter of
course. Their best interests are served when as many pack members
as possible feed and care for them. Understandably, therefore, their
most typical attitude toward adults is submissive, care- and food-
eliciting behavior. In their mutual relationships, they only must be
certain to acquire an adequate share of the food delivered by older
pack members. Serious conflict, however, is limited, as the dominance issue is still remote, pack membership is unquestioned and reproduction is not possible.

Pack Stability and Inbreeding

Inbreeding in wolves is favored by:

> the relatively stable pair bond of the alpha pair and ex-
> clusive breeding by the alpha pair in successive years,

> recruitment of pack members from pups raised in the
> pack,

> the high aversion to pack strangers, preventing their inte-
> gration into the pack.

Thus, a wolf pack is potentially an exclusive reproductive unit
within the population. This arrangement favors reduction of gene
variability among pack members and an increase of gene variability
among packs (Woolpy and Eckstrand, 1979). A population divided
into such partially isolated subunits provides the most favorable
conditions for evolutionary advance (Wright, 1939). Social behavior,
as a result of evolutionary processes, is simultaneously a determinant
of the same processes. Therefore, the social and reproductive organ-

ization of the wolf pack, obvious adaptations to specific ecological conditions, may have enhanced the tremendous adaptability of the species. On the other hand, inbreeding favors recessive traits and could ultimately lead to extermination of the entire reproductive unit. It is therefore important to know how long a wolf pack has been reproductively isolated.

Unfortunately, little information is available describing free-living populations and pack stability. On Isle Royale where the wolf population has been under observation since 1959, wolves were not individually marked so only a few could be recognized during the annual winter research periods. Once, however, the splitting of a pack was observed after a change in the alpha male position (Peterson, 1974). Observation of my captive wolf pack also indicates that a change in one of the alpha positions often occurs at a time of great instability and unrest within the pack. It need not go so far as to complete pack disorganization, but such periods are also likely to be the times when major changes occur in pack composition; i.e. the emigration of whole subgroups or the immigration of strange wolves into the pack. The time a wolf stays in one of the dominant positions may therefore indicate a period of pack stability and of reproductive isolation.

Between 1970 and 1978 my wolf pack suffered relatively few external disturbances. In these eight years, six aggression-related changes occurred in the alpha male position and two in the alpha female position, involving five males and three females (Table 7.2). The much longer reigns of the dominant females and the rather rapid changeover in the dominant male position could be purely circumstantial or the result of confined living conditions, but the few continuous observations of free-living packs show the same tendency. From 1968 to 1975, four different males, but only two females, occupied the top positions in the so-called West Pack on Isle Royale. In the East Pack, one female even remained in the alpha position for six successive breeding seasons, from 1972 to 1977, during which time three males occupied the counterpart position (Peterson, pers. comm.). In northern Minnesota, Mech (pers. comm.) repeatedly recaptured and radio-marked the breeding female of one pack over 10 years.

These data, as well as observations of social behavior, indicate the alpha female is actually the most prominent member of the pack. She is its real center despite all the attention afforded the alpha male by the pups. Depending on prey size and availability, she attracts varying numbers of juveniles and adults, preferably males, to herself and her offspring to maximize her hunting and, ultimately, reproductive success. Although the tendency is only slight, the wolf pack is one of the few polyandrous systems known in mammals. Large-prey hunting and the enormous food requirements of pups seem to have been the selective forces responsible for the evolution of the system. In any case, a reproductive pair does not usually remain together for many seasons. Since all top-rank changeovers are during periods of increased instability with the possibility of drastic changes in pack composition, it can be assumed that total reproductive isolation and consequent inbreeding over many generations do not normally occur in wolf packs.

SUMMARY

The subject of this paper is who does what, when and to whom in a wolf pack. Data from 10 years of observation of a captive wolf pack were presented. Forty-eight different social behavioral patterns were identified. The observed behavioral interactions between duads in the pack are grouped for each behavior pattern and month of observation. Rank relationships were registered independently for each month from the wolves' expressive behavior during encounters. Thus, matrices were calculated for each behavior pattern relating the behavior of the wolves to their sex, age and rank within the pack. The matrices of the 13 most common social behavioral patterns were presented.

Social interactions without agonistic components such as muzzle-to-fur and muzzle-to-muzzle contacts were observed mainly among core wolves, the alpha pair, the beta male, the juveniles and the pups. Pack activities appeared centered around the alpha male for most of the year: 40 percent of all submissive behavior was directed toward him. Play behavior in adults was identified as a tactic to redirect incipient aggression.

Agonistic behavior of low intensity was observed between all core pack members. Aggression of moderate intensity was typical of male suppression of peripheral males. High-intensity aggression was restricted almost exclusively to female interactions. The incumbent of the alpha female position, often supported by her juvenile and pup offspring, was highly aggressive during winter breeding seasons, so that most low-ranking adult females were either expelled from the pack, or their sexual activities temporarily suppressed. Thus, breeding was restricted to the alpha female. In most years the alpha male sired the pups, although the alpha female's copulation with other males was possible. Territorial behavior was most intense in high-ranking adults, whereas low-ranking pack members were extremely tolerant of pack strangers, often seeking contact with them. Social bonding (expressed in distances between individuals) was strong between core members of the pack. Older pups, juveniles and low-ranking adults exhibited stronger tendencies to isolate themselves from the pack temporarily, resulting ultimately in emigration.

The behavioral strategies typically attached to each position in the pack are discussed in relation to the strategist's reproductive success. It emerges that the alpha female is the most prominent pack member, bonding to herself and her offspring a varying number of juveniles and male adults.

Part IV

Conservation

Introduction

Previous sections described the wolf throughout much of its present range in the northern hemisphere. In several areas, the wolf and its prey continue their age-old drama, little affected by man. In other regions, however, man's influence is extreme and pockets of wolves cling tenuously to a pittance of their former habitat. In many cases, wolves have become beggars, surviving on the offal of civilization. Over the next decade, it seems inevitable that the wolf's range will continue to shrink. Some populations will disappear completely while others will be only sporadically infused with life from more fortunate neighboring groups.

There is perhaps a smugness in the attitude of most North Americans concerning "their" wolves. After all, North American wolves continue to thrive throughout vast portions of the continent, a condition not found in Europe with the exception of parts of the USSR. However that smugness appears to be unjustified. Historically, Europe has maintained a far denser human population than North America, yet wolves have not been completely eliminated. North America, on the other hand, has lost virtually all its wolves wherever moderate levels of human activity have occurred. With proper management, European wolves could be ready for a rebound.

This section addresses problems encountered in attempts to conserve the wolf. L. David Mech provides a worldwide summary of the wolf's current status, the probable reasons for local declines and immediate actions deemed necessary to ensure continued survival or to regenerate what once was. What is immediately apparent from this survey is that the North American wolf is represented by a single, continent-wide population, whereas the Eurasian wolf exists in a number of isolated regional populations, especially in southern Europe. Some of these wolves may have survived for decades or longer with little or no contact with other populations. These tenacious remnant populations could provide valuable insights into methods of preserving this increasingly endangered species. Unfortunately, most of these populations remain unstudied. We must be concerned with how many will disappear before we learn even their most basic characteristics.

Rolf Peterson and James Woolington recount the history of the Kenai wolf, a small Alaskan population which was apparently exterminated in the early 1900s and then reappeared nearly half a century later. Their account suggests that humankind's direct influences (hunting, trapping and poisoning) may be more deleterious to the wolf's survival than our indirect influences (prey reduction, habitat change, etc.). Sverre Pedersen adds an interesting and timely companion study to Peterson's. Taxonomists will long argue over the details or necessity of subspecific classification. However, there were once 32 recognized wolf subspecies, a number which has been seriously diminished over the past century. Pedersen's analysis of morphological features suggests that the four subspecies once recognized in Alaska are now reduced to three. Apparently, the Kenai wolf, which has now returned, may not be the Kenai wolf of centuries past. In our efforts to conserve wolves, should we focus on the species as a whole, or on individual populations or subspecies?

Although the sparsely populated regions of the western U.S. seem to possess adequate room and abundant prey to support viable wolf populations, few wolves are thought to inhabit these areas. Yet reports of wolves are common. One area where sightings have been frequent and consistent is the northern Rocky Mountains near the Canadian border. Robert Ream and Ursula Mattson conducted a long-term study to catalog and evaluate the authenticity of these sightings. The project eventually led to the capture and radio-collaring of one of these rare, elusive animals. One wonders why the Rocky Mountain wolf and other western populations have not become re-established. Robert Hook and William Robinson present a reasonable explanation. After observing the failure of an attempted reintroduction of wolves in northern Michigan, Hook and Robinson decided to identify the reasons for that failure. Since all these wolves were killed by humans, they concentrated on attitudes toward the wolf. Their results were somewhat surprising: only a handful of Michigan residents harbor deep-seated resentment toward wolves but, unfortunately, this group is in a position to exercise its resentment. The section concludes with Robert Henshaw's analysis of factors to be considered prior to wolf reintroduction, and specifically to Adirondack Park in northern New York. He concludes that the successful return of wolves to one of the largest wilderness areas east of the Mississippi River rests with the reaction of two resident groups—the Adirondackers and the eastern coyote.

Knowledge of the wolf has been influenced by contributions from various disciplines, each of which emphasized specific conceptual

approaches in their collection of information. For many years, wildlife biologists followed a population approach. Individual wolves became lost in populations which interacted with prey populations
which, in turn, were influenced by climate, habitat quality and human
exploitation. The sum of these interactions was stability, decline or
increase. The search was for the key factors which regulated these
interactions so wildlife managers could intervene, when needed, to
ensure that all populations remained where the manager's constitu-
ency wanted them. The behavior of individual wolves was immaterial
as long as the final result could be reliably predicted, or dictated.

The ethologist operated at the opposite extreme. Studies were
focused on individuals and the multitude of factors impinging upon
them. Perhaps the sociobiologist and behavioral ecologist, who have
extended the focus from individuals through kinship groups to populations, may successfully span the gap.

The final two papers in this section present other approaches to
our study of the wolf. Henry Sharp discusses the methods of the
cultural anthropologist, who neither stresses the individual nor the
population, but rather the non-living, yet alive 'thing' that wolves
create—society. He argues that although wolf social behavior may
emerge from wolf biology, it stands on its own and should be studied
in its own right. Robert Stephenson then advises us to step back and
take a lesson from the Nunamiut Eskimo. We often are so enamoured
with our theories and conclusions that we hurry our observations and
miss small details which are often so important. The Nunamiut,
with no theories to vindicate or complexities to simplify, have for
centuries recorded details of wolf life in the same manner as you or
I describe our own lives. Each observation makes for interesting conversation, and may some day prove important to modern science.

The IUCN-SSC Wolf Specialist Group

L. David Mech

The "International Union for Conservation of Nature and Natural Resources" (IUCN) is the foremost international conservation organization, having representatives from 101 countries including 50 governments, 103 government agencies and 261 citizens' organizations. Its administration is funded by membership fees, and many of its conservation activities are financed by the World Wildlife Fund. The organization functions primarily by setting conservation policies and urging member governments to adhere to them.

The main activities of IUCN are carried out by six commissions including the Species Survival Commission (formerly the Survival Service Commission or "SSC"). The SSC is composed of 40 to 50 "Specialist Groups," usually dealing with a particular species or group of animals or plants. The Wolf Specialist Group is one of these. Dr. Douglas Pimlott was the first chairman of the Wolf Specialist Group and I succeeded him in 1978.

The SSC Wolf Specialist Group includes members from Canada, the U.S., Italy, Sweden, Finland, Spain, Portugal, Greece, Israel, India, the Soviet Union and Poland, and attempts are being made to recruit members from several other east European countries.

Following, are the members of the Wolf Specialist Group who met in 1979 in Portland, Oregon, U.S.A., and assembled the draft, "Table of Information for World Conservation Strategy for the Wolf" (Table 2.1): Dr. Anders Bjarvall – Sweden, Dr. Luigi Boitani – Italy, Dr. Ludwig Carbyn – Canada, Mr. Curtis Carley – U.S., Dr. Javier Castroviejo – Spain, Dr. L. David Mech – U.S., Dr. H. Mendelssohn – Israel, Dr. Jon C. Ondrias – Greece, Dr. Erkki Pulliainen – Finland, Mr. S.P. Shahi – India, Dr. Ronald Skoog – U.S., and Dr. Erik Zimen – West Germany.

Table 2.1: Draft Table of Information for World Conservation Strategy for the Wolf, 1979

Region	Wolf Population Status*	Approx. No.	% of Former Range	Subspecies	Reasons for Reduction	Legal Status	Actions Needed (in order of priority)	Main Prey
Alaska	I	10-15,000	100	*ligoni, pambasileus, tundrarum, alces*	—	hunted, trapped, seasons, bag limits, some control work, enforcement	research, management, education	moose, caribou, sheep, deer, beaver, goat
British Columbia, Yukon	I	10,000?	80	*crassodon, fuscus, columbianus, pambasileus, mackenzii, occidentalis, tundrarum, ligoni, irremotus*	— —	game species, furbearer (BC), no closed season (Y)	research, management, education	moose, caribou, sheep, deer, beaver, goat, elk
Alberta	I	?	80	*occidentalis, griseoalbus, irremotus, nubilus*	— —	furbearer	research, management, education	moose, caribou, sheep, deer, beaver, goat, elk, bison
Saskatchewan, Manitoba	I	?	70	*hudsonicus, griseoalbus, nubilus, lycaon*	—	furbearer	research, management, education	moose, elk, deer, beaver, bison, caribou
Ontario, Quebec	I	?	80	*hudsonicus, lycaon, labrodorius*	—	furbearer?/predator (Q)	research, management, education	moose, deer, caribou, beaver

(continued)

Table 2.1: (continued)

Region	Wolf Population Status*	Approx. No.	% of Former Range	Subspecies	Reasons for Reduction	Legal Status	Actions Needed (in order of priority)	Main Prey
Newfoundland, NWT	I	?	95	arctos, bernardi, hudsonicus, mackenzii, manningi, labrodorius, orion, griseoalbus, occidentalis	—	furbearer	research, management, education	moose, caribou, beaver, bison, musk-oxen, hares
Minnesota	I	1,200	40	lycaon, nubilus	persecution, habitat destruction	full protection	research, management, education	deer, moose, beaver
Michigan, Wisconsin	III/IV	60	10	lycaon	persecution, habitat destruction	full protection	enforcement, survey, education	deer, beaver, moose
Northern Rockies Northwestern U.S.	IV	10	—	irremotus	persecution, habitat destruction	full protection	survey, research, education	deer, elk, moose sheep, goats, beaver
Southeastern U.S.	IV	25	<0.5	C. rufus	persecution, habitat destruction, hybridization with C. latrans	full protection	captive breeding, reintroduction, management	swamp rabbits, nutria
Southwestern U.S.	IV	<6	—	baileyi	persecution, habitat destruction	full protection	captive breeding, reintroduction, management	deer, livestock

(continued)

Table 2.1: (continued)

Region	Wolf Population Status*	Approx. No.	% of Former Range	Subspecies	Reasons for Reduction	Legal Status	Actions Needed (in order of priority)	Main Prey
Mexico	IV	<50	<10	baileyi	persecution, habitat destruction	unenforced full protection	captive breeding, reintroduction, law enforcement	livestock
Norway	IV	<10	?	lupus	persecution	full protection	survey, law enforcement, management	moose, reindeer, roe deer
Sweden	IV	10	?	lupus	persecution	full protection	survey, law enforcement, management	moose, reindeer
Finland	III/IV	100	<1	lupus	persecution	no protection (north), game status (east), protected (south)	protection, education, survey	moose, reindeer, white-tailed deer
Greenland	?	?	?	orion	?	?	?	musk-oxen, caribou
Turkey	I/II	?	?	lupus, pallipes	?	no protection	survey, protection, education	livestock, ?
Syria	?	?	?	lupus, pallipes	?	?	survey	?
Jordan	III	200?	?	–	?	no protection	survey	?
Israel	III	100	60	pallipes, arabs	habitat destruction	full protection	research, management, education	hares, livestock, ?

(continued)

Table 2.1: (continued)

Region	Wolf Population Status*	Approx. No.	% of Former Range	Subspecies	Reasons for Reduction	Legal Status	Actions Needed (in order of priority)	Main Prey
Egypt (Sinai)	IV	30	?	arabs	persecution	no protection	protection, survey, education	hares, livestock, ?
Lebanon	V	0	?	—	?	?	?	?
Arabian Peninsula	II	?	?	pallipes, arabs	?	no protection	survey, protection, education	?
Iran	I	1,500?	80	pallipes, campestris	persecution	game species	survey, research, education	gazelle, mountain sheep, livestock, wild boar, deer, capra
Iraq	?	?	?	?	?	?	?	?
Afghanistan	?	?	?	pallipes, chanco	?	?	?	?
Pakistan	?	?	?	pallipes, chanco	?	?	?	?
Bhutan	?	?	?	pallipes, chanco	?	?	?	?
Nepal	?	?	?	pallipes, chanco	?	?	?	?
India	III/IV	?	?	pallipes	habitat destruction	full protection (not enforced)	law enforcement, survey, research	livestock, hare, deer, antelope
Mongolia	?	?	?	chanco	?	extermination efforts	protection, survey, management	livestock
China	?	?	?	chanco	?	extermination efforts	protection, survey, management	?

(continued)

Table 2.1: (continued)

Region	Wolf Population Status*	Approx. No.	% of Former Range	Subspecies	Reasons for Reduction	Legal Status	Actions Needed (in order of priority)	Main Prey
USSR (Europe) (Asia)	I I	10,000? 50,000?	50 } 70 }	*lupus, albus, campestris, chanco*	persecution, habitat destruction	protected in nature preserves only	management, education, research	ungulates, livestock
Poland	III/IV	200	10	*lupus, campestris*	persecution, habitat destruction	partial protection	protection, research, education	(moose) roe deer, red deer, wild boar
CSSR	III/IV	100	10	*lupus*	?	no protection	protection, survey, research	(moose) roe deer, red deer, wild boar, mufflon
Romania	II	2,000	20	*lupus*	persecution, habitat destruction	no protection?	protection, survey, education	(moose) roe deer, red deer, wild boar, mufflon
Bulgaria	III?	100	?	*lupus*	persecution, habitat destruction	?	survey, protection, education	(moose) roe deer, red deer, wild boar, mufflon
Greece	I/II	2,000?	60	*lupus*	persecution, habitat destruction	partial protection, bountied	management, research, education	deer, wild boar, chamois, livestock
Yugoslavia	I	3,000-4,000	55	*lupus*	persecution, habitat destruction	none, no poisoning, bountied	management, research, education	livestock, roe deer
Albania	II/III?	?	?	*lupus*	?	?	?	?
Hungary	IV	?	?	*lupus*	?	?	?	?

(continued)

Table 2.1: (continued)

Region	Wolf Population Status*	Approx. No.	% of Former Range	Subspecies	Reasons for Reduction	Legal Status	Actions Needed (in order of priority)	Main Prey
Italy	III/IV	100	5	*lupus*	habitat destruction, prey extermination	full protection	law enforcement, prey reintroduction, education	livestock
Spain	III/IV	200	10	*signatus (lupus)*	persecution, habitat destruction	game status	survey, research, education	livestock, red deer, roe deer, chamois, wild boar
Portugal	IV	100	20	*signatus (lupus)*	persecution, habitat destruction	no protection	protection, survey, research	livestock, red deer, roe deer, chamois, wild boar
Central Europe	V	0	0	–	persecution, habitat destruction	no protection	protection, law enforcement, education	livestock, red deer, roe deer, chamois, wild boar

*Population Status:

I = fully viable
II = phase of steep decline
III = lingering, low density population, highly threatened
IV = lone wolves or pairs only, highly endangered
V = extinct

The Apparent Extirpation and Reappearance of Wolves on the Kenai Peninsula, Alaska

Rolf O. Peterson and James D. Woolington

INTRODUCTION

A most significant reappearance of wolves (*Canis lupus*) in formerly occupied habitat occurred on Alaska's Kenai Peninsula during the 1960s. During the 1970s, the wolf population expanded rapidly and became established throughout the Peninsula. In 1976, a wolf research program was initiated as part of a comprehensive predator-prey study sponsored jointly by the U.S. Fish and Wildlife Service and the Alaska Department of Fish and Game. Field research included a basic ecological study of wolves on the northern Kenai Peninsula, concentrating on relationships between wolves and moose (*Alces alces*) on the Kenai National Moose Range.

Soon after the research program was initiated, it became clear that there was no consensus in the recent literature or among biologists familiar with the Kenai as to the reasons for the disappearance of the original wolf population early in this century. Likewise, there was some uncertainty regarding the origin of the newly established population: Did it arise from a remnant population, or did wolves recolonize the Peninsula from south-central Alaska? These questions prompted a historical review of Kenai wolves and a search for causes for the initial disappearance, long period of absence, and sudden reappearance of wolves on the Peninsula.

THE KENAI PENINSULA AS WOLF HABITAT

Located between Prince William Sound and Cook Inlet in south-central Alaska, the Kenai Peninsula lies just south of Anchorage (Figure 3.1). Though 26,000 km² in area, it is connected to mainland Alaska by a narrow neck of land and ice only 16 km wide. Two major landforms characterize the Peninsula: The rugged Kenai

334

mountains, rising to 1,500 m (with two major icefields), dominate the eastern half, while the Kenai lowlands on the western half consist of a rolling plateau ranging from sea level to about 500 m near the southern end.

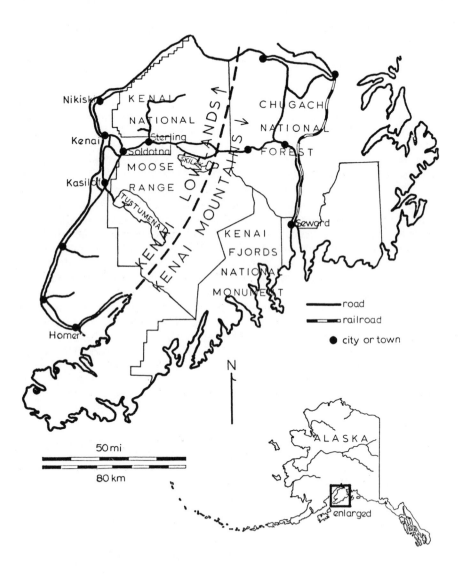

Figure 3.1: Map showing location of Kenai Peninsula in Alaska, with local details indicated in enlarged area.

Forest vegetation includes white and black spruce (*Picea abies* and *P. mariana*), white birch (*Betula papyrifera*), aspen (*Populus tremuloides*) in stream bottoms, and Sitka spruce (*Picea sitchensis*) in coastal areas near ocean influence. The altitudinal limit of trees in the mountains is approximately 500 m. Much of the Peninsula has been burned in the last 100 years, with recent large fires on the Moose Range in 1947 and 1969. Fire has been primarily responsible for the creation of large areas of good moose habitat (Spencer and Hakala, 1964; LeResche et al., 1974; Oldemeyer et al., 1977).

Wildlife has been a characteristic feature of the Kenai Peninsula throughout recorded history. Currently, the most prominent species are moose, wolves, brown and black bear (*Ursus arctos* and *U. americanus*), Dall's sheep (*Ovis dalli*), mountain goat (*Oreamnos americanus*), and a small population of reintroduced caribou (*Rangifer tarandus*). The "giant Kenai moose" and highly accessible Dall's sheep were highly prized by early sportsmen, and concern for the future of these game animals led to the establishment of the Kenai National Moose Range in 1941.

In 1978, the human population of the Kenai Peninsula was 25,281 (1978 census, Kenai Peninsula Borough). Excluding icefields and glaciers (20% of the Peninsula), the Kenai contains about 1.2 people/km^2, which is comparable to human density in wolf inhabited northern Ontario and northeastern Minnesota (Wiese et al., 1975). The human population is concentrated in major towns and distributed along the road system, leaving large blocks of land with no human residents. Until major oil and gas discoveries were made on the Kenai lowlands and adjacent Cook Inlet in the late 1950s, most communities were small and fishing-oriented. Oil development prompted rapid human population growth in the early 1960s and sustained it in the 1970s. The 1978 Borough census report provided the following statistics. The human population on the Kenai Peninsula has increased threefold since 1960. At current growth rates, the population will double in another 15 to 20 years and oil-related development could sharply increase the rate of human population growth. Over half (13,660) the current population lives in the central peninsula lowlands, including the settlements of Sterling, Soldotna, Kenai, Kasilof and Nikiski (Figure 3.1). Other concentrations include the cities of Homer and Seward, with small settlements along two major state highways and the Cook Inlet coast. About 56 percent of the population lives in unincorporated areas and can be considered rural. Only a few hundred head of cattle are present in settled areas, primarily along the coastline of the southern Kenai lowlands.

Land management patterns will obviously affect the future status of Kenai wolves. Three federal agencies are responsible for land management on the majority of the Kenai and contain 75 percent of the available wolf habitat: Chugach National Forest (U.S. Forest Service) occupies roughly the northern half of the Kenai mountains, the Kenai National Moose Range (U.S. Fish and

Wildlife Service) covers about half of the Kenai lowlands plus adjacent mountains, and the newly established Kenai Fjords National Monument (U.S. National Park Service), not important as wolf habitat, lies along the south-central coast of the Peninsula. State, borough and native land settlements are currently being decided under the Alaska Statehood Act and the Alaska Native Claims Settlement Act of 1972; these transactions will affect some federal holdings as well as the type of future development that is likely to occur on nonfederal lands.

THE ORIGINAL KENAI WOLF POPULATION

In the 1890s, wolves were considered "numerous" (Lutz, 1960) on the Kenai Peninsula and "rather common" in the Cook Inlet region (Osgood, 1901). After a hunting trip on the Kenai in 1898, Studley (1912:279) wrote: "I bought a nice lot of prime fur just before leaving Kenai, which consisted of three grey wolves' skins, 18 magnificent beaver pelts, six marten skins and two beautiful silver fox skins."

Wolves were undoubtedly a prominent furbearer on the Kenai prior to the arrival of large numbers of white men. Throughout the 19th century, the Peninsula was regarded as a rich source of furs for the local Indians. The Russians established three forts in the Kenai before 1800 for the purpose of trading for furs with the Indians (Victoria, Sister, 1974). Two of these forts were located at the present sites of Kasilof (Fort St. George) and Kenai (Fort St. Nicholas). Principal furbearers sought on the Kenai, in addition to wolves, were beaver (*Castor canadensis*), fox (*Vulpes vulpes*), sea otter (*Enhydra lutris*), land otter (*Lutra canadensis*), bear, wolverine (*Gulo luscus*), marten (*Martes americana*) and mink (*Mustela vison*) (Osgood, 1901; Moffitt, 1906).

Osgood (1901) regarded the Kenai wolf as *C. occidentalis* (Richardson), evidently because that was the closest formal designation in use at that time. Goldman (1944:423–424) undertook a complete revision of wolf taxonomy in North America and assigned the original Kenai wolves to a new subspecies, *C. l. alces*. He based his classification on skulls from five wolves (only two were adults) killed on the southern Kenai lowlands around 1904:

> This peninsular race reaches the maximum size attained by the species in North America. The skulls of two adult females are longer than those of any others examined, and present other peculiarities pointed out. Skulls of three immature males are not widely different from those of *pambasileus* (interior Alaska wolves) of comparable age, but differ uniformly in the greater width of the supra-occipital shield. The new subspecies may range throughout the Kenai Peninsula, which at its base is narrowly connected with the mainland

of Alaska. Specimens from north of Turnagain Arm
of Cook Inlet are assignable to *pambasileus*. The principal natural prey of the Kenai wolf is doubtless the giant
moose of the region. The large size of the wolf may be
the result of adaptation enabling it to cope with so large
an animal.

While the two adult skulls examined by Goldman were obviously
larger than those of adjacent subspecies, the subspecific assignment
may logically be questioned on the basis of the small sample size.
However, it is possible that Kenai wolves were sufficiently isolated
and subjected to different selection pressures to the extent that they
developed unique characteristics (Pedersen, this volume).

ELIMINATION OF THE KENAI WOLF

There are several consistent misconceptions concerning the
causes of historical fluctuations in certain wildlife species on the
Kenai Peninsula (Davis and Franzmann, 1979): (1) Moose are a
recent arrival, moving in after forest fires in the late 1800s. Lutz
(1960) refuted this idea with considerable evidence that moose
were present throughout the previous century. (2) Caribou disappeared early in the century because fire destroyed their habitat.
Davis and Franzmann (1979) reviewed a number of studies which
indicate that fires were probably not responsible for widespread
declines of caribou in North America, and indicated overhunting as
the likely cause of caribou disappearance on the Kenai Peninsula.
(3) Wolves disappeared with the caribou, implying a cause and effect relationship between decline of prey and predator alike. Alternative prey for wolves at this time included both moose and Dall's
sheep, however, and there is no reason to think that early Kenai
wolves depended heavily on caribou. The assertion that Kenai wolves
were eliminated by human overexploitation will be developed in
this section.

Human influence caused dramatic changes in abundance of
many wildlife species on the Kenai early in this century, and a
historical context is helpful in understanding these fluctuations.
Only a few white traders, trappers and prospectors lived on the
Kenai in the early 1890s, although there was a substantial native
population which dated back to prehistoric times (Reger, 1974;
Workman, 1974). Active staking of gold claims began in the Kenai
mountains in 1893, and the gold that was discovered in the next
two years brought a "rush" of thousands of prospectors. Towns
of natives were by this time well established on the western side
of the Kenai Peninsula and human activity now pervaded the Kenai
mountains.

Moose, caribou and Dall's sheep on the Kenai were slaughtered
around 1900 due to a high market value for meat and trophy heads

(Davis and Franzmann, 1979; Barry, 1973; Studley, 1912, Anonymous, 1908:1988,3407). Caribou disappeared completely in little more than a decade. Palmer (1938) reported a sighting of a bull caribou by surveyors south of Skilak Lake in 1910.

It should be noted that between 1900 and 1902, there were no game laws in Alaska and, when laws were passed to protect wildlife, enforcement was certainly inadequate. Prior to 1900, the laws of the state of Oregon, including the game laws, had been extended by Congress to the territory of Alaska (Anonymous, 1902:3842). The Alaska Code of Laws was developed in 1900 but no game law was passed until 1902. Continued concern over game animals on the Kenai Peninsula and traffic in deer hides from southeast Alaska prompted a more comprehensive game law in 1908 which provided belated protection for Kenai caribou and provided for the hiring of game wardens (Anonymous, 1908).

At the turn of the century, wolves were universally regarded as undesirable (Langille, 1904; Shiras, 1935) and little time was lost in attempting to eradicate them. Prospectors over-wintering on the Kenai often turned to trapping furbearers (Barry, 1973), modern firearms and ammunition became readily available (Davis and Franzmann, 1979) and whites introduced the practice of poisoning furbearers, an important technique used in wolf control at that time in the western United States (Young and Goldman, 1944). While traps and firearms were undoubtedly used to kill many wolves, we believe the extensive use of poison was instrumental in the complete elimination of Kenai wolves.

Trapper George Coon of Hope (located on the northern coastline of the Peninsula) took 14 wolves with poison in 1899 (Osgood, 1901) and is said to have poisoned the last nine wolves in the Hope area around 1900 (Carl Clark, pers. comm.). In 1904, English trophy hunter Radyclyffe (1904:216) wrote:

> It is surprising how few noises one hears in the woods on the Kenai Peninsula. Poison and traps have wrought havoc with the wolves and such noisy denizens of the forest at night.

Two years later, Moffitt (1906:51) expressed a similar opinion: "The number of wolves has been greatly reduced through the use of poison". Miners may have also set out poison baits in an effort to eliminate the possibility of rabies (Rearden, 1972). This was a genuine fear during the gold rushes of the 1890s; a rabies outbreak in Dawson, Yukon, prompted painful vaccination of all local residents around 1900 (Inglis, 1978). Poison was the only technique used to kill wolves that was commonly mentioned in the literature; Bennett (1916) and Culver (1923) provide additional references to the use of poison to eliminate wolves from the Kenai Peninsula.

The effectiveness of early efforts to eliminate wolves on the Kenai can be readily seen in the writings of hunters, naturalists and other scientists who visited the Kenai. Wolves were among the furbearers Mendenhall (1900:338) found to be in "limited and constantly decreasing numbers" on the Kenai in 1898. A dramatic reduction in wolf abundance was noted by 1905 (Radyclyffe, 1904; Moffitt, 1906). In 1911, the "practical extermination of the wolf" was hailed as an important step toward increasing game stocks on the Kenai (Shiras, 1935:417). Wolves were said to be "scarce" in 1914 (Bennett and Rice, 1914:135), "practically extinct" in 1915 (Martin et al., 1915:29) and were believed "exterminated by poison some years since" in 1916 (Bennett, 1916:150), a point confirmed by Culver (1923). From these accounts it is apparent that the Kenai wolf population was greatly reduced by 1905 and very likely eliminated by 1915.

Ironically, Kenai wolves taken for a museum collection in 1904 were the last individuals to be specifically mentioned in the literature we reviewed. The virtual disappearance of wolves on the Kenai went unheralded; indeed, when Goldman (1944) declared Kenai wolves a separate subspecies, he apparently did not realize that wolf presence on the Kenai had not been documented for over 30 years.

Wolf absence from 1915 to about 1960 was probably maintained by extensive trapping (prompted by government bounties) and, beginning in 1948, by formal predator control by the federal government. The first territorial legislation in Alaska in 1915 established a $10 bounty on wolves and, in 1929, soon after the first coyotes (*C. latrans*) appeared on the Kenai, a $5 bounty on this species was added (McKnight, 1970). In 1935 the bounties on both wolves and coyotes were raised to $20, and by 1949 the wolf bounty was $50 and coyotes brought $30. Federal predator control was conducted with the cooperation and financial support of the territorial government, and relied heavily on aerial shooting plus strychnine stations and cyanide guns ("getters") (Anonymous, 1950). Predator control, while not centered on the Kenai, probably eliminated the likelihood of natural reestablishment of wolves on the Peninsula. Wolf control virtually surrounded the mainland at the base of the Kenai and was most intensive in the Matanuska Valley, Nelchina Basin and Copper River valley north of Cordova, including some mountains west of Cordova, according to M.W. Kelly (pers. comm.), Head of the Division of Predator Control of the U.S. Fish and Wildlife Service from 1948 to 1960. It is clear that government wolf control efforts substantially reduced the wolf population in many areas of south-central Alaska (Rausch, 1969) and we believe this prevented effective recolonization of the Kenai Peninsula by wolves. Government predator control on the Kenai itself seems to have been limited to killing coyotes in Dall's sheep habitat (Anonymous, 1951).

During the period 1915 to 1960, very few wolves were reported on the Kenai Peninsula, and apparently only one was killed. Hjalmar Anderson shot a "small wolf" on Skilak Lake in 1928 (coyotes had just colonized the Kenai according to C. Clark, pers. comm.) and, in the winter of 1937-38, several residents of Cooper Landing reported seeing a few wolf tracks on the north side of Skilak Lake (Palmer, 1938). Tracks of a large male wolf that was missing a toe were seen regularly in the vicinity of Tustumena Lake around 1950, and this wolf was finally shot in February 1951 by local resident, Jes Willard (pers. comm.). There was no evidence of a breeding pack on the Peninsula for at least 45 years (1915-60) in spite of accelerated human activity during World War II and subsequent development. The Kenai National Moose Range was under active management by the late 1940s, and frequent aerial moose surveys failed to reveal any evidence of a remnant wolf population (David L. Spencer, pers. comm.). This was also the case during the 1950s, a period of intensive oil exploration. We conclude, therefore, that the original Kenai wolf population was completely eliminated by approximately 1915, and that during the period 1915-60 only a few transient wolves immigrated to the Peninsula.

Since wolf populations elsewhere have shown great tenacity in the face of intensive human persecution (Boitani, this volume), it seems improbable that the original Kenai wolf population could have been completely eliminated. Yet we hypothesize that this was indeed the case, basing our argument largely upon the following:

(1) Exploration for gold around 1900 brought to the Kenai an unprecedented degree of human presence, not duplicated before or since. Virtually no game protection laws were enforced and there is ample evidence of the use of poison to reduce the Kenai wolf population. This unique combination of circumstances provided a potential for elimination of wolves that may not have been equalled elsewhere.

(2) The avenue for passage from mainland Alaska to the Kenai is very narrow and historically, has been an area of high human activity, so the Peninsula can be practically regarded as an island, subjected to greatly reduced rates of emigration from adjacent areas. Wolves reappeared on the Kenai shortly after cessation of government wolf control in the 1950s, suggesting that reduced wolf populations in south-central Alaska further reduced the potential for wolf immigration to the Kenai.

(3) Only one record of a wolf on the Kenai between 1915 and the 1960s has been confirmed, and only a handful of unconfirmed reports exist from this period. Human activity on the Kenai declined between 1915 and 1940 (between the gold rush and World War II), and should have provided an opportunity for a remnant wolf population to expand, if any existed. Extensive aerial surveys and human activity associated with development of the Kenai National Moose Range and oil resources on the Kenai lowlands failed to uncover evidence of wolves during the 1950s. A few isolated immigrant wolves could explain all reports of wolves between 1915 and 1960.

It should be emphasized that our evidence is largely negative. Thus, our hypothesis may eventually be disproven if additional records of wolves appear. However, all records we know of have been reviewed and our hypothesis is consistent with that evidence.

KENAI WOLF RE-ESTABLISHMENT

Wolf control by the federal government in south-central Alaska, which reached a peak in the early 1950s, subsided by the time of statehood in 1958 and ended in 1960. The newly-created State Department of Fish and Game sought to upgrade the image of the wolf by discouraging wolf control programs and wolf bounties and encouraging classification of wolves as furbearers and big game animals (Rausch and Hinman, 1977). As government control was reduced, wolf populations throughout the south-central Alaskan mainland increased and wolves eventually appeared on the outskirts of Anchorage, near the base of the Kenai Peninsula (Rausch, 1969). Wolf re-establishment on the Kenai followed closely after the recovery of wolf populations on the mainland, further suggesting that effective colonization of the Peninsula by wolves had been prevented by low mainland populations.

Wolf hunting and trapping on the Kenai were prohibited by the State in 1962 following a confirmed sighting of a wolf on the Peninsula. Wolf hunting and trapping remained closed until 1974 when it was apparent that the wolf population was well established.

After initial sightings of single wolves, a pair of wolves was first observed in 1967, and a large pack was sighted the following year. While most early sightings were on the Kenai lowlands between Skilak and Tustumena Lakes, by the late 1960s large packs had been observed in both the northern and southern parts of the lowlands. Records kept by the Alaska Department of Fish and Game and the U.S. Fish and Wildlife Service indicate a dozen reliable wolf observations on the Kenai between 1960 and 1969, and over 100 sightings between 1970 and 1975 (Figure 3.2). Both agencies desired to see wolves re-established on the Kenai and there was apparently little or no public opposition. The wolf population expanded rapidly in the early 1970s and, by 1975, wolves had probably recolonized most suitable habitat on the Peninsula.

We estimate that the Kenai Peninsula currently provides 13,700 km^2 of suitable wolf habitat, i.e. land supporting wild prey populations with relatively few human inhabitants. Since 1976, wolf research has been conducted on the northern Kenai, primarily on the lowlands where wolf density is higher. Based on a consideration of available habitat and wolf density in the study area (Peterson, unpubl.), we estimated the wolf population on the Kenai Peninsula in 1979 at about 185 animals in early winter, before any human harvest. The Kenai National Moose Range probably contains half of the Peninsula's wolf population.

There is considerable current interest in reintroducing wolves to formerly occupied habitat in North America (Klinghammer, 1979). The natural re-establishment of wolves on the Kenai Peninsula indicates several key factors that should be considered in any attempt at wolf reintroduction: (1) large blocks of land uninhabited by people, mostly roadless wilderness; (2) low human population concentrated away from wolf habitat; (3) ample prey populations capable of providing year-around food supply; (4) no formal predator control programs and little livestock; and (5) adequate legal protection and favorable public opinion. All of the above conditions were met on the Kenai Peninsula and, after recovery of wolf populations on the adjacent mainland, natural wolf re-establishment on the Kenai was finally possible.

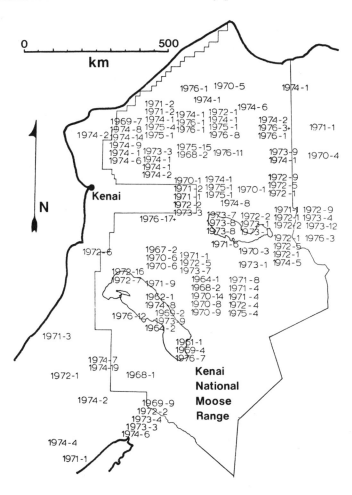

Figure 3.2: Observations of wolves on the Kenai Peninsula from 1961 until June 1976. Approximate location of sighting is indicated by placement on map. Year of sighting is followed by the number of wolves seen. Reliable counts of fresh wolf tracks are included. Data from files of Alaska Department of Fish and Game and Kenai National Moose Range (U.S. Fish and Wildlife Service).

In 1904, forester W.A. Langille considered wildlife to be a primary value of Kenai lands and believed that, if properly cared for, wildlife would be a "source of revenue to the inhabitants and pleasure to the world for many years to come." Seventy-five years later, enriched by the addition of a major carnivore such as the wolf, this vision is still alive.

Geographical Variation in Alaskan Wolves

Sverre Pedersen

INTRODUCTION

The taxonomic arrangement of North American wolves (*Canis lupus* L.) currently in use is largely due to the efforts of E.A. Goldman (1944). His classification of North American wolves recognized 23 subspecies, based largely on cranial characteristics, of which four occurred in Alaska (Figure 4.1): *Canis lupus ligoni* Goldman — Alexander Archipelago wolf (Goldman, 1937), *Canis lupus alces* Goldman — Kenai Peninsula wolf (Goldman, 1941), *Canis lupus pambasileus* Elliot — Interior Alaskan wolf (Goldman, 1944), and *Canis lupus tundrarum* Miller — Alaskan Tundra wolf (Goldman, 1944).

A number of authors (Rausch, 1953; Jolicoeur, 1959; Manning and Macpherson, 1958; Rausch, 1967; Kelsall, 1968; Standfield, 1970; Mech, 1970; Kolenosky and Standfield, 1975) have questioned the validity of some of the northern subspecies described by Goldman (1944). For instance, R.L. Rausch (1953) doubted that the subspecies *C. l. tundrarum* and *C. l. pambasileus* could be separated by the 15 cranial measurements listed by Goldman (1944), and further considered the validity of *C. l. alces* questionable. Stephenson (pers. comm.) has reiterated R.L. Rausch's (1953:110) conclusion that "restudy of the various subspecies of wolves seems necessary to determine whether the existence of so many named forms is justified." To date, the validity of the Alaskan subspecies remains unresolved.

Clearly, taxonomic questions must be resolved in order to: (1) settle doubts regarding the validity of the four Alaskan subspecies, and (2) provide information important to wolf management in Alaska. This study, based on a section from my Master's thesis (Pedersen, 1978) using multi-variate statistical tools (Jolicoeur, 1959; Lawrence and Bossert, 1967; Nowak, 1973; Sneath and Sokal, 1973; Gipson et al., 1974; Jolicoeur, 1975; Kolenosky and Standfield, 1975; Elder and Hayden, 1977), will attempt to clarify this situation.

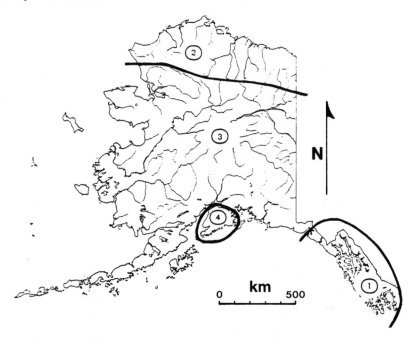

Figure 4.1: Subspecies distribution, transect target group specimen locations and sequence of analysis:

Subspecies (and Distribution After Goldman, 1944)	Transect Target Group Specimen Locations (Small Circles)
Canis lupus ligoni	(1) Southeast Alaska
Canis lupus tundrarum	(2) Arctic Alaska
Canis lupus pambasileus	(3) Interior Alaska
Canis lupus alces	(4) Kenai Peninsula

MATERIALS AND METHODS

Specimens

Wolf craniae for this study came from the University of Alaska Museum, Fairbanks; the Alaska Department of Fish and Game (ADFG), Fairbanks; the Kenai National Moose Range, Kenai; and the U.S. National Museum (USNM), Washington, D.C. Published craniologic mensural data on Alaskan wolves were also utilized (Goldman, 1944).

Methodology

Only adult specimens (using the same age criteria as Nowak, 1973) were used in the analysis. Full reliance was placed on the Museums', ADFG's and published records indicating the sex, age and capture locality of specimens used. Measurements used in this analysis were drawn from a large number of references (Pedersen, 1978), and only cranial characteristics known to be diagnostic in detecting morphological variation among canids were considered.

Multi-variate statistical tools, particularly discriminant analysis, were employed. Both sexes gave essentially similar results in all important analyses so, for brevity, I will only present the male results. The 15 variables selected by Goldman (Table 4.1) were utilized to analyze his published data (Goldman, 1944). Twelve selected variables, reduced from an original 22 (Table 4.2 and Figure 4.2), were used to analyze the recent data. (For a detailed discussion of this matter, see Pedersen, 1978.) Measurements were taken with calipers on the left side of the specimen, whenever possible; those under 155 mm recorded to the nearest 0.10 mm, and those over 155 mm to the nearest mm.

Table 4.1: Description of Measurements Taken by Goldman

(1) *Greatest length* — Length from anterior tip of premaxillae to posterior point of inion in median line over foramen magnum.

(2) *Condylobasal length* — Length from anterior tip of premaxillae to posterior plane of occipital condyles.

(3) *Zygomatic breadth* — Greatest distance across zygomata.

(4) *Squamosal constriction* — Distance across squamosals at constriction behind zygomata.

(5) *Width of rostrum* — Width of rostrum at constriction behind canines.

(6) *Interorbital breadth* — Least distance between orbits.

(7) *Postorbital constriction* — Least width of frontals at constriction behind postorbital processes.

(8) *Length of mandible* — Distance from anterior end of mandible to plane of posterior ends of angles, the right and left sides measured together.

(9) *Height of coronoid process* — Vertical height from lower border of angle.

(10) *Maxillary tooth row, crown length* — Greatest distance from curved front of canine to back of cingulum of posterior upper molar.

(11), (12) *Upper carnassial, crown length (outer side) and crown width* — Antero-posterior diameter of crown on outer side, and transverse diameter at widest point anteriorly.

(13), (14) *First upper molar, antero-posterior diameter and transverse diameter* — Greatest antero-posterior diameter of crown on outer side, and greatest transverse diameter.

(15) *Lower carnassial (crown length)* — Antero-posterior diameter at cingulum.

Table 4.2: Description of Measurements Taken by Pedersen

(1)   *Greatest length*

(2)* *Condylobasal length*

(3)* *Palatal length* — Distance from the alveolus of the median upper incisor on one side to the notch of the posterior edge of the palatal shelf on the same side.

(4)   *Post-palatal length* — Distance from the notch of the posterior edge of the palatal shelf on one side to posterior face of the ventral lip of the foramen magnum on the median line.

(5)* *Facial length* — (Nasion to prosthion) from junction, on the median plane of the right and left nasofrontal sutures to the anterior end of the intermaxillary suture.

(6)* *Maxillary tooth row, alveolar length (P^1-M^2)* — Distance from anterior edge of alveolus of P^1 to posterior edge of alveolus of M^2.

(7)   *Maxillary tooth row, crown length (C-M^2)*

(8)* *P^4 length, crown*

(9)   *Mandible length*

(10) *P_4 length, crown*

(11) *Width of nasals* — Distance between the extreme points of anterior lateral processes.

(12)* *Width of piriform aperture* — Maximum anterior lateral width of the piriform aperture (incisive foramen).

(13)* *Width of rostrum*

(14) *Width between retroglenoid foraminae* — Least distance between the retroglenoid foraminae.

(15) *Width across outer edges of M^1 or P^4 (greatest crown)* — Greatest breadth between outer sides of most widely separated upper teeth (P^4 or M^1).

(16) *Width of P^4; greatest transverse, crown*

(17)* *Width of M^2; greatest transverse, crown* — Maximum transverse diameter from outermost point to innermost point of crown.

(18) *Zygomatic breadth*

(19)* *Interorbital breadth*

(20)* *Squamosal constriction*

(21)* *Postorbital constriction*

(22)* *Rostrum height (between P^1 and P^2)* — From lower edge of skull in region between P^1 and P^2 to upper edge of skull directly above P^1 and P^2.

*The diagnostic measurements used in this analysis.

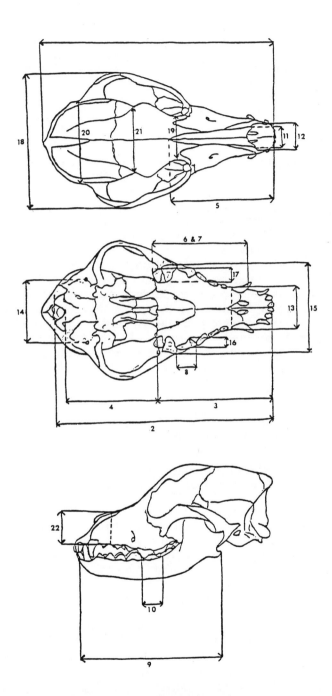

Figure 4.2: Illustration of measurements, described in Table 4.2

Statistical Analyses

The following computer programs, available at the University of Alaska Computer Center, were employed: (1) BMD 01D — Simple Data Description (Dixon, 1973), (2) BMD 05M — Discriminant Analysis for Several Groups (Dixon, 1973), (3) BMD 07M — Stepwise Discriminant Analysis (Dixon, 1973), and (4) Discriminant Analysis (SPSS), based on the D^2 statistic (Nie et al., 1975).

Systematics

The systematic relationships of wolves in Alaska were investigated with two methods. In Method 1, the assumption was made that the four subspecies described by Goldman (1944) were valid. To assess this assumption, both Goldman's original published mensural data and the recent data (arranged into four groups according to Goldman's 1944 range descriptions) were subjected to a series of discriminant analyses.

The second method was based on the assumption that a gradual clinal variation could be expected (Jolicoeur, 1959) since no major geographical barriers presently isolate portions of wolf range in Alaska. To assess this assumption, specimens from two widely separated areas approximately 2,000 km apart with widely different habitats were selected: Population 1, the Alexander Archipelago wolf population in southeastern Alaska, and population 2, the Alaska tundra wolf population north of the Brooks Range (Figure 4.1). The two groups were compared, using stepwise discriminant analysis (SPSS and BMD), to determine the following:

(1) the number of variables necessary to correctly identify the specimens above the 75 percent level, and

(2) the assigned probability of specimens belonging to one or neither of the tested groups.

Next, discriminant analyses were performed to determine the morphological affinities of specimens from areas intermediate to the two target populations. This was accomplished by weighing them as unknowns in the discriminant analysis, and requesting the program to assign them a probability of belonging to one or neither of the two groups. As a result, two additional probable areas of morphologically distinct populations were defined (3 and 4, Figure 4.1).

Finally, a number of discriminant analyses were conducted to establish the affinities of specimens from the four designated populations to determine the existence of isophenes, marked morphological and/or geographical boundaries or aberrant specimens in the total sample. These data are referred to as the transect data in later sections.

RESULTS

Goldman's Data

Multiple discriminant analysis of Goldman's (1944) Alaskan data suggests that his four subspecific groups were morphologically distinct. The four group male classification matrix, based on 15 variables, correctly assigned 100 percent of the reported sample (Figure 4.3). The maximum generalized distance (D), based on 15 variables, was 480520.6, an exaggerated value, suggesting mainly that the sample size was inadequate.

Recent Data

The recent Alaskan sample, when grouped according to Goldman's (1944) subspecific breakdown and analyzed by multiple discriminant analyses, suggests the presence of two, and possibly three, distinct subspecific groups rather than four, as proposed by Goldman. Seventy-three percent of the sample was correctly assigned (Figure 4.4), and the maximum generalized distance (D) was 13.86 for the four subspecific groups. Misclassification of specimens in analysis of population pairs occurred.

Transect Data

The "transect" data analyzed by multi-variate discriminant analysis, utilizing 20 or more specimens from the central portion of each region, where possible, suggest a clinal size variation over much of Alaska with a decreasing size cline from the Interior to the north, south and east. The maximum generalized distance (D) was 21.54 and indicated some differences among groups.

Interregional variation was attributed to: (1) a cline of decreasing size from the Interior to the Arctic; (2) possible geographic isolation on the Kenai Peninsula; and (3) strong indication of marked morphological differences between wolves from southeastern and Interior Alaska. The size cline from southeastern Alaska to the Interior appeared to be sharply delineated but, due to lack of specimens from the area between the Alexander Archipelago and Interior Alaska, the possibility that the cline may actually be gradual cannot be discounted.

In the final four interregional analyses, the classification matrix correctly assigned 100 percent of the specimens (Figure 4.5).

DISCUSSION

The multi-variate discriminant analyses of specimens examined by Goldman, and recently collected specimens, resulted in a different arrangement from that proposed by Goldman (1944). Only two local areas were identified in Alaska (Southeast and Interior) where wolves differed markedly from each other on the multi-variate level (Figure 4.4).

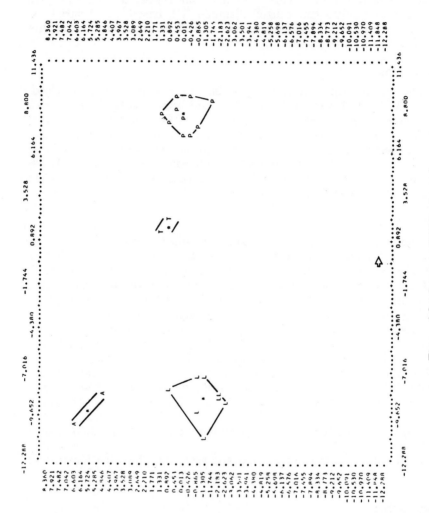

Figure 4.3: See page 355 for legend.

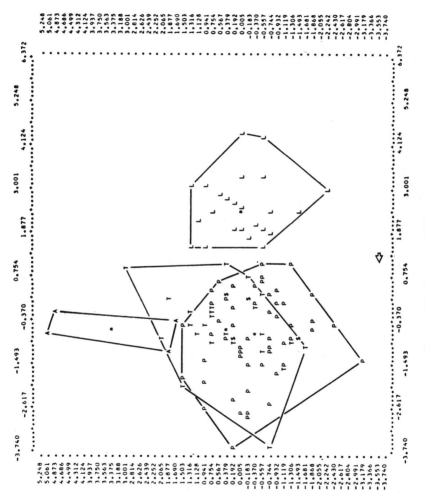

Figure 4.4: See page 355 for legend.

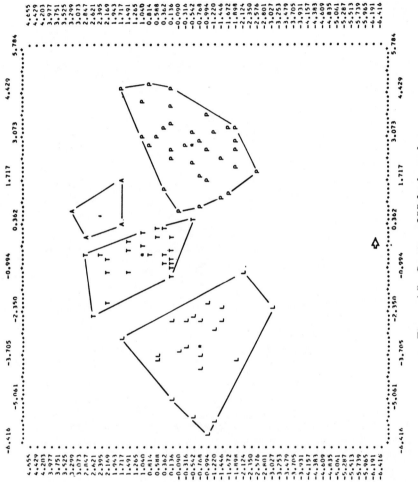

Figure 4.5: See page 355 for legend.

Figure 4.3: Multivariate comparison of Goldman's data for the four groups of males. Classification matrix and plot of the two first canonical variables. The horizontal axis (canonical variable 1) represents the general size trend (increasing from left to right, indicated by direction of arrow in the figure).

Classification Matrix

	Group	L	A	P	T	Sample Size
L	(C. l. ligoni)	9	0	0	0	9
A	(C. l. alces)	0	2	0	0	2
P	(C. l. pambasileus)	0	0	11	0	11
T	(C. l. tundrarum)	0	0	0	2	2

Figure 4.4: Multivariate comparison of the recent data for the four groups of males, using Goldman's (1944) subspecific grouping. Classification matrix and plot of the two first canonical variables. The horizontal axis (canonical variable 1) represents the general size trend (increasing from right to left, indicated by direction of arrow in the figure).

Classification Matrix

	Group	L	A	P	T	Sample Size
L	(C. l. ligoni)	24	0	1	0	25
A	(C. l. alces)	0	3	0	1	4
P	(C. l. pambasileus)	2	1	43	14	60
T	(C. l. tundrarum)	1	1	10	15	27

Figure 4.5: Multivariate comparison of the transect data for the four groups of males. Classification matrix and plot of the two first canonical variables. The horizontal axis (canonical variable 1) represents the general size trend (increasing from left to right, indicated by direction of arrow in the figure).

Classification Matrix

	Group	L	A	P	T	Sample Size
L	(C. l. ligoni)	24	0	0	0	24
A	(C. l. alces)	0	4	0	0	4
P	(C. l. pambasileus)	0	0	34	0	34
T	(C. l. tundrarum)	0	0	0	24	24

The question of whether or not the groups warrant subspecific status was explored, and findings suggested that: (1) even though Goldman's data suggested the four subspecific groups, his small sample did not support the four subspecies statistically; (2) when the recent specimens were analyzed according to the geographic limits of Goldman's four subspecies, much misclassification between subspecies resulted; and (3) the transect analysis suggests clinal size variation in the 12 skull characters over much of Alaska (Figure 4.5), with the exception of southeastern Alaska (and possibly the Kenai Peninsula). In the analysis, only one sharp phenotypic boundary was clearly evident, that between southeastern Alaska and Interior Alaska. There were distinct morphological differences between the two wolf populations with little misclassification or overlap of characters. The possibility exists that a more gradual phenotypic boundary actually exists and that the phenotypic gradient between Interior and southeastern wolves is actually more clinal than indicated in this analysis, since few specimens were available from the region separating Interior Alaska from the Alexander Archipelago.

The fact that wolves from the Kenai Peninsula were infrequently misclassified as members of the other three populations could result from their being geographically isolated from other wolf populations. It is also possible that the limited number of specimens examined (four) gives a misleading view of the Kenai population, and that additional specimens could bring the subspecific characterization closer to that of Interior Alaska wolves.

The misclassification and intermediacy of specimens between the Interior and Arctic target groups, and consequent lack of a narrow hybrid belt suggests relatively unrestricted gene flow from Interior Alaska north through the Brooks Range, as suggested by Rausch (1953). Nevertheless, Interior wolves are generally larger in skull dimensions than those from the Arctic. Jolicoeur (1959) and Skeel and Carbyn (1977) reported a similar relationship between northern and central Canadian wolves.

In this analysis, results of the transect and associated search for subspecific boundaries suggest a greater phenetic similarity between southeastern and Arctic wolves than between the southeastern and Interior populations (Table 4.3).

The source of this phenetic similarity is likely to be an adaptive process, and evidence to support the adaptive nature of the variations between regions ranges from probable Pleistocene distribution of wolves to the relative size of their principal prey. The glacial and postglacial distribution of Alaskan wolves probably influenced the phenotyic variations observed in modern Alaskan wolves. After the late Wisconsin glacial period, Southeast Alaska was mainly repopulated by mammals from the present northwestern states since the glaciated coast range to the north and northeast presented a physical barrier to access from the Interior of Alaska. Klein (1965:

16) suggested that "the wolf probably followed the deer (which came from the southern refugium) into coastal southeastern Alaska."

Table 4.3: Morphological Similarities Between Alaskan Wolf Populations Based on Maximum Generalized D, A Measure of Multivariate Phenetic Distance

Base Area	Area Compared to	D (Distance)
Area 1 (Southeast)	4 (Kenai)	8.44
	2 (Arctic)	15.31
	3 (Interior)	21.74
Area 2 (Arctic)	4 (Kenai)	7.35
	1 (Southeast)	15.31
	3 (Interior)	15.54
Area 3 (Interior)	4 (Kenai)	9.46
	2 (Arctic)	15.54
	1 (Southeast)	21.74
Area 4 (Kenai)	2 (Arctic)	7.35
	1 (Southeast)	8.44
	3 (Interior)	9.46

Wolves inhabiting the present north slope of the Brooks Range likely originated from wolves inhabiting the Interior-Bering Sea refugium during the late Wisconsin glacial period (Guthrie, 1968; Nowak, 1973). Thus, following the Wisconsin Age, most of Alaska was occupied by wolves originating from either the southern refugium or the Interior Alaska-Bering Sea refugium.

As noted, a distinct difference between Interior and southeastern wolves was evident in this analysis. The observed difference may be due, in part, to the postglacial invasion of southeastern Alaska by wolves from a southern refugium, while the remainder of the State was populated by wolves from the Interior refugium. Maintenance of genetic separation beween the two groups has perhaps been aided by the Coast Mountains, which probably provide a relatively strong geographic barrier between the two groups.

Detailed analyses of the transect data show trends in phenotypic similarities among wolf populations that go beyond the postglacial distribution theory (Table 4.3). The overall general size trend basically follows Bergmann's Rule with an increase in overall skull size from the south (southeastern Alaska) to the west and north (Kenai and Interior) (Figure 4.5). However, wolves from the Arctic present an anomaly as their skulls are smaller than those from the Interior. Jolicoeur (1959) and Skeel and Carbyn (1977) also found skulls of arctic wolf populations to be smaller than boreal populations. Mayr (1974:199) pointed out that, "in many species of warm-blooded vertebrates, the northernmost populations are exposed to conditions tending to neutralize the advantage of increased body size" (which he calls "latitude effect"). McNab (1971:849) found that the largest species of canid members, such as wolves, normally have "a size that does not vary with latitude. Sometimes, however, it conforms to Bergmann's Rule or becomes smaller at high latitudes."

In Alaskan wolves, there may be other selective factors that strongly influence the size trend from the Interior to the Arctic type. Mayr suggested that, "the latitude effect is due to the shortness of the Arctic winter day, which depresses size by reducing daily food intake," and Jolicoeur (1959) hypothesized that the anomaly may be due to a "metabolic and/or hormonal imbalance" during the long Arctic winter. Possibly the short face of the Arctic wolves is a genetically-based adaptive trait, or perhaps it is a result of seasonal nutritional limitations.

Another ecological parameter, size of principal prey, must be considered in attempting to explain the overall size difference between, and variability of some dimensions among wolves in the four areas. This study provides evidence (Table 4.4) to support associations between the size of wolves and the size of their major prey, as suggested by other authors (Goldman, 1944; McNab, 1971; Kolenosky and Standfield, 1975). The physical size of wolves may be a compromise between: (1) the pack size required for wolves to locate, approach and kill the most common ungulate prey, and (2) the weight and stature required of individual pack members to kill the most common ungulate prey. Rausch (1967:262) found that, "pack-size composition is similar in Southcentral and Interior Alaska. Southeastern and Arctic Alaska also are similar, but pack sizes in the Interior and southcentral regions are larger than in the Arctic and Southeast regions." This supports findings in the present study of similarities between Arctic and southeastern Alaskan wolf populations (Table 4.4).

The smallest major prey of Alaskan wolves are deer (*Odocoileus hemionus sitkensis*) in southeastern Alaska (Rausch, 1967), where the wolves are also smallest. The next largest major prey are caribou (*Rangifer tarandus*) in the Arctic (Rausch, 1967; Stephenson and Johnson, 1972) (disregarding the Kenai Peninsula for the present), where the wolves are decidedly larger than those in southeastern Alaska but smaller than those of the Interior, where moose (*Alces alces gigas*) and caribou predominate as the main prey (Rausch, 1967; Stephenson, 1975) (Table 4.4).

Kolenosky and Standfield (1975) hypothesized a wolf and prey size relationship similar to Table 4.4 for wolves in Ontario. They found that large wolves prey on the larger ungulates, and that small wolves depend mainly upon deer and smaller mammals. These researchers related wolf and prey types to invasion paths that wolves and ungulates may have followed during the retreat of the Wisconsin glacier. They hypothesized that larger boreal wolves probably came from the west or northeast, whereas the smaller Algonquin type wolves probably invaded from the south following deer. I propose an analogous explanation for the size relationships between southeastern and Interior/southcentral/Arctic wolves.

Table 4.4: Size Relationship Between Wolves and Their Main Prey in Four Areas in Alaska

Area	Overall Main Prey	References	... Relative Size of ...	
			Main Prey	Wolf Skull*
Southeast (1)	deer	Rausch, 1967	small	small
Arctic (2)	caribou	Rausch, 1967; Stephenson and Johnson, 1972	medium	medium
Interior (3)	moose, caribou	Rausch, 1967; Stephenson, 1975	large and medium	large
Kenai Peninsula (4)	moose, caribou, sheep	LeRoux, 1975	large and medium	medium and large

*Using the overall skull size trend from Figure 4.5.

Variations between the Interior and Arctic wolf populations of Alaska have possibly been erased by man's recent activities in the Alaskan Arctic. From the late 1940s and throughout the 1960s, wolves were subjected to extensive poisoning programs and both legal and illegal aerial hunting, particularly north of the Brooks Range (Stephenson and Johnson, 1972; Stephenson and Sexton, 1974). In addition, during the same period and continuing today, moose invaded the large river valleys of the north slope of the Brooks Range, possibly attracting wolves from the central and south slopes of the Range. Since the wolf population north of the Brooks Range was drastically reduced, wolves dispersing from the south could have become established and minor differences noted in this study between wolves from the Arctic and the Interior (2 and 3) may prove to be the result of increasing interbreeding between the two populations.

Wolves from the Kenai Peninsula present an interesting problem, as Goldman (1944) classified them as a separate subspecies and considered them to be the largest North American wolf subspecies. Due to the small sample size available to this study (4 males) it was not possible to draw firm conclusions regarding their current genetic or phenotypic relationships. The Kenai group was intermediate between the Interior and Arctic populations on the multivariate size gradient (Figure 4.5). Phenotypically, wolves from the Kenai Peninsula most closely resemble the Arctic group (Table 4.3), but the sample is too small to remain confident about this relationship.

Several factors should be considered in assessing the morphology of the wolves currently inhabiting the Kenai Peninsula. These wolves probably originated from wolves in the Interior-Bering Sea refugium. Their main prey, which include caribou, moose and sheep, apparently invaded the Kenai from the same refugium. Prior to the demise of caribou on the Kenai Peninsula (early 1900s), they may have been the most important prey of wolves. With the decline of the caribou and an increase in moose populations, wolves may have responded to the new selective pressures with an increase in size and adjusted social organization.

Apparently wolves became quite scarce on the Kenai Peninsula some time between 1910 and 1920, and their numbers remained low until the early 1960s (Peterson, this volume). Today the population is estimated at roughly 160 animals (Stephenson, personal communication). Wolves may have persisted in very low numbers in remote sections of the Kenai, but a breeding population was probably absent until the early 1960s, a period of 20 to 40 years.

SUMMARY

Two phenotypically distinct subspecific groups were identified (Southeast and Interior/Southcentral).

The observed differences between subspecies *C. l. pambasileus* and *C. l. ligoni* warrant continued recognition as subspecific. This decision is supported by data available on Pleistocene and post-Pleistocene distribution and dispersal of wolves, eco-geographical influences (formulated loosely as "Bergmann's Rule," "latitude effect"), and phenotypic/social adaptations to attain a physical size appropriate to successfully pursue the major ungulate prey of each area.

The difference between Interior and Arctic Alaskan wolves, although noticeable, is manifested as a cline in which the two forms imperceptibly grade into each other. A distinct population may have inhabited the northern foothills of the Brooks Range and the Coastal Plain at one time, but it is probable that these wolves have since interbred with wolves from the southern foothills and central Brooks Range, as their population size was diminished by heavy aerial hunting in the late 1950s through the 1960s. No clear hybrid belt or distinct phenotypic break was found between wolves from these two regions that would warrant subspecific designation of the northern population. Wolves north of the crest of the Brooks Range best fit the definition of an ecological race, as per Mayr (1974:212).

Wolves currently found on the Kenai Peninsula do not fit Goldman's original description of the subspecies, but do show phenotypic characteristics that may warrant the subspecific designation *C. l. alces*. At present, an insufficient number of wolf skulls from the Kenai Peninsula have been analyzed to permit a conclusive statistical analysis.

Finally, I realize that future measures of biochemical, chromosomal or behavioral traits could lead to an understanding of geographic variation in Alaskan wolves different than that presented in this paper.

Acknowledgements

This study of Alaskan wolf systematics was conceived in cooperation with Mr. R.O. Stephenson, Game Biologist, Alaska Department of Fish and Game; Dr. R.D. Guthrie, Department of Biology, University of Alaska; and Dr. P.S. Gipson, Assistant Unit Leader, Alaska Cooperative Wildlife Research Unit, University of Alaska.

Special thanks go to my wife, Fran, a steady source of understanding and encouragement throughout this study.

Wolf Status in the Northern Rockies

Robert R. Ream and Ursula I. Mattson

HISTORICAL PERSPECTIVE (pre-1860 to 1930)

To better understand the current status of wolves (*Canis lupus*) in the northern Rockies, we must review the human attitudes and historical events leading to the present. Throughout this paper, unless indicated otherwise, we discuss only the U.S. portion of the northern Rockies.

Prior to the 1860s, there was no organized effort to eradicate wolves in the west. Lewis and Clark referred to wolves several times, calling them "shepherds of the buffalo" (De Voto, 1953). Early trappers were more interested in valuable beaver (*Castor canadensis*) hides than wolf pelts. Undoubtedly, a few wolves were shot when they raided food caches, but the wolves offered no real threat to the trapper's livelihood.

The discovery of gold in the 1860s initiated a change in land usage and native inhabitants, both human and animal. Permanent settlers arrived in great numbers with their families and livestock. Eradication of bison (*Bison bison*) began. During the 1870s, wolves thrived on the abundant bison carrion left to rot on the plains. During this period, professional "wolfers," many of whom were former buffalo hunters, were common in Montana. These men were usually employed to haul supplies to mining settlements during the snow-free months. During winter, they poisoned wolves using bison carcasses laced with strychnine (Curnow, 1969). A wolfer could average $1,000-1,500 per winter selling wolf hides (Ludlow, 1876).

Bison in Montana were exterminated by 1884 and wolves turned to killing cattle and sheep, outraging ranchers and settlers. Their "Indian problem" was in the government's hands; droughts, floods, bitter winters and natural catastrophes could not be changed; but wolves were an enemy the settlers could directly deal with. They pressured state and territorial governments to pass bounty laws. In some cases, state, county and private bounties were paid on a single wolf (Day, 1977). Every available means of destroying wolves was used: poisoning, shooting, trapping, burning pups in their dens. One scheme was to infect and release wolves with mange, hoping they would infect others (Curnow, 1969). Between 1883 and 1918, bounties were paid on 80,730 wolves (Curnow, 1969).

In 1915, the U.S. Bureau of Biological Survey began predator control on federal lands using professional trappers to exterminate the remaining wolves. A few renegade wolves continued to take livestock until the 1930s, and efforts to eliminate them continued.

Even in the national parks, wolves and other predators were sought out and destroyed. An official policy of predator control was actively carried out in Yellowstone and Glacier National Parks from 1914 to 1926 (Singer, 1975a; Weaver, 1978). Similarly, wolves were eradicated by 1921 in southwestern Alberta near Waterton Lakes National Park because of livestock losses on the eastern periphery of the Park (Cowan, 1947).

Thus, between 1870 and 1930 wolves were all but exterminated in the western U.S. Large scale eradication began when wolves were an economic threat, but later continued with seemingly irrational fervor. Fear and hatred of wolves remains. Hook and Robinson (this volume) partly explain our persistent persecution of them and their current endangered status throughout most of the U.S.

RECENT DEVELOPMENTS

Following the Depression of the 1930s, agricultural expansion resumed. More efficient methods of predator control were developed, including new poisons and aerial hunting. Although records indicate a few lone wolves or pairs may have survived the major predator control period, there is no indication that packs survived and re-established themselves. Very few reports of wolves in the western states exist from 1940 through the early 1960s. Aulerich (1964) states that any wolves left probably survived on large tracts of national forest land. One or two small packs in Yellowstone Park may have survived the major control period (1914-1926), but did not persist for long (Weaver, 1978). Regular sightings and an occasional dead wolf in the North Fork of the Flathead River Valley in Glacier Park indicate that wolves were present during this time, but only in small numbers (Singer, 1975a).

During the late 1960s, a new environmental consciousness led many people to re-evaluate their attitudes toward wildlife. The Endangered Species Act resulted from concern for species on the verge of extinction. The Northern Rocky Mountain wolf (*C. lupus irremotus*) was one of three subspecies listed as endangered in the U.S. in 1973.

PRESENT RESEARCH

It was during this time that the Wolf Ecology Project (WEP) began. Through lectures, workshops and interviews with interested people, a network for gathering wolf information was established. Each report was followed with a telephone conversation or a personal interview. When a series of reports came in from one area, or even a single promising report, a field assistant was sent to the area to look for wolf sign, try to elicit howling responses, and interview residents of the area.

Each report was classified as either a wolf sighting or wolf sign. Sign reports included tracks, scats, howling, dens, kills or any other type of wolf sign. Standard information forms were developed for both types of reports. Each completed information form will hereafter be referred to as a "wolf report" unless specified as a sighting or some type of sign.

Certain biases are inherent in observational data of an elusive animal such as the wolf. Reliability of the observer is perhaps the most critical factor in obtaining accurate information. Additional biases include differences in vegetation and topography, varying degrees of accessibility to areas, amount of time spent by the researcher in each area, distance of the observer from the animal, weather conditions and lighting, similarity in appearance of wolves, dogs and coyotes under certain conditions, and similarity in track size of large dogs and wolves.

Day (1977), working with the WEP, developed a system of ranking wolf reports to compensate for these biases. Each report was rated as *very good*, *good*, *fair*, or *questionable*. Reports of *questionable* value are filed separately and not included in any analysis or discussion of reports. Criteria used to rank reports were the observer's credibility, back-country experience, occupation, experience with wolves and coyotes, circumstances and details of the observation, and correlation of the report with other reports in the area. This subjective rating system is most consistent when the same person rates all reports. A more objective rating system has recently been developed by the WEP. It assigns points using the same criteria, with the final rating based on the total number of points.

Nearly 400 wolf reports were compiled between 1972 and June 1979 from western Montana, Idaho and northwestern Wyoming (Figure 5.1). These reports had a north-south bimodal distribution in western Montana between 1974 and 1977 (Day 1977). No wolf reports occurred within a 50 km strip (T3N-T7N) in west-central Montana, and only five reports within a 145 km strip (T3S-T12N). More reports were collected from northwestern than southwestern Montana. Wolf reports for the seven and a half year period being discussed here closely correspond with this geographical distribution pattern and will be presented as two groups of reports—northwestern and southwestern Montana.

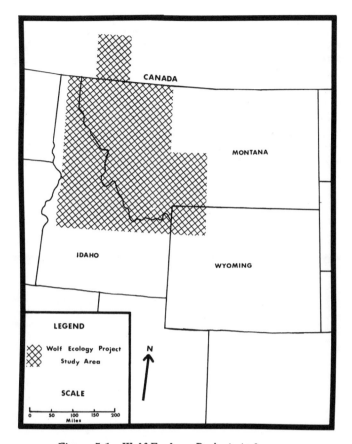

Figure 5.1: Wolf Ecology Project study area.

RESULTS AND DISCUSSION

Southwestern Montana

Wolf reports from southwestern Montana were concentrated in an area bounded on the north by Township 3S and on the east by Range 5E (Figure 5.2). The southern and western boundaries are formed by the Continental Divide. One major mountain chain, the Bitterroot Range, forms the Continental Divide along the western boundary, and numerous smaller ranges of mountains are scattered throughout the area. Public forest and range lands occupy much of the area, but large expanses of private ranchland are found along the major river valleys. Livestock grazing and mining are the major land uses.

Predator control efforts in Yellowstone Park and adjoining areas of Montana, Wyoming and Idaho reduced the wolf population to

such a low level by the 1930s that it never recovered. The last confirmed wolf killed in Yellowstone Park was in 1926 (Weaver, 1978),
and in southwestern Montana in 1941 (Flath, 1979). Sporadic sightings of wolves or wolf-like canids have continued to the present (Day,
1977; Flath, 1979; Weaver, 1978, 1979). However, there is no indication of a successful, natural re-establishment of a viable population
of wolves in the area.

Seventy-four wolf reports were collected between 1972-79 in
southwestern Montana (Figure 5.2). Additional reports of wolves
were collected by Weaver (1978, 1979) in and adjacent to Yellowstone Park during this time. Three main areas of wolf activity were
identified by Day (1977): 1) the Sheep Creek area along the Continental Divide, 2) the Gravelly Range west of the Madison River, and
3) the Big Hole-Pioneer area east of the Continental Divide.

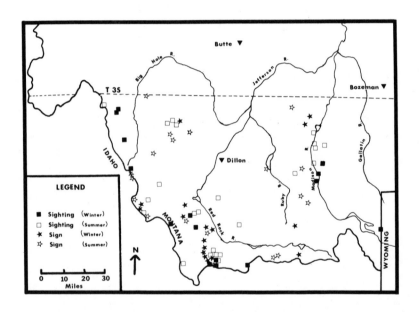

Figure 5.2: Wolf reports from southwestern Montana, 1972-1979.

During 1974-76, the number of wolf reports averaged 18 per year
and then dropped to the previous level of five reports per year (Figure 3). The sharp increase in the number of reports may be due to:
1) a concerted effort by WEP field assistants to gather reports in
the area from 1974-76, 2) an increased public awareness of the plight
of the wolf, 3) a general northwesterly movement of wolves or
wolf-like canids from Yellowstone Park (Flath, 1979), or 4) an
actual increase in the number of wolves occupying the area due to
successful reproduction.

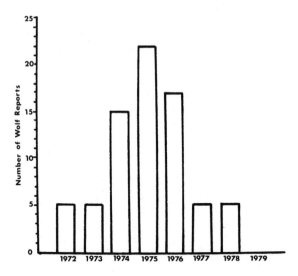

Figure 5.3: Number of wolf reports from southwestern Montana, 1972-1979.

Adults with pups were reported in 1973, 1974 and 1975. Re-peated sightings of adults with young and a den in the Gravelly Mountains in 1974 present good evidence that a litter of pups may have been produced that year (Day, 1977). Group sizes ranged from two to 10 animals. Of the 38 sightings, 61 percent were of single animals.

The sharp drop in number of reports after 1976 may reflect less time spent in the area by WEP field assistants. However, Flath (1979) increased his efforts to document wolf activity in the same area after 1976 with the same results. Flath (1979) noted that average reported group size has been decreasing steadily since 1974, and reports have been irregular, both temporally and geographically.

The identity and origin of wolves reported in southwestern Montana has been the topic of much speculation (Day, 1977; Flath, 1979; Weaver, 1978). No specimens have been taken from this area since 1941; thus, no skulls are available to confirm their identity as wolves and to determine their taxonomic status. Weaver (1979) suggests the possibility, among others, that these may be a wolf-coyote hybrid and refers to "wolf-like canids." This interpretation is certainly possible, but no skeletal or other evidence supports it at this time.

As Weaver (1978) and Flath (1979) suggest, it seems very un-likely that a small population of remnant wolves has persisted in southwestern Montana. The reports have been too irregular, tempo-rally and geographically, to support such a hypothesis.

The possibility exists that deliberate reintroduction(s) have taken place in the area (Flath, 1979; Weaver, 1979), and that their success has faded. Although no documented evidence supports this possibility, its controversial nature would certainly warrant maximum security with a minimum of people involved.

Because of increases in reports of wolves from the Beaverhead-Madison area, with corresponding decreases in reports from Yellowstone National Park, Flath (1979) suggests that the wolf-like canids reported from Beaverhead-Madison may have originated in or near Yellowstone National Park.

Finally, it is possible that some or all of the reports have resulted from immigration of wolves to this area from the north. Long-distance movements have been reported for radio-tagged wolves (VanCamp and Gluckie, 1979; Berg and Kuehn, this volume; Scott and Shackleton, this volume). During the past year, a wolf was found killed near Glasgow in the plains of eastern Montana, and another was killed in southern Idaho 80 km north of Boise. Both were at least 300-400 km from known wolf populations.

Northwestern Montana

The second area of concentrated wolf reports is northwestern Montana. This area extends north from Township 13N to the Canadian border, and from the western boundary of the state east to Range 5W (Figures 5.4 and 5.5).

There were a total of 270 wolf reports in northwestern Montana from 1972 through June 1979. Two hundred thirty-seven (88%) of the reports are within a "wilderness corridor" that straddles the Continental Divide from the Canadian border south 225 km. This corridor includes Glacier National Park, the Blackfeet Indian Reservation, the Great Bear, Bob Marshall and Scapegoat Wilderness areas, and surrounding national forest lands. Much of this land is pristine wilderness with recreation the major type of land use. Only one major highway bisects this corridor from east to west. Even though the Continental Divide is not a barrier to wolf movements, it is biologically significant for discussing wolf reports east and west of the Continental Divide. Winter snow depths along the Divide undoubtedly affect prey species distribution for six to seven months each year.

West of the Continental Divide: One hundred forty-eight wolf reports were collected west of the Continental Divide. In comparison with the 122 reports collected east of the Divide, the reports west of the Divide were more widely dispersed both temporally and geographically (Figure 5.4). Despite their scattered nature, most reports west of the Divide fall within two large areas: the Kootenai National Forest in the northwestern corner of Montana and the western portion of the "wilderness corridor," previously discussed.

Logging, mining and recreation are three important land uses on the Kootenai National Forest. The area is sparsely populated, with the largest town having fewer than 5,000 residents. White-tailed deer (*Odocoileus virginianus*) are the most abundant potential prey species for wolves with up to 52 deer/km² in some of the major wintering yards (Firebaugh et al., 1975). Mule deer (*Odocoileus hemionus*), elk (*Cervus elaphus*) and moose (*Alces alces*) are also present, but in lower numbers than white-tailed deer.

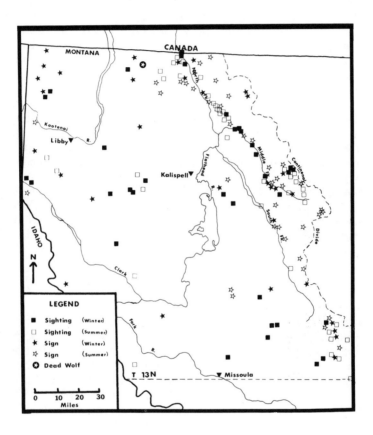

Figure 5.4: Wolf reports west of the Continental Divide, 1972-1979.

Half of 28 wolf reports collected in the area were sightings. Ten of the 14 sightings were of single animals, and four sightings were of a pair. Ten of the 18 wolves involved in the sightings were described as black or dark gray. The frequency of winter reports (November through April) was almost four times that of summer (May through October) reports. There were no reports of wolf pups or dens in the area. In January of 1974, a trapper caught and killed a light gray male wolf in a coyote trap 13 km south of the Canadian border. The skull and carcass were lost, but the pelt was examined.

Considering the scattered nature of these reports, small group sizes and the incidence of dark animals, wolves in the Kootenai Forest are most likely transients from British Columbia that range south more frequently during winter than summer. There is no evidence that these wolves are reproducing, or that they are year-around residents of that area.

Eighty percent of all reports west of the Continental Divide came from the western strip of the "wilderness corridor" previously described. This area includes the western portion of the Lolo Forest. Much of the land is de facto or designated wilderness. Elk, mule deer, white-tailed deer and moose are all present in the area and locally abundant where adequate winter range exists.

The 115 wolf reports from this area are scattered along the entire 225 km length of the wilderness corridor. There were 59 wolf sightings and 56 wolf sign reports. Reports were divided nearly equally between summer (58) and winter (57). Of the 59 sightings, 48 (81%) were of single animals and 11 (19%) were of groups of two and three wolves. There were no reports of pups or dens in the area.

Two areas that consistently produce wolf reports are the Middle and North Forks of the Flathead River. Both rivers are classified as wild and scenic, and the lands surrounding the Middle Fork were recently designated as the Great Bear Wilderness. An elk herd of approximately 200 animals (Smith, 1978) winters at the Middle Fork and deer are plentiful. All recent reports from the area have been of single wolves, but the color descriptions are not consistent. Reports of wolf tracks, scats, howling and sightings in the area indicate that at least one, and possibly two, wolves have utilized the Middle Fork drainage.

The North Fork of the Flathead River originates in British Columbia and flows south into Montana. Thirty-six reports were collected from this area. Singer (1975a) and Kaley (1976) compiled additional information on the occurrence of wolves in the North Fork Valley and adjacent lands. Singer (1975a) estimated a population of five to 15 wolves in his study area from 1973 to 1975. Day (1977) gives a more conservative estimate of one wolf in the North Fork area. He was considering reports exclusively from Montana, whereas Singer included reports from B.C. and areas east of the Continental Divide. Singer reports breeding activity in 1973 and 1975; however, no pups have been reported since that time. It is very likely that wolves reported from this part of Montana are transients from Canada. In 1979, a wolf was trapped and radio-tagged by the WEP 10 km north of the U.S.-Canadian border in the North Fork Valley. More detailed information on that individual is presented later in this paper.

East of the Continental Divide: One hundred twenty-two reports were collected east of the Continental Divide (Figure 5.5). These were all located within the eastern portion of the wilderness corridor including portions of Glacier National Park, the Blackfeet

Indian Reservation and the Lewis and Clark National Forest. Topography of the area is largely influenced by overthrust faulting of sedimentary layers of bedrock meeting abruptly with the Great Plains on the east. The climate is drier with higher winds and more extreme temperatures than west of the Divide. A herd of approximately 1,000 bighorn sheep (*Ovis canadensis*) winters along the eastern slope of the Rockies. Two thousand to 3,000 elk and even larger numbers of mule deer also winter along this interface between the mountains and Great Plains (Mattson and Ream, 1978). All 122 wolf reports are concentrated along a 200 km by 50-km strip with no scattered reports to the east. Any wolves dispersing from this wilderness corridor would probably not survive very long on the plains to the east where cover is scarce, aerial coyote hunting is a common practice and the land is under intensive agricultural management. To help adequately examine the nature of wolf reports from this area, two time periods will be discussed: 1972-76 and 1977-79.

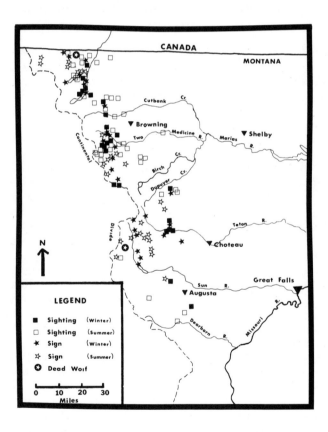

Figure 5.5: Wolf reports east of the Continental Divide, 1972-1979.

1972-1976: The first five years of the WEP were spent gathering wolf information from a large region to determine the most promising areas for intensive field surveys. The area east of the Continental Divide was chosen for intensive field work beginning in 1977 because of an abundance of reports from this region prior to 1977. First, it is necessary to discuss the nature of reports prior to 1977 and compare them to other areas.

Thirty-nine (60%) of the 65 wolf reports collected during 1972-76 were sightings, 25 (38.5%) were wolf signs and one (1.5%) was a dead wolf. Reports were slightly more common in summer than winter (55% versus 45%). Group sizes were generally larger than west of the Divide (Table 5.1). Adults with pups were reported in 1972, 1974 and 1975. A freshly dug possible wolf den was reported in 1972, but was apparently never used. In November 1974, a dark gray female wolf was found shot near the North Fork of the Sun River 140 km south of the Canadian border.

Table 5.1: Group Sizes East and West of the Continental Divide, 1972-1976

Group Size	West (n = 47)	East (n = 39)
1	38 (81%)	26 (67%)
2	6 (13%)	7 (18%)
3-5	3 (6%)	6 (15%)

The wilderness corridor east of the Continental Divide was thought to have the best potential for wolf recovery for several reasons: 1) extensive areas of winter range for deer, elk and bighorn sheep are included in the area, 2) reports of adults with pups suggested that some limited reproduction was taking place, 3) the clumped distribution of wolf reports in the area gave a measure of protection for the few remaining wolves, and 4) beside the wolf killed in 1974, three others were killed in, or very close to, this area in 1964, 1968 and 1977. In January 1977, a trapper caught and killed a black female wolf in the northwest corner of the Blackfeet Indian Reservation four km south of the Canadian border. There was good evidence of additional wolves in the area. Thus, the corridor east of the Continental Divide was the only place in the northern Rockies where actual specimens, including skulls, provided solid, recent evidence of wolf presence in 1977.

1977-1979: A full-time field assistant was stationed in the area just outside Glacier Park early in 1977, and intensive field surveys were initiated. The objective was to radio-collar one or more wolves for the purpose of gathering ecological information on wolves in a low-density area. To do this, it was necessary to first identify wolf activity centers or travel routes, and secondly, to begin trapping in these areas.

Because of substantial evidence of wolves in the area, the WEP requested and was subsequently denied permission to trap wolves on the Blackfeet Indian Reservation. Several months later, we were ordered to stop all work on the Reservation, including field surveys and interviews of local people. Work continued along the eastern slope of the Rockies on Park Service, Forest Service, Bureau of Land Management and private lands.

Fifty-seven wolf reports were collected during the two and a half year period from 1977 to mid-1979 compared to 65 reports collected during the previous five year period. The number of reports increased from 1974 to 1977, and sharply declined after 1977 (Figure 5.6). The low number of reports during 1972-1973 may be the result of less time spent in the area interviewing local residents and searching for wolf sign, although this same reasoning cannot be applied to the sharp decline in the number of reports during 1978 and 1979. One hundred fifty days were spent searching for wolves and sign, and over 200 local residents were interviewed concerning recent evidence of wolves during those two years.

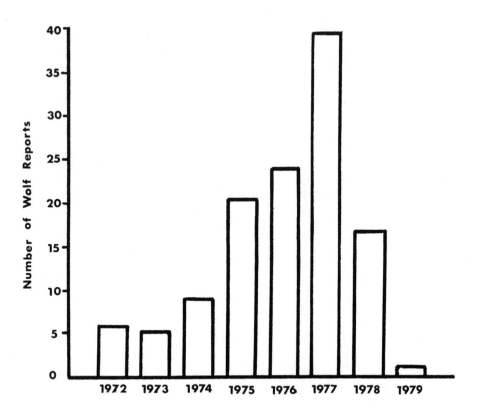

Figure 5.6: Number of wolf reports east of the Continental Divide, 1972-1979.

Not only did the number of reports decline during 1978 and 1979, but the average group size decreased from 1.6 to 1.2 and observations of lone wolves increased from 67 to 84 percent from 1972-76 to 1977-79. Groups of three, four and five wolves were observed during the earlier period, but there were no groups larger than two wolves observed after 1976.

The sharp drop in total number of wolf reports, the decrease in average group size and the increase in percentage of lone wolf sightings may be due to several factors: 1) the Blackfeet Indian Reservation, an area that had provided 50 percent of all wolf reports east of the Continental Divide from 1972 to 1977, was closed to further study by the WEP in mid-1977, 2) extreme weather conditions and blowing, drifting snow during the winter of 1978-79 caused very poor tracking conditions, 3) wolves from Canada may have travelled south of the border less frequently during these years, and 4) the adults with pups that had been reported previously, may have succumbed to natural or human-caused mortality.

Wolves reported south of Highway 2 may have been permanent residents. A few wolves probably immigrated from Canada into this area during the early to mid-1970s, and may have produced one or more litters of pups from 1972 through 1975. If so, the litters were victims of natural or man-caused mortality as there were no groups larger than two wolves seen after 1976. According to Mattson and Ream (1978), approximately five to eight wolves inhabited this area along the eastern slope of the Rockies and areas farther west within the Great Bear-Bob Marshall-Scapegoat Wilderness complex.

Wolves reported north of Highway 2 on the Blackfeet Indian Reservation and Glacier Park were probably transients from Canada. Wolves in southwestern Alberta have expanded their range southward during the 1970s (John Gunson, pers. comm.). In response to livestock depredations, a pack of wolves was removed from the Porcupine Hills area 100 km north of the U.S. border in 1976-77. The frequency of wolf reports has increased during the past few years between the Porcupine Hills and Waterton Park on the U.S.-Canadian border. In the winter of 1976-77, several wolves were observed in the vicinity of an elk herd near the border. Their tracks were seen repeatedly. One of the wolves killed a yearling bull elk in Waterton Park (Keith Brady, pers. comm.). It was at this same time that the female wolf was trapped and killed just south of the border on the Blackfeet Reservation. If wolves from Alberta continue to expand their range southward, they may eventually breed and establish home ranges in the U.S. The area east of the Continental Divide is ecologically ideal for wolf recovery, but an active management program will be required to assure recovery. Such management must include means for reducing man-caused mortality of wolves. This would require full cooperation of all resource management agencies in the area.

Other Areas

Idaho: Forty-two wolf reports were collected from Idaho between 1972-79 (Figure 5.7). Two areas in Boise and Clearwater National Forests produced well documented reports and will be discussed
separately. The remaining 16 reports are scattered throughout the
state with very little temporal or geographic consistency. Several
reports in east-central Idaho near the Montana border correspond
with, and add credibility to, a group of Montana wolf reports.

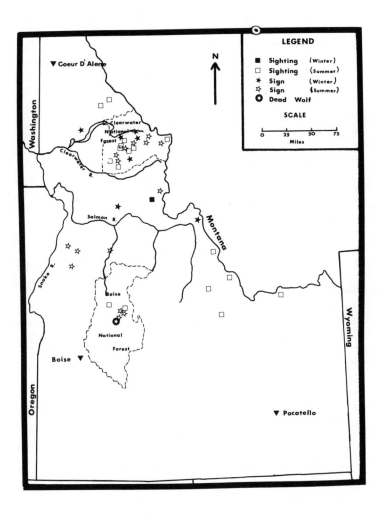

Figure 5.7: Wolf reports from Idaho, 1972-1979.

Clearwater Forest: Twenty wolf reports were collected from the Clearwater Forest, mostly from experienced outdoorsmen including outfitters, taxidermists and biologists. Seven were documented with photographs, plaster casts and/or photos of tracks or a fur sample.

A series of excellent reports occurred during summer 1978. On 4 June Mike Schlegel, an Idaho Department of Fish and Game biologist, photographed a wolf from a helicopter during an elk survey. Less than one mile from this sighting and within a period of three days, he found the remains of two previously radio-collared elk calves. Wolf tracks, a scat and wolf fur were near the carcasses. Three months later, an Idaho Department of Fish and Game conservation officer spotted four wolves (apparently an adult and three young) 11 km away. The adult was light gray whereas the wolf photographed in June was dark, suggesting that a pair of wolves with a litter were in the area. Shortly thereafter, members of WEP made several short trips into the area to search for wolf sign and interview local residents.

A WEP field assistant spent summer 1979 in the Clearwater area searching for wolf sign in cooperation with the Forest Service and Idaho Department of Fish and Game. The areas where 1978 reports originated were investigated intensively and many local contacts were made, but only two possible sets of wolf tracks were found.

The Clearwater National Forest has excellent potential for wolf recovery for several reasons: 1) a series of well documented reports in 1978-79 indicate that several wolves occupied the area; 2) large portions of unroaded land still exist, making human access difficult; and 3) there is substantial prey base available, including several hundred elk and fewer deer.

Boise National Forest: In October 1978, a wolf was shot in Bear Valley in south-central Idaho. The carcass was confiscated by the U.S. Fish and Wildlife Service and the skull was confirmed as wolf by the Smithsonian Institution. Five additional wolf reports were collected within a 19 km radius of where the wolf was killed. These include a sighting of an adult wolf with five or six pups one week prior to the wolf-shooting incident. Wolf howling, a scat and four wolf pups were reported during the summer of 1979. The origin of the dead wolf is unknown. The area should be examined more thoroughly for evidence of additional wolves.

North Fork of the Flathead Ecological Studies

On 8 April 1979, a female wolf was trapped and radio-tagged at the North Fork of the Flathead River Valley. This was the first wolf ever radio-collared in the northern Rocky Mountains. The North Fork of the Flathead River originates in southeastern British Columbia and flows 125 km south into Montana to its confluence with the Middle Fork of the Flathead River (Figure 5.8). The Montana portion of the river is bounded on the east by Glacier Park and on the west by the Flathead National Forest and private lands. Major land uses in this area include recreation, logging, oil and gas exploration, and a limited amount of livestock grazing. The recent popularity of summer and permanent homes in rustic settings has led to increased "wilderness subdivision" development. By contrast, in British Columbia there is very little private land in the valley and few year-around or summer residents.

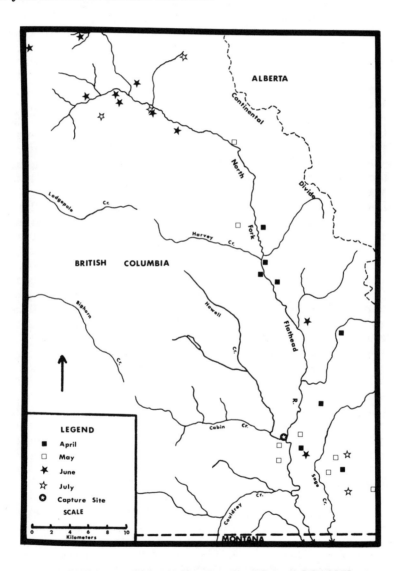

Figure 5.8: Locations of radio-tagged wolf, April-July 1979.

During the past three years, an outbreak of mountain pine beetle (*Dendroctonus ponderosae*) throughout the entire valley has accelerated logging of seral lodgepole pine (*Pinus contorta*) and has increased access to some areas by upgrading existing roads and building new ones. Oil development and coal strip mining are proposed to begin north of the border in the near future.

The light-gray female wolf was captured 10 km north of the U.S.-Canadian border using trapping techniques similar to those described by Mech (1974b) in northeastern Minnesota. The wolf was captured after six days of trapping (23 trap nights). At the time of her capture, she weighed 36 kg and her age was estimated to be seven or eight years. She did not appear to be pregnant at the time, and subsequent tracking revealed that she did not occupy a den or produce pups that season.

Aerial radio-locations have been made twice weekly since her capture. In addition, some ground tracking was done from roads when feasible. During the first three and a half months of radio-tracking (8 April-20 July), the wolf ranged north-south along the main river valley with very limited east-westerly movements. Her range during this time was 44 km long by 8-10 km wide. The southern-most end was 5 km north of the border and the northernmost end was at the headwaters of the Flathead River.

Locations during the first three weeks of April and the last three weeks of May were concentrated in the southern portion of her range. The intervening time was spent travelling to and from the northern portion of her range. Beginning 1 June, she again set out for the headwaters of the Flathead and moved farther west and north than on her previous excursion. She stayed in this general area for six weeks and then moved south again in mid-July only to return to the northern portion by 31 July. The farthest distance moved within a 24 hour period was 13 km. Her locations were concentrated in these northern and southern portions of her range with repeated visits to some of the same sites and very little time spent in the area between the two portions. There has been considerably more human activity and development in the southern end of her range than in the north. New roads are being built, the area has been intensively logged and human settlement includes a logging camp, a Forest Service complex, an outfitters' ranch and several occupied cabins. Plans are being made to mine coal in the area. In contrast, the northern portion has a few low-grade roads and very little development.

She apparently used roads extensively for travelling. She was captured in a trap set in the middle of a snow-covered road. Her tracks have been seen several times from the air along snow-covered roads. Her average distance from the nearest road was one km (n = 32). She crossed the river frequently, even during spring run-off.

She has not been seen during any of the 27 flights as she has consistently been in heavily timbered areas. Twenty-five percent of the locations were on forested benches above the river. The remaining 75 percent were divided evenly between heavily timbered sites adjacent to bogs or wet meadows, and timbered lower slopes often close to a stream bed. Elevations at location sites range from 1,300-1,850 m.

Major potential prey in the area include white-tailed deer, mule deer, elk, moose and beaver. South of the border, elk and white-tailed deer are relatively more abundant than the remaining species (Singer, 1975b). Initial observations indicate that moose and beaver may be comparatively more abundant north of the border than in areas farther south. An investigation of the relative numbers and distribution of ungulates and beaver within the wolf's range is currently underway.

Scat collection and analysis are currently underway to identify the wolf's prey items using criteria established by Weaver and Fritts (1979) to separate wolf from coyote (*C. latrans*) scats. Other supporting evidence such as tracks or proximity of wolf locations are being used to further separate wolf scats from those of coyote and mountain lion (*Felis concolor*). To date, no kills have been located. Singer (1975a) recorded wolf kills of white-tailed deer, moose, elk, beaver and snowshoe hares (*Lepus americanus*) in his study area which overlaps the southern portion of the radio-tagged wolf's range. Collecting scats and locating kills should be easier by snow tracking during the winter.

Wolf reports within the radio-tagged wolf's range and areas further south indicate that there are additional wolves in the North Fork Valley, including a black wolf seen several times by loggers. However, judging by her tracks when she was captured, she was travelling alone. Since she has not been seen since capture, it is unknown whether she is alone or with other wolves.

All radio-location sites are being sampled this summer (1979) to determine habitat characteristics and to look for scats, tracks or food items. This information and the movement data should begin to give us information on ecological requirements of wolves in a sparsely occupied range in the Northern Rockies. It should also prove useful in assisting survey efforts in other areas of the Northern Rockies.

SUMMARY AND CONCLUSIONS

The Northern Rockies region has experienced an increase in wolf activity in the 1970s. This is substantiated not only through reports, but also by an increase in wolves killed. It follows 40 years of very sparse activity which followed intensive eradication programs. Although this increase leads to optimism regarding wolf recovery, we still cannot point to a single area in the U.S. Rockies and say, "Here is a population of wolves that should be managed and protected and these are the critical elements of their habitat." However, there are several areas which appear to have the best potential for wolf recovery. One problem in initiating intensive field work in these areas, however, is that reported activity in them is not consistent from year to year. We have observed declines in reports from Yellowstone National Park (Weaver, 1979), southwestern Montana, the Rocky Mountain Front and, apparently, now in Idaho. Even the North Fork of the Flathead seemed to show a decline in activity between the field work of Singer (1975a) and our recent wolf capture and radio-tagging effort.

What does this mean in terms of wolf status, distribution and potential for recovery? In the first few years of our survey work, we felt that there must be remnant packs or populations of wolves that have persisted in the most remote portions of the Northern Rockies. This possibility still exists, particularly in the "wilderness corridor" extending for 225 km (135 miles) down the Continental Divide from Glacier National Park through the Scapegoat Wilderness. However, the sporadic nature of these reports, particularly from year to year, leads us to conclude that most of the reports obtained represent wolves from sources other than remnant packs. In northwestern Montana and northern Idaho, the simplest explanation is immigration from Canada. Wolves have increased in southern Alberta and parts of British Columbia in recent years, so relatively short immigration distances would be required. The high incidence of single wolf sightings suggest "loners" which might also be dispersing wolves.

A similar situation exists in Finland. For most of Finland except Lapland and Karelia, 90 percent of wolves observed were lone wolves and 10 percent were pairs (Pulliainen, 1965). Wolves are dispersing from the USSR where populations are healthy, to Finland where wolf numbers are lower (Pulliainen, this volume).

So far, we have no solid evidence of viable packs producing pups in Montana in recent years, so dispersers and/or residents apparently have not been reproducing successfully. In southwestern Montana and Yellowstone National Park, the situation is even less clear. As discussed earlier, there is no skeletal evidence yet to support an interpretation that wolf-like canids have developed through hybridization with coyotes (Weaver, 1979). Skeletal material from coyote control programs around Yellowstone should be investigated for evidence of such hybrids. Distances appear to be great for immigration from Canada, yet some recent evidence points to this as a possibility to consider and to watch for in the future. Finally, release of wolves by humans, either captive or wild, cannot be summarily discounted. Whatever the source, the number of wolf reports that were coming in around 1970 has sharply dropped in the last five years so that evidence for wolves being present is now minimal.

The future of wolves in the northern Rockies remains uncertain. For Yellowstone National Park and vicinity, it now appears that if recovery is to occur, at least in the near future, a reintroduction program may be necessary (Weaver, 1978, 1979). In northwestern Montana, recovery can occur through protection of any existing wolves and through natural immigration. For biological and political reasons, this is obviously preferred over reintroduction attempts to achieve wolf recovery. In both cases, a good program of public education is needed. Protection of the wolves, their prey and their habitat is critical to recovery of viable wolf populations. Enforcement is one tool, but efforts by resource management agencies to reduce human-wolf contacts are at least as important. Land use planning for our remaining wildlands must take this into account if the present potential for wolf recovery in the northern Rocky Mountains is to be realized.

Acknowledgements

The Wolf Ecology Project has been funded by the U.S. Fish and Wildlife Service (Office of Endangered Species), the World Wildlife Fund and the University of Montana Forest and Conservation Experiment Station. We also wish to thank those employees of the U.S. National Park Service, U.S. Forest Service and U.S. Fish and Wildlife Service who gave their time, logistical support and cooperation to the Project. Without the diligent and often unpaid efforts of students Gary Day, Rich Harris, Dennis Daneke, Rick Johnson, John Stern, Randy Rogers, Brian Giddings, Carol Schmidt and Sharon Gaughan, this work would not have been possible.

Attitudes of Michigan Citizens
Toward Predators

Richard A. Hook and William L. Robinson

INTRODUCTION

Since the late 1950s, the eastern timber wolf *(Canis lupus lycaon)* has been near extinction in the Upper Peninsula of Michigan (Hendrickson et al., 1975). A bounty paid by the state until 1960 encouraged the killing of wolves and was likely an important factor in their near extermination. In 1974, an experimental effort was made to re-establish the wolf in the Upper Peninsula by releasing four radio-collared wolves from Minnesota in northern Marquette County. There was considerable controversy surrounding the experimental wolf release, with support coming in the form of letters, resolutions and a few contributions from some groups, and opposition coming from others. In eight months all four wolves had been killed: two shot, one trapped and shot, and one struck by an automobile (Weise et al., 1975).

After the experimentally released wolves had been killed, three other wolves killed in the Upper Peninsula between November 1974 and March 1976 were examined. One had been shot twice prior to its third and fatal shooting, one had been trapped once and shot once prior to its fatal bullet wound, and one had escaped from a trap minus a couple of toes once before its final trapping (Robinson and Smith, 1977). Nearly all wolves examined were in good nutritional condition, indicating that prey populations in Michigan are adequate to sustain wolves (Weise et al., 1975).

From these observations, it appears that the primary reason why wolves have failed to thrive in Michigan despite state legal protection since 1965, despite their endangered status under the Federal Endangered Species Act of 1974, despite natural immigration of animals of both sexes, and despite an experimental effort to re-establish them, is direct persecution by humans. It seems that the wolf's future in Michigan depends upon the attitudes of Michigan residents toward this animal, and upon the number of encounters between these citizens and wolves. A continuation of antagonistic human attitudes and overt hostile behavior toward wolves will likely restrict population growth or ensure total extinction of wolves on the Michigan mainland. Conversely, restoration of the wolf in suitable range in the Upper Peninsula will depend primarily upon changes in human attitudes.

Until this study (cf. Hook, 1981), the extent of opposition to wolves among Michigan citizens had not been measured, nor had the bases for anti-wolf attitudes and behavior been examined. The major objectives of our study were to assess the extent of anti-wolf attitudes in Michigan and to determine their underlying causes.

METHODS

Since wolves probably share a common persecution with other predators, we addressed the broader question of attitudes toward all predators. Our model tested the following hypotheses:

A negative attitude toward predators is caused by: (1) lower educational level, (2) rural childhood background, (3) poor understanding of the ecological role of predators, (4) concern for economic losses caused by predators, (5) fear of predators, (6) participation in hunting, (7) participation in hunting by family elders.

A positive attitude toward predators is caused by: (1) higher educational level, (2) nonconsumptive interest in wildlife, (3) non-rural childhood background.

In addition, we included questions to assess the attitude types described and named by Kellert (1976): naturalistic, moralistic, negativistic, dominionistic, scientistic, ecologistic, humanistic, aesthetic and utilitarian.

We developed a questionnaire which contained several questions pertaining to each hypothesis, tested the questionnaire through one mailing of 300 copies, then extensively revised it, discarding questions which were found to be ambiguous, and adding questions which we felt were more clear and which more adequately tested our hypotheses. Most questions were in the form of statements with the respondent asked to indicate one of five choices: strongly agree, agree, undecided, disagree, strongly disagree.

The final questionnaire (available from Hook) containing 122 questions was mailed to 3,382 Michigan residents chosen randomly from Michigan drivers' license files from six counties, three from the Upper Peninsula and three from the Lower Peninsula (Figure 6.1). Counties were chosen on the basis of urban-rural ratios. In the Upper Peninsula, Marquette County is the most urban. Iron County is a wooded, rural county and Menominee County is the most agricultural county in the Upper Peninsula.

Figure 6.1: Map of Michigan showing counties sampled (diagonally hatched).
Map from U.S. Bureau of the Census (1973).

In the Lower Peninsula, Wexford County has a similar urban-rural ratio and density to Marquette County, and Charlevoix County has similar rural-urban characteristics to Menominee County. Bay County, which includes the cities of Saginaw and Bay City, was randomly chosen from 10 Michigan urban areas. It has no equivalent in the Upper Peninsula in terms of population density, and likewise, no Lower Peninsula county was found to have the wooded-rural characteristics of Iron County. Table 6.1 lists characteristics of the sampled counties.

Table 6.1: Some Summary Statistics of the Six Sampled Counties and of the State[1]

County	Percent Urban Population	Median Years of Education	Median Age	Percent Female	Median Family Income	Density (#/km^2)
Bay	66.8	11.7	25.4	51.1	10,408	101
Charlevoix	39.2	12.1	28.8	50.7	8,535	15
Wexford	50.7	12.0	28.9	51.7	8,024	14
Iron	19.4	11.7	42.8	50.7	7,443	5
Marquette	65.1	12.2	24.2	48.1	8,562	14
Menominee	43.7	11.2	30.5	50.7	7,703	9
Entire state	73.8	12.1	26.3	51.0	11,032	60

[1] From U.S. Bureau of the Census (1973).

Two reminders were sent to tardy respondents. A total of 1,664 questionnaires were returned, a response rate of 49.2 percent. Of these, 1,290 returns, or 38.1 percent, were deemed useable. Non-useable returns include a large number from Midland County, which was inadvertently included through an error in zip code translation, and another large number from people who had moved to a different county. No questionnaires were discarded because of content.

RESULTS AND DISCUSSION

Bias

Bias was tested for by comparing some characteristics of the respondents with those of the population of the state as a whole. More hunters (45.7%) answered the questionnaire than in the population of the state as a whole (about 10.7%); 56.4 percent of the respondents were male compared with a statewide percentage of 49.0 percent, and educational levels of respondents averaged 12.8 years, while the state median among adults is 12.1 years (Bouchard and Lerg, 1977; U.S. Bureau of Census, 1973). Other unmeasured biases in responses undoubtedly exist and, therefore, appropriate caution should be exercised in interpreting results. We felt, however, that many of the relationships revealed in our study are strong enough so that the effects of sample bias will not cause serious misinterpretation of the results, although some results should not be extrapolated to the general population.

Factor Analysis

The results were subjected to factor analysis using the Statistical

Package for the Social Sciences (SPSS) (Nie et al., 1975). Factor
analysis consists of a system of matrix rotation by which patterns of
responses to questions are identified. In our study, the answers to
over 100 questions were found to fall into 11 meaningful factors
which may be thought of as representing various attitudes.

Once the factors were identified, the contribution, or "loading,"
of each question to each factor was statistically evaluated on a scale
ranging from -1.0 to 1.0. Based upon the SPSS "reliability" sub-
program and the advice of experienced social analysts, a question
which had a loading of about 0.4 or better on a factor was regarded
as contributing significantly to that factor. This was checked with
the SPSS subroutine "reliability" to determine which questions went
into the scales.

Anti-Predator Attitude

One factor clearly and strongly segregated itself out of the
questionnaire. Eighteen questions loaded onto this factor with a
value of about 0.4 or greater. The leading questions in this factor are
listed in Table 6.2. The tone of all these questions is anti-predator,
and answers reflect the respondents' feelings on such anti-predator
questions.

Table 6.2: Questions Loading Strongly Onto
Anti-Predator Attitude Scale

	Question	Loading
1.	Predators should be eliminated.	+0.76
2.	A wolf is a varmint and should be eliminated.	+0.76
3.	Predators are generally not much good.	+0.74
4.	Since they kill animals useful to man, coyotes should be shot.	+0.60
5.	Since foxes carry rabies, it is a good idea to eliminate foxes.	+0.70
6.	If someone told me a coyote was a useful animal in nature, I'd think he was crazy.	+0.67
7.	An animal that eats another animal is bad.	+0.64
8.	When a wolf kills a moose, it is murder.	+0.65
9.	When not controlled by man, coyotes seriously reduce wildlife.	+0.57

Each respondent's answers to all of the anti-predator questions
were then coded, with "strongly agree" given a value of 1.0, and
"strongly disagree" a value of 5.0. Thus, a response average of 1.0
would indicate that the respondent strongly agreed with all of the
statements with an anti-predator tone. The anti-predator scale mean
for the entire sample was 3.72, indicating a general disagreement
with the anti-predator statements. Table 6.3 shows the frequencies
of attitude, with answer categories subdivided into high and low
"strongly agree," high and low "agree," etc.

Table 6.3: Frequencies of Scores on Anti-Predator Scale

Scale Category	Anti-Predator Scale Mean	Number in Group	Percent in Group
High "Strongly agree"	(1.0 -1.50)	1	0.1
Low "Strongly agree"	(1.51-2.00)	9	0.8
High "Agree"	(2.01-2.50)	23	1.9
Low "Agree"	(2.51-2.99)	65	5.5
"Uncertain"	(3.00)	3	0.3
Low "Disagree"	(3.01-3.50)	202	17.1
High "Disagree"	(3.51-4.00)	415	35.2
Low "Strongly disagree"	(4.01-4.49)	319	27.0
High "Strongly disagree"	(4.50-5.00)	143	12.1
Total sample	(3.72)	1,180	100.0

The anti-predator scale average suggests a generally favorable outlook, but may be an artifact of the wording of the questions. We also asked two direct questions pertaining to the future of wolves in Michigan, as follows: (1) "Wolves should be restored in the Upper Peninsula," with the usual scale of responses, and (2) "If plans were made to reintroduce wolves into the Upper Peninsula, would you: (a) oppose the plan?, (b) hope that the wolves were killed?, (c) not pay much attention to the wolves?, (d) be favorably inclined toward the wolves?, (e) actively support the effort?" The answers (Tables 6.4 and 6.5) also indicate a favorable inclination among the respondents.

Table 6.4: Response Frequencies to Question 95; "Wolves Should be Restored in the Upper Peninsula"

Category	Number in Group	Percent in Group
Strongly agree	187	14.7
Agree	502	39.4
Uncertain	428	33.6
Disagree	101	7.9
Strongly disagree	56	4.4
Total sample	1,274	100.0

Table 6.5: Response Frequencies to Question 112 and the Anti-Predator Scale Means for Each Category

Ques. 112: If plans were made to reintroduce wolves in MI, would you: (√ one)
   A. Oppose the plan?
   B. Hope that the wolves were killed?
   C. Not pay much attention to the wolves?
   D. Be favorably inclined toward the wolves?
   E. Actively support the effort?

Category	Number in Group	Percent in Group
Oppose the plan	164	15.1
Hope the wolves were killed	7	0.6
Not pay much attention	266	24.4
Be favorably inclined	489	44.9
Actively support the effort	163	15.0
Total sample	1,089	100.0

Thus, according to the anti-predator scale, a "pro-wolf" scale developed in response to other questions, and direct questions pertaining to wolf reintroduction, the general attitude of the public toward predators appears favorable, although some people do harbor anti-predator sentiment.

Other factors may contribute to these anti-predator attitudes. To examine our original hypotheses:

(1) Educational level. There was a linear relationship ($p < 0.01$) between anti-predator attitude and educational level, with anti-predator feeling decreasing with increasing education (Table 6.6).

Table 6.6: Anti-Predator Scale Means for Various Educational Levels

Category	Anti-Predator Scale Mean	Number in Group	Percent in Group
Grade school	3.33	96	8.2
1–3 years' high school	3.67	133	11.4
High school diploma	3.82	513	44.0
1–3 years' college	4.00	248	21.3
College diploma	4.14	122	10.5
Advanced degree	4.08	54	4.6
Total sample	3.85	1,166	100.0

(2) Rural background. People who were reared on a farm tended ($p < 0.01$) to have a less favorable attitude toward predators than the population as a whole (Table 6.7).

Table 6.7: Anti-Predator Scale Means of Respondents and Home Location at Age 16

Home Location	Anti-Predator Scale Mean	Number in Group	Percent in Group
In wooded country	3.91	149	13.6
In open country	3.81	69	6.3
On a farm	3.77	238	21.7
In a small city	3.87	490	44.7
In a medium city	3.88	64	5.8
In a suburb	4.02	42	3.8
In a large city	3.94	45	4.1
Total sample	3.86	1,097	100.0

(3) Lack of knowledge. We tested each respondent's knowledge of predators by asking them to identify predators from a list (Table 6.8). There was a significant relationship ($p < 0.01$) between anti-predator attitude and the ability to identify predators (Table 6.8).

Table 6.8: Anti-Predator Scale Means and Ability to Identify Predators[1]

Number of Correctly Identified Predators	Anti-Predator Mean	Number in Group
1	3.66	65
2	3.60	127
3	3.72	208
4	3.86	317
5	3.99	202
6	3.99	149
7	4.22	28
8	4.30	23
All	3.86	1,119

[1] *Predator Identity Question:* The word predator means different things to different people. Usually a predator is defined as an animal that kills and eats other animals for food. Please consider the following list and place a check by each animal that you would think of as a predator.

a. ___ Brown bat
b. ___ Striped skunk
c. ___ Black bear
d. ___ Opossum
e. ___ River otter
f. ___ Red fox
g. ___ Cottontail rabbit

h. ___ Woodchuck
i. ___ Coyote
j. ___ Song sparrow
k. ___ Chipmunk
l. ___ Meadow mouse
m. ___ Robin
n. ___ Bobcat

o. ___ White-tailed deer
p. ___ Mink
q. ___ Marsh hawk
r. ___ Porcupine
s. ___ Raccoon
t. ___ Beaver

(4) Concern for economic loss. Concern for economic loss contributed to anti-predator attitude. Those who showed concern for economic loss to predators had increasing anti-predator feelings ($p < 0.01$) (Table 6.9).

Table 6.9: Response Frequencies to Question 59; "Farmers in Michigan Lose a Lot of Livestock Because of Predators"

Category	Anti-Predator Mean	Number in Group	Percent in Group
Strongly agree	2.54	9	0.8
Agree	3.18	100	8.5
Uncertain	3.65	318	26.9
Disagree	3.89	553	46.9
Strongly disagree	4.42	200	16.9
Total sample	3.85	1,180	100.0

. . . . Farmers' Answers to Question #59 . . . .

Category	Anti-Predator Mean	Number in Group	Percent in Group
Strongly agree	—	0	0.0
Agree	1.91	2	9.5
Uncertain	3.47	4	19.0
Disagree	3.53	11	52.4
Strongly disagree	4.64	4	19.0
Total sample	3.58	21	99.9

(5) Fear of wolves. Five questions contributed to what we termed the "fear-of-wolf" scale (Table 6.10). There was a significant correlation between the fear-of-wolf and anti-predator scales ($r = 0.72$; $p < 0.01$). As fear of predators increased, so did general anti-predator feeling.

Table 6.10: Comparison of Scores on Fear-of-Wolves Scale[1] and Anti-Predator Scale

	Score on Fear-of-Wolf Scale	Number in Group	Percent in Group	Score on Anti-Predator Scale
High "Strongly agree"	(1.00–1.50)	3	0.3	2.56
Low "Strongly agree"	(1.51–2.00)	31	2.7	2.93
High "Agree"	(2.01–2.50)	97	8.5	3.24
Low "Agree"	(2.51–2.99)	198	17.4	3.55
"Uncertain"	(3.00)	118	10.3	3.69
Low "Disagree"	(3.01–3.50)	224	19.6	3.86
High "Disagree"	(3.51–4.00)	274	24.0	4.00
Low "Strongly disagree"	(4.01–4.49)	117	10.3	4.37
High "Strongly disagree"	(4.50–5.00)	79	6.9	4.56
Total sample		1,141	100.0	3.84

[1] *Questions contributing to fear-of-wolves scale:* (1) Since there are many wolves on Isle Royale, one must be careful while hiking on the Island. (2) Wolves in the woods can often be dangerous to humans. (3) We are lucky no one was killed by one of the transplanted wolves. (4) Wolves are especially dangerous when they are hungry. (5) Many people have been attacked by wolves in North America.

(6) Participation in hunting. Our hypothesis that hunting activity increases anti-predator attitude was not borne out. Increased participation in hunting was associated with an increase in the participant's sympathy for predators ($p < 0.01$) (Table 6.11). Those who seldom or never hunted had less favorable attitudes toward predators than those who hunted more frequently.

Table 6.11: Anti-Predator Scale Means for the Various Hunting Activity Level Categories

Category	Anti-Predator Mean	Number in Group	Percent in Group
Never	3.78	601	52.4
Seldom	3.81	87	7.6
Sometimes	3.90	193	16.8
Often	3.98	266	23.2
Total sample	3.85	1,147	100.0

(7) Hunting as a family tradition. Contrary to our original hypothesis, a family hunting tradition was associated with a more

favorable outlook toward predators (Table 6.12). People who engaged in another consumptive activity—fishing—also had increased positive attitudes toward predators, but those who trapped and bountied animals did not (Table 6.13).

Table 6.12: Anti-Predator Scale Means for the Categories of Participation in Hunting by Family Elders

Category	Anti-Predator Mean	Number in Group	Percent in Group
Only respondent hunts	3.73	118	18.2
Father and respondent hunt	3.81	118	18.2
Grandfather, father and respondent hunt	3.95	412	63.6
Sample	3.88	648	100.0

(8) Nonconsumptive recreation. The hypothesis that participation in nonconsumptive outdoor activities is associated with higher appreciation of predators was confirmed to some extent (Table 6.13). Ardent backpackers had the highest scores, but there was not much difference between attitudes toward predators among avid hunters, fishermen and park visitors. Increased participation in bird watching did not linearly increase positive feelings toward predators, an unexplainable result. We conclude that, with the exception of trapping and possibly bird watching, increased outdoor activity is associated with an increasingly favorable attitude toward predators, whether that activity is hunting, fishing, backpacking, canoeing or camping.

Table 6.13: Anti-Predator Scale Means for Various Outdoor Activities (sample mean = 3.74)

	Never	Seldom	Sometimes	Often
Consumptive activities				
Hunting	3.78	3.81	3.90	3.98
Fishing	3.73	3.94	3.89	3.98
Trapping*	3.86	3.85	3.80	3.82
Bountying an animal	3.86	3.87	3.65	3.96
Nonconsumptive activities				
Hiking	3.62	3.85	3.99	4.03
Camping	3.77	3.87	3.97	4.00
Bird watching	3.73	3.88	3.85	3.94
Visiting a park	3.73	3.84	3.90	4.03
Backpacking	3.80	4.02	4.24	4.30

... Scale Means for Activity Levels ...

*Not statistically different from the population

We tested whether anti-predator attitudes of members of various wildlife oriented organizations differed from those of the sampled population. Members of most organizations, both of a consumptive and nonconsumptive nature, generally had more favorable attitudes than nonmembers (Table 6.14).

Table 6.14: Anti-Predator Scale Means for Members of Various Organizations

Organizations	Anti-Predator Mean	Number in Group	Percent in Group
Consumptively-oriented organizations			
Sportsmen's Club	3.97	80	7.6
Michigan United Conservation Clubs	4.16	42	4.0
Michigan Deer Hunters' Association	3.83*	10	1.0
Ducks Unlimited	4.30	18	1.7
National Rifle Association	4.27	45	4.3
Nonconsumptively-oriented organizations			
Audubon Society	4.35	25	2.4
National Wildlife Federation	4.27	45	4.3
Fund for Animals	3.92*	17	1.6
Sierra Club	4.23*	5	0.5
None	3.82	763	72.7
Total sample	3.86	1,050	100.1

*Not statistically different from the population.

Using multiple regression, we identified the factors contributing most heavily toward anti-predator attitude. The most important factor, surprisingly, was fear of the wolf, accounting for 35 percent of the variability. This was followed by negativistic attitude toward all animals (8.7%), tested using questions from Kellert's (1976) scales, and the third factor was the respondent's age (3.9%), with older people having less favorable attitudes toward predators. The fourth was anti-Department of Natural Resources (DNR) feeling (2.6%). The other factors–education, dominism attitude, positive ecological feelings, rural childhood, residency and income–each contributed less than one percent.

Factors such as consumptive or nonconsumptive wildlife interests, hunting, hunting by family elders, ecological awareness, knowledge of predators, home location and other socio-economic indicators contributed little in predicting negative attitudes in this group.

If, as we have shown, the average respondent is favorably inclined toward predators, why are so many killed? To answer this question, we examined the responses from individuals most likely to kill predators: hunters with negative attitudes toward predators (<3.0 on the anti-predator scale). This included 41 people, seven percent of the people sampled. Their average age was 51.8, compared with 41.4 years of all hunters who responded. Their average educational level was 11.0 years, compared with a statewide median of 12.1 years and an average among all sampled hunters of 12.6 years. Their sentiment was distinctly anti-DNR and their fear of the wolf was much higher than average.

A discouraging feature about this group of hunters with anti-predator attitudes appears to be their apparently low receptivity to information (Table 6.15). They indicated they received less information about predators from all sources than hunters in general, and most of the general public as well.

Table 6.15: Sources of Information About Predators

Source	Total Sample (1,135)	Persons with Anti-Predator Attitudes (103)	Hunters (589)	Hunters with Anti-Predator Attitudes (41)
Television	41.2%	31.1%	41.0%	34.1%
Discussion	30.6%	20.3%	39.2%	29.3%
Magazine	27.8%	11.6%	32.6%	12.2%
Book	10.4%	2.9%	11.5%	0.0%
Radio	6.1%	0.9%	8.3%	0.0%
Lecture or display	3.9%	0.9%	8.3%	0.0%

About 8.5 percent of the hunters in the Upper Peninsula and 4.3 percent of the Lower Peninsula hunters had anti-predator attitudes, although differences among counties were greater than between peninsulas (Table 6.16).

Table 6.16: Comparison of Frequency of Anti-Predator Attitudes Between Residents of the Upper and Lower Peninsulas

Peninsula	County	Number of Hunters Sampled	Number of Hunters with Anti-Predator Attitude	Percent
Lower	Bay	56	1	1.8
	Charlevoix	90	8	8.9
	Wexford	87	2	2.3
	Lower Peninsula total	233	11	4.7
Upper	Iron	143	15	10.5
	Marquette	120	5	4.2
	Menominee	93	10	10.8
	Upper Peninsula total	356	30	8.4

Educational material to reduce anti-predator sentiment among Michigan citizens should be designed to take into account the results we have described. The content of the material should portray the wolf in a realistic way, as an interesting predator and a part of the natural community. Information that wolves rarely, if ever, attack humans should be incorporated.

Television appears to be the most received form of communication, although new approaches should be considered. For example, free informational brochures placed in taverns and general stores in rural communities may reach a substantial proportion of the audience which may be most in need of education about predators. If this relatively small group of people can be favorably influenced, there may be hope for the restoration of the wolf in Upper Michigan and elsewhere.

Acknowledgements

We acknowledge the financial support of the Michigan Department of Natural Resources through state and federal Endangered Species Programs, The Huron Mountain Wildlife Foundation, the National Audubon Society and the National Rifle Association; and the professional counsel of sociologists Thomas Sullivan and Richard Wright of Northern Michigan University. We also received valuable advice from Stephen Kellert of Yale University and much technical assistance from Larry Ryel of the Michigan Department of Natural Resources.

Can the Wolf Be Returned to New York?

Robert E. Henshaw

INTRODUCTION

Wolves *(Canis lupus)* once inhabited all of New York State. Not until the arrival of the white man in the early 1600s did the wolf conflict with human enterprise. Van der Donck (1655) documented that as Long Island was settled in the early 1600s and overland trade routes were established, the wolf was quickly extirpated from the Island. As the interior of the New Amsterdam colony was settled, forests were cleared and agriculture established. By the early 1700s, sheep farming was well established in settled areas and competition with the wolf was intense. An eradication program was begun with the early goal to extirpate wolves from areas where domestic livestock were present. However, the program gained momentum. By the mid-1800s, the wolf was forced out of all of the state but the northern mountainous areas (DeKay, 1842). The early utilitarian extermination program had been corrupted over time into a bountied varmit eradication program. The last wild wolf in New York State was taken in the Adirondack Mountains in 1897 (Fowler, 1974).

Recent interest in the wolf coupled with a new sympathetic attitude toward predators, has spawned inquiries as to whether the wolf can be returned to its former range in New York State (Jolly, 1975; Engelhart and Hazard, 1975). Wildlife professionals inside New York have shown interest (The Wildlife Society, 1975), and the Eastern Timber Wolf Recovery Team has asked whether sites can be found in New York for possible reintroduction of the species (Mech, pers. comm.). While it may be concluded *a priori* that on ethical grounds, the wolf deserves access to its former range, no such *a priori* assumption can be made of the feasibility of reintroduction to its former range. In this paper I examine some of the questions which should be asked before attempting to reestablish wolves in the state of New York.

395

NICHE PARAMETERS OF THE WOLF

The wolf's propensity to select the easiest prey available (Mech, 1970) means that it may kill livestock when vulnerable individuals of wild species are less available. Some wolves become nuisances by entering inhabited areas, killing dogs and considering man as a provider of food. As the largest and most social of the wild canids, the wolf is the top carnivore in its ecosystem. It hunts in packs and therefore must kill prey, deer *Odocoileus sp* for instance, at least every several days on the average (Mech, 1970). This means its home range must be very large and it may travel great distances. In regions with high human density, or along frequently travelled roads, wolves are likely to be seen, and thereby to become targets of local ire. Therefore, in this paper I will discuss the following selected relevant parameters of the wolf's ecological niche: density of humans (chance of conflict with landowners), designated land uses in wolf range (assurance of suitable habitat in the future), prevailing local attitudes about wolves, wild prey availability, and interspecific competition.

CANDIDATE REGIONS OF NEW YORK STATE

With 128,400 km^2, New York is the largest state in the northeast. Its eight physiographic regions (Figure 7.1) differ strikingly in topography, climate, density of human population, and in the potential of each region to support wolves. The four uplands are sparsely populated, hilly to mountainous and mostly forested, whereas the four interspaced lowlands contain virtually all the major cities, commerce, transportation routes and prime agricultural land (Thompson, 1966). Clearly, candidate regions for reintroduction of wolves would be found only in the uplands.

The Appalachian Upland includes the rolling hilly Southern Tier throughout the western and central parts of the state, and the high rugged Catskill Mountains on the east. The Southern Tier is sprinkled with small villages, local businesses, marginal or abandoned farms, and three small industrial cities. The Catskills, including the state-designated Catskill Forest Preserve, are mostly enclosed in Catskill Park. Resorts and villages seasonally accommodate very large numbers of visitors and residents. The New England Upland contains second-home communities throughout.

The Adirondack Upland and the Catskill Mountains were recognized early as important natural areas. In 1892, the New York Legislature created Adirondack Park, an area larger than the State of Massachusetts, and the smaller Catskill Park. In 1894, the state constitution was amended to declare that the associated state-owned forest preserve lands "will be forever kept as wild forest lands." Despite this land use designation, development and growth has not been greatly restricted. The Adirondack Upland is bordered on the south, west and north by the agriculturally productive lowlands of the Mohawk, Black and St. Lawrence Rivers, respectively. Only the Tug Hill Upland is uniquely roadless and uninhabited, but like the Adirondack Upland, is surrounded by important dairylands. The acceptability of these areas for reintroduction of the wolf can be interpreted by analysis of pertinent parameters of the wolf's ecological niche.

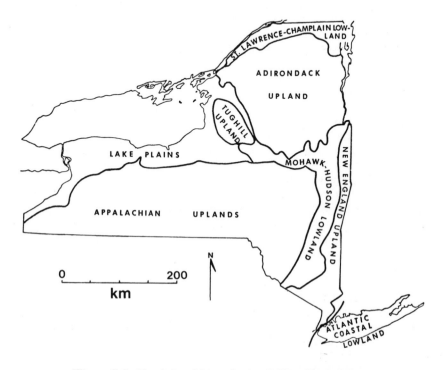

Figure 7.1: Physiographic regions of New York State. Within each region, climate, topography, vegetation, fauna, demography and commerce are remarkably homogenous [based on Thompson (1966)].

COMPATIBLE DENSITY OF DOMESTIC LIVESTOCK

If domestic animals are an easy and likely prey for the wolf, regions containing large numbers of livestock would not make good reintroduction sites. Cattle, horses, sheep, hogs and chickens are all potential prey; however, their vulnerability varies. Although very large numbers of chickens are raised in New York, modern methods maintain the birds entirely indoors. Sheep and hogs are raised in open farm yards and would be potentially vulnerable; however, sheep and hogs are kept mainly in the lowlands on a relatively small percentage of the farms (NYCRS, 1972). Further, because many farms keep only a few animals for home slaughter, and only a few keep large numbers, the distribution of sheep and hogs is uneven in the lowlands and cannot be judged a significant industry in any of the uplands except the Appalachian. Horses are alert, fast animals which would be expected to defend themselves and to signal alarm if approached by wolves. Further, horses are raised as pets and for recreational use in New York. Therefore, their distribution is likely to be even more uneven than sheep and hogs, their numbers probably are lower, and they often are housed indoors. Because of this, chickens, sheep, hogs and horses do not appear to be highly vulnerable prey; their presence need not influence decisions to reintroduce wolves to New York.

Cattle are raised commonly throughout New York. Regions differ in total number of cattle, and in the numbers of farms containing cattle (NYCRS, 1972). Since they are often grazed continuously on open, often remote pasture, and because they are slow, passive animals, cattle may readily become a favored prey of the wolf. Therefore, presence of cattle appears to be an important factor determining whether wolves should be reintroduced into a region.

Ubiquity, rather than density of cattle or number of cattle per farm, probably will determine their vulnerability to wolf predation. Wolves may be as likely to kill single cattle on isolated farms as they would be to single out one animal from a herd on a larger farm. Therefore, density of farms containing cattle was calculated for each county and examined with respect to the large physiographic regions. Size of farms was not so different among regions as to have biased this calculation for areas with large farms. [This calculation is appropriate also because each head of cattle killed by a wolf would be considered of equal (and high) value whether on a large or small farm, and would be as likely to generate adverse publicity.]

Density of farms with cattle varies strikingly among regions (Figure 7.2). Therefore, the highest density of farms within Adirondack Park (AP), 0.12 farms per km^2, was taken as a reference density. Areas containing less than 0.12 farms per km^2 occur in the southeastern portion of the state, the Tug Hill Plateau, and AP. I would not propose Manhattan and Long Island for reintroduction of wolves, even if the wolf had first claim to the region; however, the Tug Hill and Adirondacks have a uniformly low enough density of farms to bear further examination. Both areas contain wild forest and virtually no farms. In this regard, they are similar to upstate Minnesota, the major wolf range in the lower 48 United States (Mech and Frenzel, 1971), and should be acceptable wolf habitat by this criterion. To give this threshold value perspective, areas of up to twice the threshold density are also shown in Figure 7.2. These regions contain mountainous, residential areas in the southeast, and Monroe County on Lake Ontario, which has a low density of cattle farms because of the presence of Rochester City. These areas would not be candidates for wolf reintroduction.

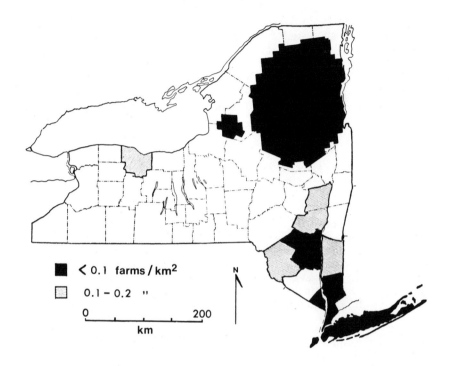

Figure 7.2: Density of farms with cattle in counties of
New York calculated from New York Crop Reporting
Service (1972).

COMPATIBLE DENSITY OF HUMANS

A human population density compatible with wolves might be
deduced by comparing present and recent wolf ranges, other factors
being considered equal. The only present extensive gray wolf range
in the lower 48 states is a small southern projection of the trans-
continental Canadian/Alaskan wolf range extending down into
northeastern Minnesota (Mech, 1973). Additionally, a vestigial population of perhaps a dozen wolves is thought to reside on the northern
peninsula of Michigan (Hendrickson et al., 1975). Wolves were reported present in northern Wisconsin until the 1950s, and have
recently recolonized the northwestern corner of that state (Thiel,
pers. comm.). Human population density in each of these regions
is certainly a major factor determining success or failure of wolf
populations in these areas (Hendrickson et al., 1975; Weise et al.,
1975).

Human population densities in northern Minnesota and Wisconsin, and in the upper peninsula of Michigan were examined using 1970 census data (U.S. Bureau of the Census, 1973). Population densities were calculated for total counties and also for each town within these counties. City and village populations were noted, and urbanized towns were distinguished. I make the assumption that human density in Wisconsin has changed little since the mid-1950s when wolves became extinct there. Patterns of distribution of people in each state, which are of diagnostic importance in determining success or failure of wolf population, are immediately apparent (Table 7.1).

Table 7.1: Mean Human Population Densities Within Wolf Ranges Compared to Two Areas in New York

	Persons/km^2
Minnesota	
Non-urban towns	1.3
Counties in primary wolf range[1]	3.7
Counties in peripheral wolf range[1,2]	5.9
All counties in ranges[2]	5.1
Michigan, Upper Peninsula	
Non-urban towns	2.8
Non-urban counties	4.4
All counties	9.3
Wisconsin, northern	
Non-urban towns	2.7
All northern counties[2]	4.4
New York, Adirondack Park	
Non-urban towns in high peaks and western regions	1.2
Eastern urbanized towns	7.5
New York, Tug Hill Upland	
Non-urban towns	1.0

[1] Ranges proposed by Mech (1977b).
[2] Except Duluth metropolitan area.

In Minnesota, Stenlund (1955) and Mech (1973, 1977b) considered the Superior National Forest and four contiguous counties to the west to be "primary wolf range." Surrounding that region, they recognized a "peripheral wolf range" approximately one county wide (about 80 km). Within those areas, people are very unevenly distributed. Essentially none reside within the entire Boundary Waters Canoe Area. In surrounding areas of the Superior National Forest and the Great Bog region to the west, people live mainly in small villages around the periphery and along the mining region of the Masabi and Vermillion Ranges. Thus, the tabulated 1.3 persons per km^2 belies the fact that there is a vast wilderness, measuring about 430 x 112 km, which is mostly uninhabited by people.

In Wisconsin, on the other hand, people are fairly uniformly distributed in low numbers in all of the counties across the northern fourth of the state. There are no large, uninhabited regions as in Minnesota. Most of these northern counties are forested. South of these counties, population density is higher. Except for occasional wolves and the recent colonizers from Minnesota, the last wolves reported in Wisconsin were observed in the 1950s in forested areas in the northern part of the state. (Thompson, 1952).

The upper peninsula of Michigan contrasts to both Minnesota and Wisconsin in that wide, north-south bands of dense human population occur in the eastern, central and western counties, alternating with bands of low population density. It is noteworthy that wolf sightings occur only within the two bands of low human population (Hendrickson et al., 1975).

Human population density within New York State should be compared to that of Minnesota, Wisconsin and Michigan. Only AP and the Tug Hill Upland deserve detailed attention because only these areas, overall, have population densities as low as Minnesota's primary wolf range. In AP, analysis is complicated by patchy distribution of population, by county boundaries overlapping Park borders, and by all counties, and even some towns, enclosing both villages and wilderness areas. The 1970 census (U.S. Bureau of the Census, 1973) showed that greatest densities of people exist throughout the eastern edge of AP, and also in towns associated with Lake Placid and surrounding all-season communities (Figure 7.3, Table 7.1).

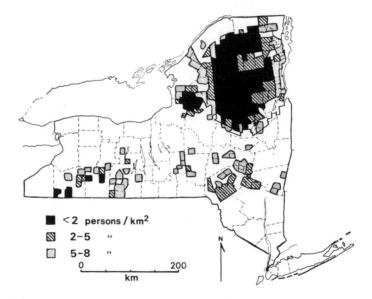

Figure 7.3: Human population density in towns of New York based on 1970 population census (U.S. Bureau of the Census, 1973). Only towns with densities less than 8 persons/km² are shown.

Throughout more than 60 percent of AP, including the uninhabited high peaks area in the eastcentral and the mostly non-urbanized western half of the Park, population density is approximately one person per km^2.

COMPATIBLE LAND USES

With the exception of AP, land use in all of New York State is governed by anthropocentric designations, e.g., residential, commercial, park, etc. Although many areas of the state contain suitable forested wolf habitat, there is no way to be assured that the habitat will be preserved in the future. It would be improper to consider reintroducing wolves where no such assurance of compatible land uses exists.

Two areas, in addition to the Adirondacks, would appear to offer long-term compatible land uses: the Tug Hill Plateau and the Catskill Mountains. The Tug Hill Upland (Figure 7.1), an outlying extension of the Appalachian Upland, rises 600 m above the encircling lowlands and, as noted above, is one of the least settled parts of the state. In all likelihood, it will remain so since poor soils and excessive snow have discouraged agriculture (Thompson, 1966), indicating that it will remain compatible with wolf cohabitation in the future.

The Catskill Mountains rise ruggedly at the eastern end of the Appalachian Upland. This region of high forested mountains is mostly enclosed in Catskill Park, and therefore is protected by the state constitution. Land uses in Catskill Park are restricted to those compatible with the scenic qualities of the region, and seemingly would be compatible with wolf habitat. However, the Catskills serve as a popular resort and second home area for tremendous numbers of residents of the nearby New York City metropolitan region. Densities of humans (Figure 7.3) at least seasonally exceed the threshold of one to two persons per square km. Therefore, it appears that wolf reintroduction into the Catskills is not advisable.

Land uses in the 26,000 km^2 AP are tightly controlled by the Adirondack Park Agency. State lands contain wilderness and wild forest designations. Most of the private lands are owned by large companies involved in forest production. Therefore, these tracts are mostly devoid of human habitation and undergo regulated logging operations. Correlating with population distribution, land uses related to commercial (except for logging) and residential use are clustered mostly in eastern parts of AP (Figure 7.4, compare to Figure 7.3). This results in at least four large areas in the east, west and north which will retain permanently their wild character. The remote parts of the eastern half contain rugged high peaks (up to 1,500 m), crisscrossed with well maintained trails popular with back country hikers and mountain climbers. There is little winter deer

(O. virginianus) habitat and few deer (Severinghaus, pers. comm.). The western half contains lower, less rugged mountains interspersed with rolling plateau, lakes and rivers, and a few small villages. In the north, topography is still more gently rolling; some of the rivers contain hydroelectric generating plants. It is instructive to compare these areas to other parks presently containing wolf packs. Isle Royale in Lake Superior, Boundary Waters Canoe Area in Minnesota and Algonquin Provincial Park in Ontario are used by thousands of campers, hikers and canoers annually, yet wolves are rarely seen by visitors (Allen, 1979; Peterson, 1977; Mech, 1970; Ontario Department of Lands and Forests, 1970). Recreational back-country use clearly is compatible with wolf presence.

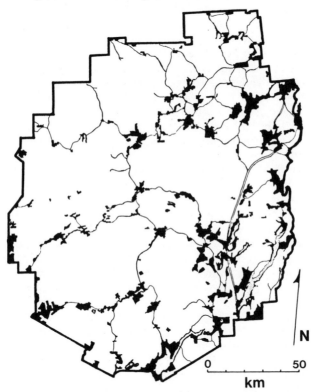

Figure 7.4: Land uses in the Adirondack Park based on the Land Use and Development Plan of the Adirondack Park Agency. Private resources management areas (forests) and all state land classifications are combined in the unshaded areas because the land uses are compatible with presence of wolves. All private land classifications, except rural use, which are incompatible to presence of wolves are shown by the shaded areas. Rural use areas containing farms with cattle in any abundance are located along the southeast, east and northeast periphery of the Park. The expressway traversing north and south would not be incompatible with wolf presence because no hunting is possible along it.

ADEQUATE AREA

Wolves normally move over great distances, whether travelling in a pack or as lone individuals. Wolf packs tend to remain within a territory, or home range, with rather established boundaries. Northeastern Minnesota, which is ecologically similar to the Adirondacks, had an essentially saturated wolf population with polygonal pack home ranges of about 200–400 km^2 (Mech, 1973, 1974b, 1977a,c). The greatest dimension of most of these territories was about 32 km.

On occasion, packs travel greater distances. The longest distance yet recorded of wolves moving with a pack in remote northern areas of North America was 296 km (Kuyt, 1962), although these particular packs traditionally migrated great distances. Lone wolves, not affiliated closely to a pack, typically move among pack territories over even greater areas than pack animals (Mech and Frenzel, 1971).

It is instructive to note the distances moved by released wolves in two previous reintroduction attempts. When five two-year-old captive-reared wolves were released 280 km from the home colony, four animals travelled 140, 160 and 280 km (Henshaw and Stephenson, 1974). Two of these were killed along a direct line between the release site and the colony, and one returned to her cage; therefore, we concluded that they had "homed." In a similar study, four wild-caught wolves from Minnesota were released in Michigan's upper peninsula. Weise et al. (1975) described three of the wolves moving over a 4,100 km^2 area, then settling in a 637 km^2 area about 88 km from the release site. After two of these were killed, the remaining female again moved over a much larger area. It appears that wolves translocated to an unfamiliar place will move over greater distances than wolves in established territories, that this distance may be greater when released where no other wolves have territories, and that the animals may move great distances due to a tendency to home. Such space requirements would have to be accommodated in any release zone.

The large areas with virtually no human habitation in AP would seem to be prime candidate areas for wolf reintroduction. These areas are dissected, however, by all-season highways connecting more than 150 villages (Figure 7.4). Not shown in Figure 7.4 are many smaller roads, travelled in summer, and open to at least snow machine traffic during winter. The result is that no point in AP is more than 32 km from the nearest major road, and many are even closer to off-road vehicular traffic. The judgment must be made whether reintroduced wolves would be imperiled by such proximity to civilization.

If released wolves are wary of civilization, they might survive near it without incident; however, the more access for vehicles in the wolves' area, the more likely wolves may be seen and fall victim to trappers and hunters, or their vehicles. The captive-reared wolves released in northern Alaska all reentered human habitation, apparently considering man to be a provider of food (Henshaw et al., 1979). All but one were killed. Of the wild-caught wolves released in Michigan, two were shot, one was trapped and shot, and one was killed by an automobile (Weise et al., 1975). If ready access to much of the back-country in AP is possible by all-terrain vehicles, safety of reintroduced wolves would depend heavily on public attitudes.

PUBLIC ACCEPTANCE OF WOLF REINTRODUCTION

Presence of humans *per se* is not the expected cause of conflict between wolves and man. Prevailing attitudes in the area either foster public acceptance or rejection of the wolf.

Two distinctly different groups of people use AP. Most are transients who seasonally use the park for many forms of recreation; many maintain second homes there. The other group are year-round residents—"Adirondackers"—who find employment in the region. Many Adirondackers are second or third generation residents. The former group would be expected to be well educated and earning more than the average New York wage. If they hunt, it is recreationally. The Adirondackers, on the other hand, are substantially less educated and earn an average income which is only 81 percent of the New York State average wage (U.S. Bureau of the Census, 1973). Some claim to "hunt for the table" out of necessity.

Public attitudes may be deduced from these facts. The former group of seasonal transients would be expected to be interested in public information regarding a proposed reintroduction of wolves. They probably would develop a good level of understanding. Possibly a large percentage do, or would, support such a venture. The Adirondackers, on the other hand, would not be expected to be educable or easily persuaded. An unknown percentage of this group probably has a basic interest and sympathy toward wolves; however, many Adirondackers have publicly and privately stated their general disapproval of wolf reintroduction into AP. One Adirondack wildlife biologist who is familiar with local opinions has predicted that "a liberated wolf hide would bring $500–1,000 on the Adirondack 'I have one of those devils' market." Adirondackers' expressed and predicted attitudes have no doubt contributed to the reticence of the Bureau of Wildlife, New York State Department of Environmental Conservation (DEC) to initiate studies of wolf transplant feasibility. (Such a study has just begun during the writing of this manuscript.)

If a political majority was all that was needed, it is possible that public acceptance might not be a limiting factor. A polling of the members of the New York Chapter of the Wildlife Society showed that 63 percent favored reintroduction of both the wolf and the lynx *(Lynx canadensis)* into AP (The Wildlife Society, 1975). Elsewhere, Hook and Robinson (in this volume) demonstrate there may be a "silent majority" of citizens, even in AP, who are sympathetic to having wolves as part of their ecosystems. Carley (1979) reported complete public support when two wolves were placed on a remote island.

But a political majority does not necessarily determine public behavior, especially if a few individuals can effectively thwart a program. The Adirondack wildlife biologist, quoted above, pointed out that a majority of Adirondackers are not trappers, yet those who trap control beaver *(Castor canadensis)* numbers throughout all areas, including deep in the interior. He feels this to be comparable for the bobcat *(Felis rufus)* population in AP. And he speculates, with good reason, that "only a very few, (perhaps) less than 500 or maybe less than 200 in opposition to wolves would control their survival." When he recounts how a good quality, all-weather road over 20 miles long (which was desired by the locals but forbidden by land use designations) mysteriously appeared through an extensive wilderness section literally overnight, we must place credence in his reasoning regarding the effect of a determined minority of citizens.

It is axiomatic that no wolf reintroduction should (could) be carried out today without full public disclosure and public involvement. It cannot be concluded that reintroduction of wolves into AP is not feasible based on factors related simply to sheer presence, numbers or distribution of people. No conclusion could be drawn prior to a complete, formal, well-planned study of public acceptance with a truly representative sample. Part of such as study should examine the degree to which public sentiment can be influenced by a well-grounded public information program, and the degree to which public cooperation might be obtained.

FOOD PREFERENCES OF THE WOLF

Prey of the wolf are mostly large (Mech, 1970). However, because wolves are opportunistic feeders, prey are taken in proportion to their convenience and availability. The most numerous large ungulate in a region usually forms the base of the wolf's food reserves, and other species are taken according to their seasonal abundance (Theberge and Cottrell, 1977). Clearly, before transplantation of wolves to AP, deer *(O. virginianus)* and beaver populations must be examined to determine if they could withstand wolf predation pressure added to that of human hunting, winter starvation and other losses.

ECOLOGY OF THE WHITE-TAILED DEER IN THE ADIRON-DACKS

The Adirondacks plus the surrounding northern counties contain 46 percent of New York's deer range, yet produce only 20 percent of New York's deer (Table 7.2). Deer productivity is less than one-fifth that of the rest of the state. The Adirondack deer population, at an average elevation of about 600 m is, in effect, ecologically at the northern limit of white-tailed deer range.

Table 7.2: Density of Deer in New York Management Units in a Typical Year—1977

Deer Management Areas	. . . . . Area			. . . . Number of Deer				
	km²	% of Deer Range	% of State	Legal Buck Take/km²	Deer/- Buck Taken[1]	Average Deer/- km²	Total Deer Popula- tion	% of NY Deer
Adirondacks and northern counties	37,967	46	31	0.2	11.5	2.2	82,090	20
Catskills and eastern counties	18,878	23	15	1.2	8.6	10.6	163,274	41
Central and western counties	25,094	30	20	1.0	6.4	6.4	155,024	39
Total	81,939	100	66				400,544	100

[1] Values calculated by Bureau of Wildlife, New York State Department of Environmental Conservation.

Deer abundance in the central Adirondacks has varied historically with the state of vegetation, but low deer productivity in the region has prevailed. The aboriginal human population was low and perhaps seasonal, subsisting not by hunting but by gathering plant materials. In the presence of a mature virgin forest and large predators, deer could not have been numerous. Despite significant reduction in wolf density by the mid-1800s, deer density did not increase significantly until about the 1870s. At that time, a new industrial process permitted use of all softwood trees for paper pulp and the forests were rapidly clear-cut (Ketchledge, 1965). Subsequent shrub and young forest communities sustained a rapidly growing deer population until about 1910, when deer increased to such densities as to completely suppress regeneration of hemlock and other hardwood species (Behrend et al., 1970). Without further regeneration of browse species, the forests again matured and the deer population declined proportionally. The deer population has irregularly decreased, limited by winter starvation due to dwindling forage stock (Severinghaus, pers. comm.).

DEC considers that 97 percent of AP is deer range. As noted, however, 90 percent of the area occurs on nutrient-poor mineralized soils and now comprises slowly maturing, second-growth northern hardwood forests. Winter snows characteristically accumulate and remain on the ground into late spring. Thus, deer are forced to feed entirely on woody vegetation of poor nutritional value for up to 180 days each winter.

William Severinghaus, who studied the Adirondack deer populations for 40 years, interpreted the combined effect of deep snow and limited food on the AP deer population. Deer mostly collect in large numbers during the winter in concentration areas or deer yards. Often these sites are in locations providing some degree of shelter, but may not contain a great abundance of browse. Deep snows make travel difficult for deer, and they tend to remain in the winter yards even after the food supply is exhausted. Severinghaus (1976) found that the energy expenditure required for fawns to move through more than 0.4 m, and for adults to move through more than 0.5 m, of snow is greater than can be sustained by available food and stored fat reserves. When snow depth exceeds 0.4 m for 60 days, or 0.5 m for 50 days, winter starvation is common and the population is adversely affected. Thus, the Adirondacks are an area of low deer population and high natural winter mortality.

The Adirondack deer herds typically sustain an annual winter starvation mortality of 12-18 percent in mild winters, and up to 42 percent in severe winters. Sources of mortality include hunters' legal and illegal takes and their crippling losses, collision with vehicles, predation (mostly by untethered dogs), and winter starvation (Table 7.3). The disproportionately high winter starvation in the Adirondacks with respect to the rest of New York (up to 10 times higher) is dramatic. Severinghaus (1976) considers winter fawn starvation so expected in AP that he proposed deer management criteria which do not "attempt to reduce the deer population to such a low level that losses of fawns from starvation during severe winters can be prevented," and also which do not "attempt to have the deer population continually increase to levels where losses of fawns from starvation occur during mild or moderate winters."

Table 7.3: Sources of Deer Mortality (Percentages) During a Severe Winter—1958[1]

Area	Legal Hunting	Crippling Loss	Illegal Hunting	Road Kills	Predation	Starvation and Other
Adirondacks and northern counties	27.0	8.9	19.5	4.7	7.1	32.8
Catskills and eastern counties	48.6	1.4	2.3	29.6	10.9	7.2
Central and western counties	69.7	5.8	5.5	11.9	3.8	3.3
Statewide	46.3	4.9	8.8	17.0	7.9	15.1

[1] Provided by Bureau of Wildlife, New York State Department of Environmental Conservation.

During a series of mild winters (1963-68), even in the presence of the greatest hunting pressure ever for any age of both sexes, deer density increased (Figure 7.5). This growth was most dramatic in AP where massive poaching does not occur to the extent it does in peripheral areas. When a series of severe winters killed a large portion of the fawn crop, reproduction and replacement in the following years was lower each year and the population declined (1968–1969; 1976–1978). This effect compounded in subsequent years (1969–1971) as adult breeding stock was not replaced. Deer productivity, reflected in fawn to doe and fawn to adult ratios in the hunters' legal take, increased during mild years (1960–1968). When antlerless hunting (does and fawns) was restricted, the population increased even during severe winters (1975-1977). As a result of Severinghaus' perseverance, the Adirondack herd was regulated until 1970 by manipulation of the antlerless harvest.

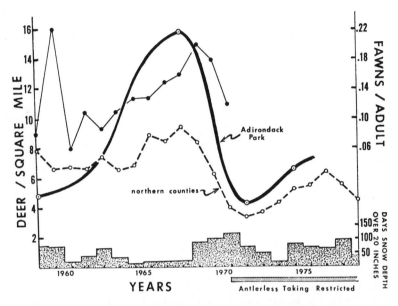

Figure 7.5: Relationship of deer density to severity of winter in the Adirondacks. Deer reporting unit includes all counties in the Adirondack Upland, Tug Hill Upland, St. Lawrence-Champlain Lowland and adjacent portions of the Lake Plains. Curve for "northern counties" describes entire deer reporting unit. The Adirondack Park (AP) contains the Adirondack Upland only. This is based on data developed by the Bureau of Wildlife, New York State Department of Environmental Conservation, and Severinghaus (1976). Snow depth data are the means of five weather reporting stations in the AP (NOAA, 1958–78).

ECOLOGY OF THE BEAVER IN THE ADIRONDACKS

Information on the density of beaver in AP is difficult to obtain and must be conjectured from aerial censuses and pelt tagging data. Aerial surveys identify all beaver colonies and estimate the number which are currently active based on the presence of fresh food cuttings. In Fulton, Hamilton and Warren Counties, the DEC experimentally closed the beaver trapping season for several years to allow populations to increase to carrying capacity. Based on these observations and the determination that regeneration of plant food species takes about three years, DEC biologists believe carrying capacity to be about 30 percent of all potential colony sites.

Excellent beaver habitat exists throughout much of AP. Very few people reside there and streams, ponds and wetlands are abundant. In areas within the central AP, beaver density is about 0.9 individuals/km^2, which is equivalent to about 10 kg/km^2 of edible tissue (Table 7.4), assuming an average body weight of 11.3 kg. In peripheral areas where human population is more dense and trapping may be more intense, beaver density is about one-fourth that of the central AP (Table 7.4).

Table 7.4: Beaver Populations (or Densities) in the Adirondacks[1]

County	Area (km^2)	Colony Sites	Percent Occupied	Number of Beavers	Individuals per km^2	Beaver Tissue Available (kg/km^2)
Fulton	1,287	457	40	1,104	0.9	10
Hamilton	590[2]	634	14	534	0.9	10
Clinton[3]	2,742	250	52	780	0.3	
Franklin[3]	4,334	409	19	466	0.1	
Warren[3]	1,150[2]	270	21[4]	336	0.3	
			15[5]	246	0.2	

[1] Data furnished by Bureau of Wildlife, N.Y.S. Department of Environmental Conservation.
[2] Censused area only.
[3] Outside of wolf range in Adirondack Park.
[4] At end of three year period closed to trapping.
[5] After first year of resumed trapping.

POTENTIAL FEEDING ECOLOGY OF THE WOLF IN THE ADIRONDACKS

To estimate potential wolf predation rates on the Adirondack deer herd, wolf consumption of deer throughout the year must be calculated. Three approaches will be used here (others are possible): (1) estimation of how many deer a wolf needs to take to meet its projected metabolic requirements; (2) an accounting of how many deer wolves have taken in other locations; and (3) projections of effects of wolf predation on the AP deer population based on analysis of a predator-prey model for AP. A combination of these methods will provide a basis for concluding whether the populations of deer and other prey species in AP would be capable of sustaining a reestablished wolf population.

The wolf is capable of gorging itself, then waiting up to a week or more before the next kill; but under some field conditions it may feed more frequently (Mech, 1970). In captivity, it may feed many times a day. Thus, the various methods of estimating the number of prey necessary to meet the wolf's physiological and ecological needs yield quite different values. Since the decision on adequacy of prey availability is contingent upon realistic projections of wolf predation, several methods of estimation are elaborated and the best estimate is proposed. This value is central among the widely varying estimates proposed by others.

Metabolic Needs of the Wolf

Comparing the estimated energy consumption of free-ranging wolves with the energy content of prey biomass will permit estimation of predation rate, if that rate is determined mostly by metabolic requirements. The average daily metabolic rate (ADMR) of free-ranging animals of many species is about two to three times the basal metabolic rate (BMR) (Gessaman, 1973). It includes the minimal maintenance metabolic cost as well as costs of digestive, thermoregulatory, and locomotory work. Because basal energy metabolism, and therefore food consumption, are functions of body weight to the three-fourth power, a 36 kg wolf may be projected to require about 725 kcal/day as its minimal energy requirement (Kleiber, 1961). Thus, an average 36 kg wolf might require 1,450–2,200 kcal/day to carry out its normal free-ranging activities. Assuming a deer contains 50 percent muscle and 50 percent viscera, then the averaged total caloric content of digestible tissue, based on a lean whole beef, is 295 kcal/kg (Geigy, 1959). (Even lean cattle may have a higher fat content than deer, and this calculated value may be high.) Thus, the 36 kg wolf may be expected to require 1.0–1.5 kg of deer/day, or 545 kg/year (Table 7.5). With about 54 kg of edible meat per deer (estimated from Severinghaus and Gottlieb, 1956), this equates to 10 deer/wolf/year.

Cowan (1947) reported, and Pimlott et al. (1969), Mech (1973), and Peterson (1977) all confirmed, that wolves tend to achieve an average maximum density of about one wolf per 26 km², even when food supply might support a greater density. Assuming that reestablished wolves in AP also would develop this density, even with 3 deer/km² available (Figure 7.5), the wolves would consume only about 12 percent of the deer population each year. But this would occur only if wolves killed prey at a rate to just meet their projected average daily (ecological) metabolic needs.

Food Availability

Ecologists commonly calculate food available to the predator as the rate of kills multiplied by the estimated weight of edible tissue in the prey. Mech (1977a) reported that the Minnesota wolf population appeared to sustain itself on eight deer-sized kills/wolf/120 days of winter, or 3.6 kg of deer/wolf/day. With a 60 percent greater food availability, they increased in number, while with 18 percent less, the wolf population declined. The average annual kill rate was estimated to be 18 deer/wolf, or 2.5 kg of deer/wolf/day (Mech and Frenzel, 1971). In the presence of an abundant moose supply on Isle Royale, estimated food availability averaged 6.3 kg/wolf/day (Peterson, 1977). During a 20 year period of observation, the wolves increased from one pack to four, and approximately doubled their density on this ration (Peterson and Scheidler, 1978) (Table 7.5). Calculating as before, if wolves were to assume a density in AP of one per 26 km² and consume 3.6–6.3 kg of meat daily, then they would take 36–64 percent of the deer present.

Table 7.5: Rates of Food Consumption Under Various Conditions and on Different Prey

	kg of Deer/ Wolf/Day	kg of Deer/kg of Wolf/Day	Food Consumption in Ratio to BMR (x BMR)	Observed By:
. . . . .Estimates of Daily Minimum Maintenance				
Wolf, 36 kg, BMR	0.5	0.014	x 1.0	(Weight-specific minimal metabolic rate)
Wolf, 36 kg, ADMR	1.5	0.04	x 3.0	(Based on BMR)
Wolves, in zoo	1.1	0.03	x 2.3	Mech (1970), quoting Fletcher
Wolves, captive	1.6	0.04	x 3.3	Kuyt (1972)
Wolves, free-ranging	1.8	0.05	x 3.6	Mech and Frenzel (1971)
Dogs, large and active	1.7	0.5	x 3.4	Mayer (1953), cited by Mech (1970)
. . . . Estimates of Ad Libitum Feeding in Free-Ranging Animals.				
Wolves, population decreasing	2.5	0.07	x 5.1	Mech and Frenzel (1971)
Wolves, on deer	2.9	0.08	x 5.8	Kolenosky (1972)
Wolves, on deer	3.6	0.10	x 7.3	Mech (1977b)
Wolves, on moose	4.4–6.3	0.12–0.19	x 8.8–12.6	Mech (1966)
Wolves, on moose	4.4–10	0.12–0.27	x 8.8–20.0	Peterson (1977)
African wild dogs (Lycaon pictus)	5.4	0.15	x 10.9	Wright (1960)

Two conclusions are possible: either these values are excessively high as estimates of deer consumption by wolves, or wolves kill from two to seven times as many deer as would meet their metabolic needs. These calculated values represent only killed food available to the wolves, but do not necessarily represent actual amounts consumed. Indeed, it seems improbable that a 36 kg wolf could pass 18 percent of its body weight through its intestinal tract each day, even if digestive efficiency is lower when large amounts of food are consumed quickly. Calculating food *available* does not account for unutilized food which scavengers such as red foxes *(Vulpes vulpes)* and ravens *(Corvus corax)* eagerly consume, or other sources of loss by the wolves.

The estimates of food consumption rates by both methods may be interpreted best in relation to rates reported for wolves and similar species under a variety of conditions, including limited activity in captivity and free ranging in the presence of limited and abundant prey populations (Table 7.5). The calculated BMR of an adult wolf likely is a close estimate of its minimal metabolism; over 60 captive adult arctic wolves, over a 10 year period at the Naval Arctic Research Laboratory, did not appear to possess unusually high or low metabolism (Henshaw, unpub.). Thus, a food consumption of 1.5 kg of flesh per wolf per day, equivalent to the projected ADMR, is grounded on a firm physiological base and the strength of comparative physiological methods based on a wide diversity of species. It is presumed that free-ranging wolves do not expend unusually large amounts of energy for locomotory work, and that they do not decrease digestive efficiency greatly when eating abundantly available food. The fact that several observations of food consumption in Table 7.5 are close to the projected ADMR appears confirmatory. On the other hand, if the metabolic limit for increasing oxygen consumption, even during brief bouts of heavy work, does not exceed 18 times the BMR (Bartholomew, 1968), then the highest estimates of food consumption, based on availability of food for Isle Royale wolves, seem to be theoretically unlikely. Indeed, Peterson (1977) documented nonconsumption of up to 50 percent of edible tissue in some killed moose.

Although Pimlott et al. (1969) concluded, and Peterson (1977) confirmed that there is no evidence of excess killing or nonconsumption of large numbers of prey, Mech and Frenzel (1971) did report "surplus killing" of deer during a severe winter. One case of massive surplus killing of reindeer *(Rangifer tarandus)* in Norway was reported by Bjarvall and Nilsson (1976). If these estimates of food consumption and prey killing can be generalized, it will be necessary to conclude that wolves may, when conditions permit, kill prey quite in excess of their needs.

Wolves do not forage or conduct other life functions at maximal rates of work. Subduing a prey usually takes a short burst of energy and is done in an energy-conserving manner. Thus, we may surmise that the observed energy consumption rates, which are about 3 x the BMR plus added increments for extensive movement through deep snow, pup production, lactation, etc., represent the best projections of the ADMR. A probable ADMR equivalent to 1.8 kg of deer/wolf/day, or 0.05 kg/kg/day, equates to an expected annual consumption of 18 percent of the Adirondack herd.

It is noteworthy that the projected annual deer consumption needed to maintain a viable wolf population at a density of 1/26 km^2 is about the number of deer lost to winter starvation in mild winters in AP. It is tempting, therefore, to conclude that re-established wolves would have little additional impact on the deer population, especially that segment hunted by man. There is no assurance, however, that wolves would restrict their predation primarily to moribund deer.

These calculations are deliberately high to be conservative. Several factors could modify this projection. Wolves might achieve an average body weight of only 22–30 kg, yielding a 19 to 38 percent reduction in food demand. Furthermore, wolves could not inhabit all parts of the deer range in AP. Areas close to human activities would not be compatible with wolves.

A PREDATOR-PREY MODEL FOR THE ADIRONDACKS

The above discussion assumes predator and prey populations interact in ways independent of population densities; they do not. The deer population at all times will be dynamically tuned to predation pressure exerted by wolves. Further, at the northern limit of deer range, density-dependent and density-independent factors operate synergistically on the deer population.

Mech and Karns (1977) projected the possible role of wolves in a long decline and eventual rapid extirpation of deer in the Boundary Waters Canoe Area (BWCA) of northeastern Minnesota. In the presence of young forest habitat, food for deer was adequate and the deer population apparently was not affected by the wolves present. However, a series of severe winters debilitated the deer, making them easier prey. An increasing wolf population exerted increasing pressure and eventually extirpated the deer herds. By simulating this interaction on a computer and inputting probable predation rates of 15 and 20 kills/wolf/year, the authors confirmed that deer populations could be extirpated within six to nine years by predation pressure.

Patterns of changing deer density in the Adirondacks over a 20 year period suggest a similar predator-prey interaction, with man as the predator (Figure 7.5). Like the BWCA, the AP forests were maturing and nutritional value for deer was declining. The deer population increased in number during a series of mild winters in spite of hunters being permitted to hunt fawns and does. However, in the presence of this hunting pressure *plus* severe winters, productivity dropped and the population declined rapidly. With doe and fawn taking prohibited, the population began a slow recovery in spite of another severe winter. A series of severe winters coupled with no doe and fawn hunting caused the population to cycle near the level considered by Severinghaus as optimal (2.6 deer/km^2). It is clear that hunting pressure on reproductive stock in deer debilitated by severe winters can exceed their reproductive potential. Continuation after 1970 of hunting at rates similar to those of the mild winters of the mid-1960s would have driven the deer population to a very low level. However, human taking is (for the most part) subject to regulation. At a strategic time, reproductive female deer were spared and allowed to reproduce.

The pattern and mechanisms of the AP deer population decline are remarkably like those described by Mech and Karns for the BWCA. The principal difference is that taking of reproductive deer by wolves is not sensitive to government regulations. In the Adirondacks, where deer density is about half that of the BWCA, re-established wolves could likely be tolerated by the deer populations except during a series of severe winters. Then, excessive killing of deer might cause the deer population to decline to a very low level until wolf numbers also declined. The AP deer population presently cycles through dramatic high and low densities. In all likelihood, these dynamics would become more exaggerated if wolves were reintroduced into the Adirondacks. However, this predator-prey model assumes that wolves and deer are uniformly distributed throughout their habitat, which would be unlikely. Further, it implies that a full annual food quota is drawn from the winter deer herd, which would not be the case (i.e., it does not account for the effect of summer release from heavy predation pressure on the deer as the wolves switch to alternate prey species). When these factors as well as the conservative nature of the above analysis are considered, we can conclude that excessive predation by wolves would occur only under extreme conditions.

THE NEW ENGLAND COYOTE

The top natural predator in the Adirondacks today is the New England coyote *(Canis latrans var.)*. Any plan to re-establish wolves must consider the niches of these two large canids, and the interactions which might occur between them. The coyote has been in New England since about the mid-1930s, and has since spread southward (Severinghaus, 1974). Physically, the New England coyote is larger than its western relatives; its mean body weight is 20 kg as compared to 14 kg for the western coyote (Severinghaus, 1974). As a result, it has been variously called wolf, wolf-dog hybrid, wolf-coyote hybrid and coyote-dog hybrid. Studies of hand-reared and inbred New England coyotes outbred with wolves and dogs provided inferential evidence of wolf ancestry, or at least little evidence for dog ancestry (Silver and Silver, 1969). Regardless of route of arrival and ancestry, the New England coyote found an unfilled top predator niche in the northeastern U.S. and has filled that niche.

The New England coyote is truly omnivorous like its Western Plains ancestors (Johnson and Hanse, 1979), taking "anything edible," including fruits, berries, corn, insects and carrion, and many species of smaller mammals and birds including mice, skunks, rabbits, woodchucks, porcupines, snowshoe rabbits, squirrels and muskrats (Fick, 1975; Chambers, 1980; Ontario Department of Lands and Forests, 1970). Consequently, its home range and range of daily movement are probably smaller than the wolf's (Andelt and Gipson, 1979). While most stomach content analyses or scat analyses divulge parts of fetal or newborn deer fawn, it is unknown whether these were killed or taken as carrion. Hamilton (1974) reported white-tailed deer occurred in about 40 percent of all winter scats analyzed in AP during the 1950s. Twenty years later, Chambers (1980) found that about 89 percent of the winter scats contained deer parts (Table 7.6). He and Hamilton agreed that the feeding niche of the New England coyote had shifted through the years to heavier dependence on deer, especially during the winter. During the summer, Chambers found evidence of every prey species discussed above, but deer remains still occurred more frequently than other species.

Table 7.6: Percent Occurrence of Prey in Scats of Adirondack Coyote[1,2]

	. . . . . .Winter.		. . . . . .June.	
	1956-61[3]	1975-77	1956-61[3]	1976
	(240)[4]	(111)	(218)	(54)
Deer	39	80	31	80
Snowshoe Hare	33	14	56	21
Red Squirrel	8	2	4	0
Mice	3	5	2	37
Birds	2	0.4	4	2

[1] After Chambers (1980).
[2] Columns do not sum to 100% because scats may have contained more than one species.
[3] From Hamilton (1974).
[4] Numbers in parentheses are numbers of scats.

Whether deer are killed frequently by Adirondack coyotes, or whether coyotes take moribund and dead deer and deer viscera left by hunters, is still not established. It was noted above that 10-40 percent of the AP deer herd dies of starvation each winter and are available to coyotes. Of inferential importance, Chambers compared food items in scats of coyotes and red foxes (Chambers, 1980). Although foxes are not known to kill deer, deer parts occurred in 56 percent of winter fox scats and 37 percent of May–June fox scats. It is possible that the entire late winter and spring consumption of deer by coyotes could be taken from moribund or dead deer.

Except for starvation-weakened adults and very young fawns, deer would probably be too formidable a target for a 20 kg canid. [Truett (1979) describes western coyotes being driven off by single mule deer does.] Cooperative hunting would probably be necessary.

Informal reports indicate several observations of two Adirondack coyotes cooperating in the taking of apparently healthy deer. Such cooperative hunting may also occur in the western coyote (Hamlin and Schweitzer, 1979; Truett, 1979). In all likelihood, then, the New England coyote has the predisposition to hunt cooperatively, enabling it to still better fill the wolf's vacant niche in the Adirondacks.

A MANAGEMENT PLAN FOR WOLVES IN THE ADIRONDACKS

If wolves were to be reintroduced into AP, a comprehensive management plan must be followed. Such a plan must provide the wolf with appropriate habitat, i.e., adequate space and food, and protection. These must be guaranteed in perpetuity. Clearly, the re-established wolves must live in remote areas where future land uses will not be conflicting. Each of these requirements could be met in the Adirondacks except, perhaps, that of adequate protection. Despite its great size, AP is small enough and so segmented that the guiding principle of a reintroduction plan would have to be *containment*. As advocated by Mech (1979), wolves would have to be maintained within designated areas.

Although vast areas of the AP could accommodate wolves, others could not. Incompatible areas ideally should remain wolf-free. Thus, it is possible to propose "wolf range" and wolf-free zones, but they could be maintained so only if the wolf's status on the U.S. Endangered Species List is modified for animals inside of AP. The Endangered Species Act of 1973 forbids the taking, harming or killing of any species having 'endangered status' (Fish and Wildlife Service, 1978). Species classified as 'threatened status' may be managed subject to specific regulations promulgated for that species in that region. Recently, the status of the wolf in part of Minnesota was modified to 'threatened.' Now the wolf receives full protection in part of its "primary range" in northern Minnesota, but south and west in its "secondary range," problem wolves may be removed or killed by the U.S. Fish and Wildlife Service (Mech, 1977b). Wolves in the Adirondacks would have to be similarly managed.

The entire central, western and northern parts of AP can be considered "wolf range" capable of meeting each of the criteria above (Figure 7.6). As was shown in Figure 7.4, areas of greater than minimal human activity are concentrated in the east and around a few communities. These areas should be designated and maintained free of wolves. Further, immediately outside the AP is prime dairy country; no wolves should locate in these areas. A wolf-free buffer zone entirely surrounding the wolf range would reduce the tendency of wolves to leave the Park (Figure 7.6).

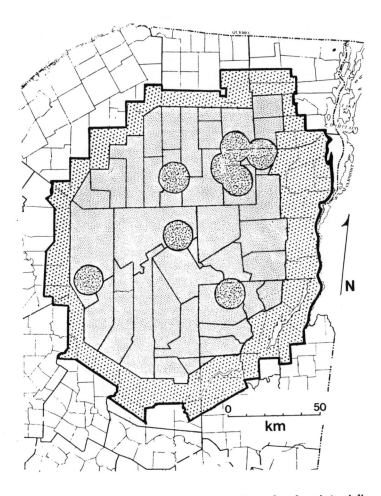

Figure 7.6: Conceptual scheme for reintroduction of wolves into Adirondack Park, incorporating designated "wolf-free" and "safe-for-wolves"
zones. Wolves would be permitted free activity within "safe-for-wolves"
zones (fine stipple). These are the high mountains zone in the east (ca.
2,330 km^2), the low mountains zone in the south (ca. 2,600 km^2), hills
and lakes zone in the west (ca. 2,600 km^2), and low mountains zone in
the north (ca. 2,850 km^2). Fewer wolves might leave AP if a wolf-free
buffer zone (dots) is maintained around the periphery of the Park. DEC
officers would be responsible for maintenance of this zone. The width of
this zone is about 8 km from the AP boundary or areas of dense human
habitation. Certain villages, for instance those which have populations of
more than 2,500 year-round, would maintain wolf-free zones (hash) in
8 km radii from the village center. These zones would be maintained by
permitted public hunting. Wolves could move unhindered near other
smaller villages unless they become nuisances, in which case conservation officers would be responsible for resolving problems. County and
town boundaries are shown to indicate basis of location of limits of zones.

Wolf-free zones should have logical and accepted boundaries. The buffer zone surrounding the wolf range should be approximately 8 km wide along the north, west and southern boundaries, and along the large lakes in the southeast (Figure 7.6). North of these lakes, the wolf-free zone should be 16–32 km wide and would encompass most of the population centers. Inner boundaries of the buffer zone should, for the most part, follow town lines for political expediency. Around villages with adequate human numbers for conflict, e.g., hamlets with a year-round population greater than 2,500 inhabitants, a wolf-free zone 8 km in radius should be maintained (Figure 7.6). This would permit peace of mind for people who harbor outdated fears of wolves, and it would reduce incidents of wolves killing or copulating with village dogs. The extensive buffer zones are designed to "protect" dairy farmers outside of AP and selected hamlets. Wolves trespassing into these zones could be trapped or killed by DEC conservation officers.

The areas among the wolf-free zones should be maintained safe for wolves. Comparison of Figures 7.4 and 7.6 indicates at least four vast wild regions in the east-central (high peaks area), southern, western and northern AP containing about 2,330, 5,200, 2,600 and 2,850 km^2, respectively. This total wolf range of about 13,000 km^2 represents 50 percent of AP. It could be expected to contain 33,500 deer at the optional density of 2.6 deer/km^2. In the unlikely event that wolves filled the entire zone to saturation (one wolf/26 km^2), about 500 wolves would reside in AP.

The scheme presented here clearly is a conceptual one. Wolf-free zones are proposed primarily because such zones would partially placate an unsympathetic public. Unpacified, these citizens could readily preclude the success of any reintroduction program either by exerting political pressure against starting a program, or by seeking out and killing reintroduced wolves. Belief in wolf-free zones might turn dairy farmers outside AP, and AP villagers inside such zones, into at least skeptical cooperators.

No assumption is made, however, that wolves could be prevented from using wolf-free zones. I recognize that these extensive wolf-free zones could not be maintained truly clear of wolves nor policed by conservation officers to prevent poaching. One approach to gaining public cooperation for "safe-for-wolves" zones in the interior might be to permit local citizens to gratify their interests in hunting wolves by permitting wolf-taking by hunters within the wolf-free zones.

The dimensions of the wolf-free zones and the minimum population of hamlets with a wolf-free zone were arbitrarily chosen. Dimensions and locations might be adjusted through public negotiation. For instance, wolf-free corridors within 1 or 2 km of selected major highways might be maintained by open hunting (much of the deer hunting occurs in these same narrow corridors). Wolf-free zones 8 km in width from the borders of all hamlets, however, would reduce these zones to unworkable dimensions (Figure 7.7). [Indeed,

during review of this manuscript, Mech (pers. comm.) and an anonymous reviewer both suggested that wolf-free zones around villages
might be reduced to a 1.6 km radius or eliminated completely without increasing problems with nuisance wolves.] Modified endangered
species status for the wolf and the use of wolf-free zones should be
considered essential for any proposed wolf re-establishment effort
in AP. Permitting Adirondackers to take trespassing wolves might
work if it could be adequately policed.

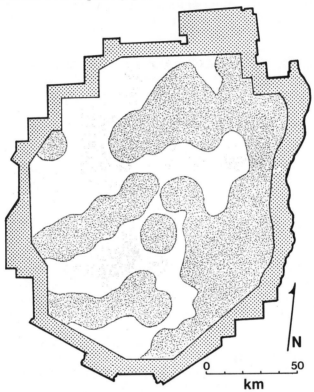

Figure 7.7: Another scheme for reintroduction of
wolves into the Adirondack Park. Shading of zones is
as in Figure 7.6. Eight km wide wolf-free zones around
all villages and areas of human habitation (see Figure 7.4)
would isolate small wolf-safe zones. Such a scheme
would be untenable.

Development of a comprehensive management plan for reintroduction of wolves will not assure success. The vagaries of climate
and prey, and the wolves' responses to disruption of established life
patterns and transplantation to an unfamiliar location at a strange
latitude and longitude, the strength of social bonds, ability to produce pups and public response will all affect survival of the trans-

planted wolves. If the decision is made to attempt such a reintroduction in AP, failure of the first pack, or even several after it, to establish themselves should not be considered a failure or allowed to cancel the effort. Much could be learned about the ecology and behavior of the wolf with each attempted reintroduction.

It is imperative that no attempts at reintroduction be attempted capriciously by amateurs. Wolves in captivity, i.e., socialized to man, may be expected to re-enter civilization and become nuisances immediately; they simply are unacceptable candidates for reintroductions. Every nuisance wolf and every failure, as interpreted by a critical public, will make more difficult the possible return of the wolf to otherwise feasible parts of its former range. An unpopular reintroduction in one location could foreclose reintroductions in other candidate regions.

NICHE ANALYSIS AND REINTRODUCTION: A SUMMARY

The factors most influencing wolf reintroduction in New York relate to the animal's size and its social hunting behavior. As the largest canid, it depends on large ungulates as its principal prey. Feeding as a pack, wolves must travel widely to obtain food for the group. In the presence of adequate prey, we could expect them to settle down to a defined home range.

The wolf is an efficient and opportunistic hunter. It may be expected to live in ecological balance with its prey. However, in AP, deer are at the northern climatic equivalent of their geographic range, and are under climatic stress even during "mild winters." With added predation pressure, even though potentially limited by intrinsic behaviors of the wolf, demands on the deer can be excessive. Under some circumstances, deer may be virtually eliminated. Wolves will then decline until the deer population recovers. Such dramatic dynamics are common in wild ecosystems. However, in managed ecosystems where hunters expect a certain deer density every year, large cycles in deer density may be quite unpopular. Administrative choice depends on the management goals and the value placed on the aesthetic knowledge that an ecosystem contains top predators.

The New England coyote may survive in the Adirondacks more easily than the wolf. Being less social, it need not hunt as widely. It may be less likely to conflict with civilization. Its public image is less ominous to those who retain outdated images of the two canids.

It could be argued that the coyote has adapted to fill the wolf's vacant niche in AP. It would follow, then, that returning the wolf to AP is not necessary, that it is ethically superfluous. An alternative view finds AP a natural laboratory for studying competition between these large and similar canids. Returning the wolf would permit extensive investigation into the niches of these animals.

WOLVES IN THE ADIRONDACK WILDERNESS?

Adirondack Park contains adequate space which has been designated to remain mostly undeveloped (although no longer wilderness). There is probably adequate isolation in the remote areas, and this isolation can be facilitated by hunting wolves in predesignated wolf-free zones near civilization. The Adirondacks, while not highly productive, contain acceptable habitat with a mix of prey species and a surplus of deer most years. That surplus could be expected to disappear following several severe winters which would make the deer unusually vulnerable to predation. The major competitor, the coyote, is there today, but might be displaced through intense competition.

In the end, man has so modified the boundary conditions that decisions do not turn on these ecological questions but, rather, on sociological questions. A conscious decision must be made whether society *as a whole* desires to manage the ecosystem for natural trophic balance or for man-facilitated stability. The former is more ethical and aesthetic; the latter may be more regular and predictable.

Acknowledgements

I thank C.W. Severinghaus and Drs. L.D. Mech and P. Sauer for critically reviewing this manuscript, and G. Rasmussen and M. Brown of the Division of Fish and Wildlife, New York State Department of Environmental Conservation, for furnishing data.

Much of the content of this paper is based on studies between 1966 and 1975 sponsored by the Arctic Institute of North America with financial support of the Office of Naval Research under contract N00014-70-A-0375-002 (subcontract ONR-439 and others).

Some Problems in Wolf Sociology

Henry S. Sharp

Social behavior, in humans as well as animals, is a puzzling phenomenon. As social animals, we intuitively understand the advantages of social life but, as the inheritors of a cultural tradition that includes the "social contract" and "individual rights," the advantages of social life are inherently inadequate as a conscious explanation of social behavior. Why should organisms live in groups? What are the advantages of social life? What impact does social life have upon the evolution of the species? How important is social life? What is social life? Most of these questions are not answerable now, but we are more able to deal with the "whats" and "hows" of social life than the "whys."

To deal with wolf social life, I suggest that it is useful to turn to the models developed by the social sciences, particularly those developed by British social anthropology and French structuralism (Kuper, 1973). It has been acceptable to use biological models to explain human social behavior since before Darwin's time—often not very useful, as the current fad for sociobiology in the social sciences illustrates—but since the work of Thorndike and Watson, it has not been considered legitimate to do the opposite. This has resulted in a one-sided view of animal social life that I am unable to justify on either theoretical or methodological grounds.

To create a wolf sociology, it is necessary to consider a few of the basic assumptions that the social anthropologist makes about human social behavior. Society is "obligatory" and "general," a multi-generational phenomenon into which each individual is born, and by whose rules the individual must abide. Society is a complex product of interaction between the individual and the social, and it is as legitimate to view society as the producer of the individual as it is to view it as the product of individuals. In fact, the former view, by not being reductionistic, is often more useful and produces more powerful explanations.

This brings us to the most important assertion, that society is a "thing." It is not tangible and can only be studied indirectly through its effects upon the behavior of organisms, but it is nonetheless real and constrains the behavior of living beings (Durkheim, 1895). The best analogy here is to language. The English language does not exist as a tangible thing; all that is tangible are individual speech acts or utterances. Yet there is such a thing as the English language and it can be studied through its constraining effects upon individual speakers. Nor is it possible to argue that the English language exists in the head of each speaker, for no speaker knows all the words, dialects or variations. It can be studied as a "thing" (Durkheim, 1895) apart from its speakers. As with language, the focus of analysis of society *is* society itself. Perhaps the single most prevalent and serious error in sociological analysis is the failure to distinguish between sociological problems and those reducible to lower levels of explanation.

Society is ". . . a number of individuals bound together in a network of social relations" (Radcliffe-Brown, 1952:140–141). Our analytical focus should be upon the social relations rather than the individuals. Individuals are a means to reach the more abstract principles of relationship; once we have adequately modelled the social system, we can begin to understand how individuals manipulate within the social system and how the system itself is generated.

Lockwood (1976) made an excellent attempt to extend social models to wolves. However, Lockwood used role theory, which is more social-psychological than sociological, because he has ultimately ". . . chosen to concentrate on the behavior of individuals" (Lockwood, 1976:126). That Lockwood and I agree on many points confirms the utility of social models but, as I argue from a position that is *purely* social, large areas of disagreement should be apparent.

Social anthropology, and hopefully wolf sociology, is very much an attempt to create accurate and useful models of social behavior. Its goal in the study of animal social behavior is to ultimately determine the point at which models of social causality collapse. By determining this point, we will gain a clearer understanding of social behavior *per se*, as well as a better understanding of those aspects of social behavior rooted in other than social causes. Separating social behavior from biologically based behavior is simply beyond our capacity at this time, yet our inability to separate them is one of our major weaknesses. Thus, any progress in this direction is worthwhile.

If these attempts prove useful, they may shed light upon a difficult problem for anthropology and zoology as well: The relationship between social behavior and evolution. Anthropology was established through the application of Darwin's principles of evolution to human social life and, for its first 50 years, was dominated by evolutionary considerations and theory. This fact is ironic because the modern discipline resulted only after the recognition that "cultural evolution" is a useless concept. Please note that I am not denying evolution or downplaying the extensive literature on the evolution of the first human societies, but I am saying that it is not possible to use evolutionary theory to explain the differences between existing human societies.[1]

This state of affairs seems to result from two conditons: (1) It is not possible to demonstrate direct links between any element of social behavior and human biochemical structure. This forces the anthropologist, trying to understand a living culture, to discard genetic, and indeed biological, factors as irrelevant because they are not specifiable. (2) There exists a basic inability to decide whether it is the culture or the population that exhibits the culture that is evolving. As we normally use the term, "culture" refers not to a pan-human phenomenon but to a subset that is separated for political, linguistic, economic or other reasons that have no *discernible* relationship to the biology of the species. I am certain that zoology is far in advance of anthropology in dealing with this problem, but I fear that this advance may be sustained at the cost of understanding *how* social behavior works.

Below, I have identified some areas where the use of sociological models may have some utility in the analysis of wolf social behavior.

DOMINANCE

Lockwood (1976) has provided a good discussion of problems in the concept of dominance. Although he concludes that "dominance" is still a useful concept, I believe that it is such a powerful concept that it obscures the subtleties of complex social behavior and, hence, should be discarded. Too often, social behavior is forced into the analytical framework rather than the analytical framework elucidating the social behavior.

Dominance theory makes a series of very powerful, and probably unwarranted, statements about the nature of social life. First of all, the concept "dominance" presupposes a "terror model" of social life. Fear becomes the mechanism by which individuals are bound together. Secondly, it implies a "zero-sum model" of social life in which there is a certain amount of power, represented by positions in the dominance hierarchy, inherent in the social group, with each animal competing for a higher position with the ultimate goal of gaining alpha status. Even if there is a deliberate analytical attempt to avoid these specific assumptions, the framework of the theory remains. There is little logic in deforming the theory beyond all recognition in order to save the word "dominance."

In the case of wolves, dominance presumably increases the opportunity to reproduce, and this is usually seen as being advantageous to the animal, which gets his/her genes passed on, and to the species, which receives the best genetic material available.[2] Aside from the implicit tautology (i.e. the best animals breed and they are the best because they are dominant, so that dominant equals best), I wonder about the relationship between genetic fitness and alpha status. Does a deposed alpha animal become any less desirable *genetically* because it is at a point in its life cycle where it cannot obtain or maintain alpha status? More importantly, dominance theory implies "strategies" or "goals" that individual animals employ throughout their lifetimes. This implication should be dealt with explicitly.

I believe it is preferable to recognize that social life is based upon *consensus* rather than upon fear.[3] According to Radcliffe-Brown (1952:140-141), "A social relation exists between two or more persons when there is some harmonization of their individual interests, by some convergence on interest and by limitation or adjustment of divergent interests...," and, "A society cannot exist except on the basis of a certain measure of similarity in the interests of its members. Putting this in terms of value, the first necessary condition of the existence of a society is that the individual members shall agree in some measure in the values that they recognize."

When considering social behavior as complex as that of the wolf, it is not adequate to say these animals are bound together by fear. Even to say that they could not survive as single animals does not explain their socialness, unless we argue that wolves "know" they cannot survive alone. It is more fruitful to recognize that wolf society is based upon some form of consensus stemming from their specialization as group-predators of large game; the social group bound together primarily by mechanical solidarity derived from hunting. In this framework, "alpha-ness" becomes a social position (perhaps "office" is a better term), carrying both rights and obligations, and reflective of organic solidarity (Durkheim, 1893).

The clearest evidence for this comes from the rather muddled relationship between alpha status and reproduction in males. From an evolutionary perspective, hence, largely from a dominance perspective, there must be a clear relationship between alpha status and reproductive status. In a consensual model, this is not the case. Alpha status becomes a social position given by the other wolves rather than something taken by two wolves. As such, there need not be any correlation between alpha status and reproductive success. This does not mean that there is a random relationship between the two, but only that no specific relationship is necessary. The relationship observed is determined by the specific history of each pack and by members that happen to be present at any given time.[4] The important point is that variability between packs is expected, and should not be a puzzle to be explained away.

Variability is crucial to a sociological analysis because only through variability are the underlying forces that produce social behavior actually observed. Behavior that is uniform throughout a social species is most likely attributed to genetic causes, whereas behavior that is variable is more likely dependent upon social factors.

It must also be stressed that the equivalence of two behaviors in different packs must be demonstrated rather than assumed. This is the simple point of cultural relativism; each society is an independent unit that must be explained in terms of its own history and social practice. To assume that "identical" behaviors in two different groups have identical causes or functions leads quickly to misleading explanations that hinder, rather than help, further investigation.

A further advantage of consensual approach is the implied recognition that other "rights and obligations" (Maine, 1861) are not associated exclusively with alpha status. Thus, wolves other than the alpha animals are expected to show initiative in specific situations rather than passively following the alpha animals. I suspect that alpha status is primarily related to what we would call political activities in human society involving relations between packs, the ritualistic centering of the pack and the maintenance of social cohesion within the pack. However, these are points for investigation rather than assumption.

HOWLING CEREMONY

An intriguing aspect of wolf social behavior is the "howling ceremony." I shall, of necessity, break the rule of cultural relativism and use this phenomenon to make a point about social behavior.

The "howling ceremony" is an example of complex social behavior, the purpose of which, in social life, is somewhat unclear. When faced with complex behavior such as this, the tendency is to explain it in terms of what it does. In this case, explanations involving the maintenance of group cohesion, territory or dominance have been offered (Mech, 1970; Harrington and Mech, 1979; Lockwood, 1976; Peterson, 1977). This approach is akin to "functional" analysis of the type advanced by B. Malinowski throughout his career (Kuper, 1973:13-51).

Functional analyses of this type are often insightful or useful as pragmatic tools, but they are inadequate as explanations of the origin or maintenance of elements of social behavior. More importantly, they are sometimes wrong, and tend to inhibit further investigation of the phenomena.

Complex social behavior such as the "howling ceremony" must be treated as "total social phenomena." Mauss (1967:1) remarked that in "*total* social phenomena, as we propose to call them, all kinds of institutions find simultaneous expression: religious, legal, moral and economic." Thus, a behavior like the "howling ceremony" cannot be explained in terms of functions because there is never any single function or cause, and "what" the behavior does is not an explanation of how or why it came to be.[5]

It seems reasonably clear that the "howling ceremony" is a ritualized definition of the pack that also reaffirms existing social relationships and generates solidarity (organic) by focusing attention on its activities and its opposition to other social groups. This, of course, does not explain how or why it exists in specific wolf packs, but leads to another problem. Social behavior is a nonreplicable phenomenon. It is necessary to recognize that we are dealing with unique events that can never be duplicated. This has largely kept social anthropology interested in explanation rather than prediction, as the latter is too crude a criterion. This is also reflected in the distinction made between social organization and social structure. Social organization has been defined as, "the working arrangements of society. It is the process of ordering of action and of relations in reference to given social ends, in terms of adjustments resulting from the exercise of choices by members of the society" (Firth, 1964:45). Social structure involves the factors that produce social organization, which means venturing into the realm of emic[6] analysis to deal with classes, categories, symbols and structures (Firth, 1964).

Social structure, as thus defined, has yet to be used with wolves. To "explain" the howling ceremony would require the analyst to leave his/her level of observation and reach the underlying structure of wolf social life. This will be difficult since wolves do not speak, but since classes are not intrinsically symbolic, it is possible. It is important that we be aware of the level of explanation we are advancing in our explanations of wolf social behavior.

NATURE OF THE PACK

Sociological models should be of utility in examining the nature of the wolf pack. Although the concept of a pack is valid, the term is used in such a variety of ways that it has no clear meaning. A pack seems to be any aggregation of more than one wolf that is identified one or more times by an observer. This problem of nomenclature may stem from a lack of interest in wolf social groups *per se*. For many studies, packs seem to be an epiphenomenon integrated into more basic examinations of ecological systems, prey selection or other variables.

It is easy to underestimate the significance of social organization on the life of a species, especially if social organization is not approached in a systematic manner. Social groups never exist in a vacuum. Simply because a group of wolves can be isolated, or appears discrete to an observer does not mean that it corresponds to a real social unit. The nature and extent of interconnections between groups is an empirical question and it should never be assumed that physical isolation or physical boundaries correspond to social boundaries (Fortes and Evans-Pritchard, 1940:1-24).

To judge from written description, the wolf pack seems to have a two-layer structure with indications that it is embedded within a third layer of organization. The core is a breeding pair that forms a domestic group and represents the minimal conditions of wolf social life. In humans, the domestic group has a definite life cycle consisting of phases of *expansion, dispersion* and *replacement* (Fortes, 1958), but the shorter lifespan of wolves coupled with reproduction by litters may make this particular model of little use.

The domestic group is not limited to reproducing animals, but is that social group "which must remain in operation over a stretch of time long enough to rear offspring to the stage of physical and social reproductivity if a society is to maintain itself" (Fortes, 1958:2). The second layer of the wolf pack may be analogous to the local band of human hunting and gathering societies; the domestic units exist within a larger group which also has corporate aspects, but not all the wolves are always in face-to-face contact and, in some circumstances, almost never together.

A key to understanding social groups in human societies has been the recognition of corporate aspects of each type of social group, and the determination of their spheres of corporateness (Radcliff-Brown, 1952:45).[7] It seems reasonable to regard the wolf pack as the equivalent of the local band. It is this group that is self-sustaining in terms of subsistence, territorial occupation and reproduction. It should be a relatively easy task to arrive at a series of characteristics that would define the wolf pack in any given area. This view also implies, or at least allows, the existence of sociological processes at a level above the wolf pack so that it is necessary to consider the relationship between packs as a part of any study.

With the pack having the appearance of a basic corporate group, the lesser groups appear to be either task groups (Helm, 1968) or domestic groups. It is to be expected, then, that the wolf pack should splinter and re-form into subsidiary groups as a part of the normal yearly cycle of exploitation. The basic corporateness of the pack does not come from continuous physical contact, but from its members' mutual support and organization for hunting and reproduction in a specific, socially bounded area.[8]

It is likely that packs and their domestic groups pass through developmental cycles, and that variations in structure between packs are simply a result of this cycle. It would be interesting to see if long-term studies might uncover cyclic patterns of pack structure.

If the ideas outlined above are applicable to wolves, several problems would deserve further attention. One task faced by any social group is its continuation. Social organization makes no sense unless it is ongoing. There exists a very impressive literature showing how wolf social groups feed and space themselves. A third aspect of the problem of continuation is how they generate a sufficient number of individuals to maintain the group. The pack is viewed by some as a means to cause an increase in inbreeding, leading to the reduction in "available genotypes (gene combinations) within a pack...and...greater variability between wolf packs" (Peterson, 1977:85, discussing Woolpy, 1968:32). Thus, "In a sense, evolution of the species would be accelerated, resulting in rapid adaptation to different environments" (Peterson, 1977:85). Therefore, wolf packs are viewed as "closed societies."

A sociological view, on the other hand, would expect to see an "open" society. As Tylor (1878) long ago pointed out, "Again and again in the world's history, savage tribes must have had plainly before their minds the simple practical alternative between marrying-out and being killed out" (quoted from Fox [1967:176]). Discarding the idea of marriage, we still have a basic fact of social life: A small group often must turn beyond its own boundaries to recruit members if it hopes to avoid extinction. For the social group, the task is recruitment rather than reproduction, although reproduction is obviously one form of recruitment.

Seen from this perspective, the wolf pack open to recruitment is in a rather different situation than one in which reproduction is the sole means of replacement. For each set of environmental circumstances, there may be a minimal number below which pack survival may be threatened, at least in its present form. When an unexpected drop in numbers occurs, the need for replacement is immediate; the pack may not survive until the next litter of pups can be raised to maturity. There would also seem to be a premium upon older, more experienced animals as replacements. A pack would seem to be at a disadvantage unless it were already in contact with adult animals beyond the boundaries of its own group.[9] This implies that packs do maintain relations with "lone wolves" and neighboring packs that are not hostile, or that some mechanism exists to selectively suspend hostility if the need arises. "Lone wolves" could play this role. Although many "lone wolves" are probably young animals or "outcasts" of other packs, it is possible, from the proposed model of the structure of wolf society, that some "lone wolves" could be animals that sometimes acquire multiple pack affiliations. Why an individual wolf becomes a "lone wolf" is not a sociological question, but the existence of "lone wolves" raises the possibility that they are a bridge between wolf packs, and through temporary (or permanent) affiliation with existing packs, minimize pack extinctions and, hence, produce greater stability in the wolf population.

Since it is fairly well established that wolf packs are generally hostile[10] toward other wolves ("closed"), the arguments above may be unlikely. However, given the importance of alpha status, and beta status where it exists, it follows that not all animals are equally important. Recruitment is most likely to occur after a disruption of the high-ranking positions, because their occupancy by experienced animals may be more crucial than the absolute number of animals in the pack.

A recruitment perspective also leads us to expect little or no decrease in genetic variability within wolf packs over a few generations. Concomitant with the exchange of wolves between packs, then, is an exchange of learned behavior patterns. These learned social behaviors, particularly regarding hunting, may be the primary adaptive mechanisms of wolf society.

Recruitment brings us to the third level of organization mentioned earlier. If wolf packs are involved with other packs in a geographical region, and if there is some exchange of members between packs, it is likely that this process is bounded by social factors at this higher level. I would expect that within units of several hundred wolves, their functioning is analogous to Athapaskan regional bands (Helm, 1968) which are so bounded that the exchange of individuals is much more frequent within the boundaries than between them. Beyond this, there is not enough data to do more than speculate, although I have dealt with a similar situation elsewhere (Sharp, 1978). This could help explain the rapidity of subspeciation in *C. lupus* in North America, because genetic diversity is more likely to be reduced at this third level of organization than at the pack level.

Finally, if evidence is found of systematic exchange of wolves among packs, we can make a dramatic shift from recruitment to alliance theory (Levi-Strauss, 1969) and the role of reciprocity, at which point the wolf will become a major focus of attention in the social sciences.

FOOTNOTES

(1) The approach I am taking dealing with wolves may seem unusual to zoologists, but is really quite simple. I am interested in social behavior and how it works in and of itself, rather than how it relates to evolutionary or ecological theory. In this approach, these two concerns become epiphenomena, whereas in most zoological studies it is the social behavior that is the epiphenomenon. There is no denial of these problems in this approach, only the recognition that they are different questions which are not amenable to explanation by a single approach or body of theory. I have attempted to keep the sociology at an elementary level so I have not taken up such promising areas as conflict, reciprocity, class or rank, and I have only mentioned structuralism, which seems the best possibility for understanding wolf social behavior and reconciling sociology with reductionistic approaches (Levi-Strauss, 1979). I have done this not because I think the models are not useful, but because there does not seem to be a body of data on wolves adequate to allow their application.

(2) I am not keeping abreast of current evolutionary theory in this paper. This is deliberate; I am not a geneticist and not qualified in this area, so I have taken much of the wolf literature as it stands rather than trying to reinterpret the evolutionary arguments in it into more modern terms.

As I am not dealing with evolutionary questions here, I think this is acceptable. A more basic reason is my rejection of sociobiological theory. This approach, whatever its utility in biological theory, is useless as a means of dealing with social behavior.

(3) A consensual model is broad enough to include punishment, terror and fear, whereas a "terror" model is not broad enough to include consensual behavior. The introduction of *Political Anthropology* (Swartz et al., 1966:1-41) provides a good introduction to this approach, but should probably be read in conjunction with the preface and introduction of *African Political Systems* (Fortes and Evans-Pritchard, 1940:X1-X23) in order to provide a context for the development of these ideas in social anthropology.

(4) To pick up a point raised by the editors I realize that this diversity is not as clear as I have made it seem. However, in the approach I am taking diversity itself is to be expected, so that any indications of it become important.

(5) Recognizing social behavior, such as the howling ceremony, as "total social phenomena" is an explicit recognition that they are symbolic even though they are nonverbal. It is through this type of social behavior that symbolism appears in nonverbal species, as it would have appeared in our own ancestral populations prior to the evolution of speech.

(6) "Emic" analysis (from "phenemic") refers to the categories of the subjects internal to a system, in contrast to "etic" analysis (from "phonetic") which refers to the categories used by the analyst which are external to the system.

(7) "Corporateness," simply put, derives from the notion of corporation. Social life is supra-individual and ongoing, as is a corporation, but it can also be hierarchial in its organization. Certain functions, e.g. reproduction, may be performed at one level of organization within the group while others, e.g. territorial maintenance, are performed at other levels of organization. It is necessary to determine the level at which various functions are performed and the way in which the various groups inter-relate. A social group may exist in isolation in one context, yet be part of a larger group in another context. What I am suggesting here is that the wolf pack is, in structure, a grouping of sets of wolves that may only rarely be together, but represents a minimal "unit" for the performance of all tasks necessary for their survival. As with humans, this structure may vary considerably and may well be compressed, e.g. a pair of wolves in isolation, according to circumstances. I must admit that a great deal of my thinking about wolves derives from my experience in the Northwest Territories in a tundra-taiga interface, and the constraints of a heavily forested area are somewhat difficult for me to deal with.

(8) In this sense, a wolf pack is defined by its organization into a hunting group, and wolves are separated from coyotes and dogs by the variations in the organization to obtain food. The differences between the three seem to me to be more ones of behavior than genetics, but I am not ready to argue this just yet as it is the topic for future work. A group of wolves in captivity does not have this organization and cannot display the corporateness resulting from this organization. Without this function for the social organization to perform, they cannot be regarded as akin to wild wolves. Captive studies are useful for individual (psychological) or small group (social-psychological) studies, and captive wolves will exhibit the

elements of behavior found in wild animals, but these elements cannot be bound into the same type of organization found in wild wolves without the activity that social organization is designed to accomplish in the wild.

(9) Let me be clear about this. I am speculating on the basis of the model, not reporting observed data. Perhaps prediction is a better word, and I am thinking, again, of the open, mobile life of the Northwest Territories, but the presence of other wolves seems to be characteristic of forest environments.

(10) In humans, hostility—including violence, fighting and killing—is often a major mechanism by which groups are integrated and bound together. This can be true both internally and externally (e.g. blood feud, crime or war).

Nunamiut Eskimos, Wildlife Biologists and Wolves

Robert O. Stephenson

INTRODUCTION

The Nunamiut Eskimos are part of the group of Inupiat-speaking people who inhabit the northern edge of the North American and Eurasian continents. The Nunamiut, "people of the land," inhabit inland northern Alaska. Prior to 1949 they were seminomadic, with seasonal movements accompanying changes in the availability of resources (Binford, 1978). Recently, their economy has changed from mobile hunting and trapping to localized and sporadic hunting, trapping and wage earning. Most Nunamiut, the remnant of what was, at times, a much larger inland population, now reside in the village of Anaktuvuk Pass in the central Brooks Range (Ingsted, 1951; Gubser, 1965). Prior to this recent change, their hunters roamed the country almost daily in search of game, their more extensive travels occurring during winter. References to the Eskimo's extensive practical knowledge and astute observations are common in the writing of early explorers, archeologists and anthropologists (Freuchen, 1915; Stefansson, 1919, 1922; Jenness, 1957; Spencer, 1959; Chance,1966; Campbell 1976). As with other aboriginal peoples, the Nunamiut gained an extensive and intimate knowledge of their physical and biological environment (Irving, 1953, 1958, 1960; Rausch, 1951, 1953), including an unusual familiarity with wolves *(Canis lupus).*

A number of factors account for their extensive knowledge of wolf habits. The Nunamiut inhabited the mountains and northern foothills of the Brooks Range, an area of open tundra where promontories and elevations allowed observers to study animals for long periods without disturbing them. Wolves were also relatively common, with densities sometimes reaching one wolf per 75 km^2 (Stephenson and Johnson, 1973). The Nunamiut relied extensively on caribou *(Rangifer tarandus)* for food and, in addition, hunted and trapped furbearers extensively for trade. From the late 1930s until

434

about 1967, a bounty on wolves provided important cash income for the Nunamiut and prompted additional efforts to take wolves (both adults and pups) during summer. Virtually every summer during this period, groups of two to five hunters searched for dens from late May until early July in an area of about 20,700 km^2. The 24 hour daylight during summer allowed them to travel and observe at night, when wolves are most active, deducing the location of active dens by tracing the movements of adult wolves through the innumerable valleys favored for denning. They knew that wolves generally began hunting in the evening, i.e. traveling away from dens, usually returning to their dens in early morning with food, where they rested during the day. More importantly, they looked for specific characteristics—age, sex, time of day, rate of travel, distension of abdomen, etc.—that together might indicate the direction of a den. Finally, their knowledge and techniques were the result of years of observation. The 18 Nunamiut I worked with had spent the equivalent of more than 30 years watching adult wolves and searching for dens in summer, in addition to their extensive contact with wolves during the rest of the year.

In early 1970, I began studying the ecology and current status of wolves in northern Alaska, and soon became acquainted with the Nunamiut. During the next three years, I spent about 16 months with Nunamiut hunters, visiting previously discovered dens, observing wolves in an attempt to find new dens, working with them on various tasks around the village, and visiting with them during many long winter evenings. From this contact, I attempted to understand their view of wolf ecology: how they interpreted the behavior of undisturbed wolves and how their interpretations compared to mine. Some technical aspects of this work have been reported previously (Stephenson and Johnson, 1972, 1973; Stephenson, 1974; Stephenson and Ahgook, 1975). My intent here is to discuss what wildlife biologists and behavioral ecologists might learn from the Nunamiut view of wolf ecology.

DISCUSSION

As one might imagine, the Nunamiut possess a great deal of knowledge about wolves collected and interpreted outside the framework of Western science. Although aspects of their knowledge are similar to those developed by Western wildlife biologists, there are differences in both theories and the specific relationships perceived between details of natural history.

The Nunamiut impressed me with their ability as observers and the basic good sense of their interpretations. For example, one morning in May 1972, Bob Ahgook and I were walking up a boulder-strewn mountainside in an area where Bob knew of two dens used in previous years. We had already spent several days there attempting to locate an active den, and observed two different wolves whose movements and behavior suggested that one of these dens was being used. When an all night watch at one den revealed no wolf activity, we decided to investigate the other den in an adjacent valley. As I followed Bob through the rocks and dwarf birch up the steep hillside, he suddenly stopped and pointed to the vague outline of a narrow trail leading up and across the slope, saying it looked like a wolf trail. Following the trail with his eyes he added, " and there's a wolf." A hundred meters away, a tan wolf hurried up the mountain, climbing almost vertically between segments of a sheer escarpment before it disappeared, after a final glance back, over the edge. Bob said the wolf was a female which had just had, or would shortly have pups. He had seen her dark abdomen where the hair was shed around her mammae. Certain that we were near a den, we retreated to avoid disturbing it further. At a safe distance, we looked back for some indication of the den location. As we watched, a robin landed about 30 m above us and quickly flitted away. Bob immediately said the den was located where the robin had landed; he had seen the bird pick up some wolf hairs for its nest. When I looked closely, I saw freshly exposed soil (evidence of digging) around a triangular boulder where the robin had landed.

As we walked back to camp, I wondered at the abilities of men like Bob. I had seen the wolf and the robin but had not noted the wolf's abdomen, which revealed its sex and breeding condition, or the robin's wolf hairs that revealed the den. I also might have crossed the wolf trail without noticing it. I thought about the difference in approach between myself, formally trained to understand wildlife, and Bob, whose familiarity with animals was based on day-to-day contact and economic dependence. I became acutely aware that my abstract, generalized knowledge was not as well suited as Bob's personal experience to the field study of an animal like the wolf.

The Nunamiut also made me aware that, by comparison, my knowledge of the country was woefully inadequate. I could not hope to appreciate, in a short time, the features of terrain, weather and prey distribution that influenced the behavior of wolves. It took considerable time to learn only some of the things that might influence their behavior. For example, while observing adult wolves hunting away from dens, we often saw them rest for varying periods. To me, these wolves were simply sleeping, but to the Nunamiut, they were also hunting. The Nunamiut pointed out that the wolves frequently slept near recently used caribou trails on the chance that caribou might pass. Where I had seen only a resting wolf, the Nunamiut perceived one that may have been hunting as well.

Their knowledge of wolves was equally impressive. They showed me how morphology, pelage and behavior could often reveal a wolf's sex and age at a considerable distance. They related the ways wolves used vocal communication to convey information and told an incredible array of stories about wolf behavior under a wide range of environmental conditions. In addition, they pointed out nearly 80 dens used during the previous 30 years and related the number of pups and adults seen at each.

For the Nunamiut, knowledge of animals is of crucial value and, thus, much sought after. I became aware of their facility in using the scientific method to evaluate this knowledge. They hypothesized steadily as we watched wolves, and were adept at testing these hypotheses. Over the years, this has resulted in some fairly refined conclusions. As an example, Justus Mekiana told me that after three days, whatever scent we had left near a den would not be detected by wolves. I asked how he knew this and he said that he had once watched a wolf cross a trail made by himself and a companion three days earlier. The wolf had not detected their scent. On other occasions he had watched even running wolves stop and investigate trails they had made the day before. In the same manner, he had deduced that wolves were best able to detect animal scent trails when the soil was dry and warm. Conclusions like these were tested and re-evaluated whenever the opportunity presented itself.

The field biologist and the Nunamiut probably react differently when each discovers a new phenomenon. For us, considerable importance is often attached to being first with a new finding. This implies, to me at least, that (1) someday we will find no more surprises in, for instance, wolf-prey interactions, and (2) that something special should be conferred upon the observer of such phenomena. The Nunamiut are less interested in these considerations than in the food for thought provided by new phenomena. They show little concern about whether an idea or fact is accepted by anyone other than themselves, nor do they have the tendency to interpret events in the same terms used by someone else.

In 1974 I spent about two weeks by myself observing a denning pack. Upon returning to the village, I shared my experiences with Bob and Justus, anxious to hear their interpretations. Justus was especially intrigued that the adults had howled extensively almost every night because, in his experience and that of recent generations of Nunamiut, wolves normally howl very little in early summer. The following day we talked again and he wondered if elements of wolf culture underwent gradual change, in the same way the Nunamiut language had metamorphosed. He thought certain habits of wolves might change over time, at least in some packs or populations, and sifted through his experiences for other things that might be explained in this manner. I found it instructive that this man, who is highly knowledgeable about wolves, could so easily alter his frame of reference. I have, at times, sensed a certain reluctance to do this on the part of wildlife biologists, although recently I think we have improved in this respect. Indeed, many of our earlier generalizations are now being examined in a more objective light.

As my association with the Nunamiut developed, I became interested in the way ecological generalizations were received. For instance, at this time, biologists were asking whether wolves killed only "inferior" prey. When I asked older Nunamiut this question, it confused them; they elaborated the many circumstances that influence the outcome of wolf-caribou encounters. While they appreciated the fact that very young, old or infirm caribou were most vulnerable, they pointed out that during winter when lakes and rivers are frozen, even the fleetest caribou is vulnerable to a determined wolf. They also stressed the importance of terrain and cover, relating cases of caribou stumbling in rough terrain, slipping on ice as they fled from wolves, or outdistancing one wolf only to be caught by another waiting in ambush. Based on their examination of thousands of wolf-killed caribou, and observation of hundreds of chases of caribou bands of all sizes during all seasons of the year and involving groups of wolves varying in size and age composition, the Nunamiut were reluctant to generalize.

The same was true for many of the classical questions posed by our discipline. They seemed naive to the Nunamiut because they assumed too little variability in animals, behavior and environmental conditions. When asked whether wolves kill healthy or infirm caribou, or whether male or female wolves are the most influential or dominant individuals in a pack, the Nunamiut broke the questions down into increasingly smaller questions which were then considered in the context of specific domains: physiology, ethology, morphology, meteorology and physiography.

Another fundamental difference between modern biologists and the Nunamiut is that the Nunamiut are willing to attribute considerably more to individual variation and to volition. The Nunamiut believe behavioral variation among wolves explains the wide range of interactions they see among wolves, other animals and the environment. The wolf is viewed as an animal whose behavior changes perceptibly with age and experience because, like people, it learns throughout its life. The Nunamiut also believe that some decisions wolves make are likely to be foolish, "inefficient," or ambiguous of interpretation. In contrast, it appears that biologists and even more so, the wildlife-oriented public, look for "adaptive" value in most details of animal behavior. The wolves I observed did many things that Western science normally refers to as anecdotal behavior, but which the Nunamiut believed contained rather significant information. I watched single adult wolves spend considerable time teasing or following grizzly bears (Ursus arctos), playing with scraps of caribou hide, scaring sleeping eagles and running in circles for no apparent reason. I also saw wolves stop chasing caribou when they appeared to have a good chance of killing them and, on the other hand, persist in chases that looked hopeless. In these things, the Nunamiut saw elements of volition, instinct and individual idiosyncrasy. We may be reluctant to recognize and speak of these things because they are difficult to quantify. Still, I think we have to recognize this side of wolf behavior and help the public understand that wolf populations are not composed of identical individuals or packs guided in their every move by ironclad laws of nature.

The Nunamiut way of thinking also suggests that besides over-simplifying complicated things, we also tend to make simple things terribly complicated. On one occasion, the Nunamiut applied "Occam's razor" quite abruptly. Bob Ahgook and I had visited a vacated wolf den and found that ground squirrels, living nearby, had moved a large amount of soil back into it, plugging the entrance. I wondered about the squirrels' motivation, but could not explain their peculiar behavior. Upon returning to the village, I related the story to another Nunamiut and asked what would compel squirrels to fill in a wolf den. Without hesitating, he replied, "too much stink."

The Nunamiut forced me to recognize that there is an important difference between a generalized knowledge of an animal obtained through reading, and a more specific working knowledge derived from field experience. I was especially concerned when I saw that my generalized knowledge interfered with my ability to comprehend the things before me. As Ghiselin (1972) noted, one deficiency that limits "the acceptance of the ideas of ecologists and their allies, and which makes the trade seem untrustworthy. . .is overgeneralization, a common failing among biology professors. In the search for general laws, we condense, we gloss over exceptions, and we are drawn into hyperbole by the approach of the final bell." The fact that wolves are controversial and that much debate occurs in the public forum creates demands for general truths that I think we must, in some measure, resist. We could use the Nunamiut as a guide. They acquired their knowledge of animals in an atmosphere free of this demand and, although this knowledge has met and continues to meet many tests, the demand for uniformity is not one of them; there is little cause for anxiety if things they see elude established categories.

I do not mean to denigrate the contribution of Western science to our understanding of wolf ecology. We have learned much and, in recent years, have sparked a more positive interest in wolves in a large segment of the public. Many of our techniques allow us to learn things that cannot be discovered with other methods and, with respect to phenomena at the population level, we can see things that are impossible for even the Nunamiut to know. However, some unknown and potentially significant aspects of wolf behavior and ecology may be obscured if we do not guard against the tendency in our discipline to blindly follow general laws.

Acknowledgements

Every family in Anaktuvuk Pass helped in some way, and I am grateful for their many gestures of kindness and criticism. Bob Ahgook and Justus Mekiana took a special interest in telling me about the Nunamiut experience with wolves, and for their companionship and insight I am extremely grateful. This work was supported, in part, by Federal Aid in Wildlife Restoration funds and by the Alaska Department of Fish and Game.

Bibliography

Adorjan, A.S. and Kolenosky, G.B. 1969. A manual for the identification of hair of selected Ontario mammals. Ontario Department of Lands and Forests, Research Report (Wildlife) No. 90.

Alcock, J. 1975. *Animal Behavior: An Evolutionary Approach*. Sinauer Associates, Sunderland, Massachusetts.

Alexander, R.D. 1974. The evolution of social behavior. *Annual Review of Ecology and Systematics*. 5:325-383.

Allen, D.L. 1979. *The Wolves of Minong: Their Vital Role in a Wild Community*. Houghton Mifflin Co., Boston.

Altmann, D. 1974. Beziehungen zwischen sozialer Rangordnung und Jungenautzucht bei *Canis lupus. Zoologisher Garten N.F.* Jena 44:235-236.

Andelt, W.F. and Gipson, P.S. 1979. Home range, activity and daily movements of coyotes. *Journal of Wildlife Management* 43:944-951.

Anonymous. 1902. Congressional Record. Vol. 35, U.S. Government Printing Office.

Anonymous. 1908. Congressional Record. Vol. 42, U.S. Government Printing Office.

Anonymous. 1950. Annual report to the Secretary of the Interior, 1948-1950. Alaska Game Commission. Copy in University of Alaska Library.

Anonymous. 1951. Annual report to the Secretary of the Interior, 1950-1951. Alaska Game Commission. Copy in University of Alaska Library.

Armstrong, G. 1965. An examination of the cementum of the teeth of Bovidae with special reference to its use in age determination. M.S. Thesis, University of Alberta.

Aulerich, R.J. 1964. Status of the wolf in North America. M.S. Thesis, Michigan State University.

Bailey, R. (ed.). 1978. *Recovery Plan for the Eastern Timber Wolf*. U.S. Fish and Wildlife Service, Washington, DC, 79pp.

Ballard, W.B. 1980. Brown bear kills gray wolf. *Canadian Field-Naturalist* 94:91.

441

Ballard, W.B., Franzmann, A.W., Taylor, K.P., Spraker, T., Schwartz, C.C. and Peterson, R.O. 1979. Comparison of techniques utilized to determine moose calf mortality in Alaska. 15th North American Moose Conference Workshop, Kenai, Alaska.

Ballard, W.B., Miller, S.D., Spraker, T.H. 1980. Moose calf mortality study, Unit 13, Alaska. Final Report, Federal Aid in Wildlife Restoration, Project W-17-9, W-17-10, W-17-11, W-21-1, Job 1.23R. Alaska Department of Fish and Game, Juneau, Alaska.

Ballard, W.B. and Spraker, T. 1979. Unit 13 Wolf Studies. Alaska Department of Fish and Game, P-R Project Report W-17-8, Jobs 14.8R, 14.9R and 14.10R, Juneau, Alaska.

Ballard, W.B., Spraker, T. and Taylor, K.P. 1981. Causes of neonatal moose calf mortality in south-central Alaska. *Journal of Wildlife Management* 45:335-342.

Ballard, W.B. and Taylor, K.P. 1978a. Moose calf mortality study, Game Management Unit 13. Alaska Department of Fish and Game. P-R Project Report W-17-9 (2nd half) and W-17-10 (1st half), Job 1.23R.

Ballard, W.B. and Taylor, K.P. 1978b. Upper Susitna River moose population study. Alaska Department of Fish and Game. P-R Project Report W-17-10, Job 1.20R.

Banfield, A.W.F. 1954. Preliminary investigation of the barren-ground caribou. Canadian Wildlife Service Wildlife Management Bulletin, Series 1, Nos. 10A, 10B.

Banfield, A.W.F. 1974. *The Mammals of Canada.* University of Toronto Press, Toronto.

Barash, D.P. 1977. *Sociobiology and Behavior.* Elsevier, New York.

Barry, M.J. 1973. *A history of mining on the Kenai Peninsula.* Alaska Northwest Publishing Company, Anchorage.

Bartholomew, G.A. 1968. Energy metabolism (Chapter 3); Body temperature and energy metabolism (Chapter 8). In M.S. Gordon, G.A. Bartholomew, A.D. Grinnell, C.B. Jorgensen and F.N. White (eds.), *Animal Physiology: Principles and Adaptations,* MacMillan Co., London, 48-65, 290-354.

Behrend, D.F., Mattfield, G.F., Tierson, W.C. and Wiley, III, J.E. 1970. Deer density control for comprehensive forest management. *Journal of Forestry* 68(11):295-700.

Bekoff, M. 1974. Social play and play-soliciting by infant canids. *American Zoologist* 14:323-340.

Bennett, H.H. 1916. Report on a reconnaissance of the soils, agriculture and other resources of the Kenai Peninsula Region of Alaska. In M. Whitney (ed.), *Field Operations of the Bureau of Soils,* U.S. Department of Agriculture, 39-174.

Bennett, H.H. and Rice, T.D. 1914. Soil reconnaissance in Alaska, with an estimate of the agricultural possibilities. In M. Whitney (ed.), *Field Operations of the Bureau of Soils,* U.S. Department of Agriculture, 43-236.

Bennett, N.L. 1979. Some aspects of co-operative pup-rearing in a pack of captive timber wolves (*Canis lupus*). M.S. Thesis, Dalhousie University.

Berg, W.E. 1981 (in press). Scent station indices in Minnesota, 1975 to 1980. *Minnesota Wildlife Research Quarterly.*

Bergerud, A.T. 1974. Decline of caribou in North America following settlement. *Journal of Wildlife Management* 38:757-770.

Bertram, B.C.R. 1976. Kin selection in lions and in evolution. In P.P.G. Bateson and R.A. Hinde (eds.), *Growing Points in Ethology.* Cambridge University Press, London.

Bibikov, D.I. 1973. The wolf in the USSR. Technical meeting on the wolf in Europe, IUCN/WWF, Stockholm, Sept. 5-6, 1973, 9pp.

Bibikov, D.I. 1975. The wolf in the USSR. In D.H. Pimlott (ed.), *Proceedings of the First Working Meeting of Wolf Specialists.* Morges, Switzerland.

*Bibikov, D.I. 1980. Certain ungulates and predators in connection with human economic activity. Influence of human economic activity on the population of game animals and their habitats. Vol. 2, Kirov.

*Bibikov, D.I. and Filimonov, A.N. 1974. The wolf: problems of population regulation. Okhota i okhotnich'e khozyaystvo, 10.

*Bibikov, D.I. and Filonov, K.P. 1980. The wolf in preserves in the USSR. Priroda, 2.

*Bibikov, D.I. and Rukovskii, N.N. 1975. Geographical peculiarities of the wolf diet in the USSR. Topical questions in zoological geography. Kishinev.

Binford, L.R. 1978. *Nunamiut Ethnoarchaeology.* Academic Press, New York.

Bishop, R.H. and Rausch, R.A. 1974. Moose population fluctuations in Alaska, 1950-1972. *Naturaliste Canadien* 101:559-593.

Bitterman, H.E., Wodinsky, J. and Candland, D.K. 1958. Some comparative psychology. *American Journal of Psychology* 71:19-110.

Bjarvall, A. (in press). An interview study of attitudes to the wolf in Sweden. In R. Soutar (ed.), *Proceedings of the International Wolf Symposium,* Edinburgh, 1978.

Bjarvall, A. and Nilsson, E. 1976. Surplus-killing of reindeer by wolves. *Journal of Mammalogy* 57:585.

Bjarvall, A., Nilsson, E. and Osterdahl, L. 1978. Jarvstammen okar — totalskydd eller avskjutning? *Forskning och Framsteg* 13:29-32.

Boitani, L. 1976. Il lupo in Italia: Censimento, distribuzione e prime ricerche eco-etologiche nell'area del Parco Nazionale d'Abruzzo. In Pedrotti (ed.), *SOS Fauna.* WWF Roma: 7-42.

Boitani, L. and Zimen, E. 1979. The role of public opinion in wolf management. In E. Klinghammer (ed.), *The Behavior and Ecology of Wolves.* Garland STPM Press, New York, 43-83.

Bouchard, M.E. and Lerg, J.M. 1977. Sex and age parameters in the population dynamics of Michigan hunters. Michigan Department of Natural Resources, Wildlife Division, Report No. 2796.

Briscoe, B.W., Carbyn, L.N. and Trottier, G.C. 1980. Applications of a large mammal system study for natural resource management in Riding Mountain National Park, Canada. Presented at 2nd Conference of Scientific Research in National Parks, San Francisco, California.

Brown, J.L. 1978. Avian communal breeding systems. *Annual Review of Ecology and Systematics* 9:123-155.

Brown, P.L. and Jenkins, H.M. 1968. Auto-shaping of the pigeon's key-peck. *Journal of the Experimental Analysis of Behavior* 11:1-8.

Bulger, A.J. 1975. The evolution of altruistic behavior in social carnivores. *The Biologist* 57:41-51.

Bunnell, F.L. 1979. Deer-forest relationships on northern Vancouver Island. *Proceedings of the Black-tailed Deer Conference*, U.S. Forest Service, Juneau, Alaska, 86-101.

Burkholder, B.L. 1959. Movements and behavior of a wolf pack in Alaska. *Journal of Wildlife Management* 23:1-11.

Burkholder, B.L. 1962. Observations concerning wolverine. *Journal of Mammalogy* 43:263-264.

Burt, W.H. 1943. Territoriality and home range concepts as applied to mammals. *Journal of Mammalogy* 24:346-352.

Burt, W.H. and Grossenheider, R.P. 1964. *A Field Guide to the Mammals*. 2nd Edition, Houghton Mifflin Company, Boston.

Cagnolaro, L., Rosso, D., Spagnesi, M. and Venturi, B. 1974. Inchiesta sulla distribuzione del lupo in Italia e nei Cantoni Ticino e Grigioni (Svizzera). *Lab. Zool. Appl. Caccia* N. 59, Bologna.

Camenzind, F.J. 1978. Behavioral ecology of coyotes on the National Elk Refuge, Jackson, Wyoming. In M. Bekoff (ed.), *Coyotes: Biology, Behavior, and Management*. Academic Press, New York, 267-294.

Campbell, J.M. 1976. The nature of Nunamiut archaeology. In E.S. Hall, Jr. (ed.), *Contributions to Anthropology: The Interior Peoples of Northern Alaska*, National Museum of Man, Mercury Series, Paper No. 49, National Museum of Canada, Ottawa.

Carbyn, L.N. 1974. Wolf predation and behavioural interactions with elk and other ungulates in an area of high prey diversity. Ph.D. Thesis, University of Toronto. Canadian Wildlife Service Report, Edmonton.

Carbyn, L.N. 1981a. Ecology and management of wolves in Riding Mountain National Park. Canadian Wildlife Service Report for Parks Canada. 182pp.

Carbyn, L.N. 1981b. Territory displacement in a wolf population with abundant prey. *Journal of Mammalogy* 62:193-195.

Carley, C. 1979. Report on the successful translocation experiment of red wolves (*Canis rufus*) to Bull's Island, South Carolina. Presented at Portland Wolf Symposium, Lewis and Clark College, Portland, Oregon.

Chambers, R.E. 1980 (in prep.). Food habits of the red fox and the coyote in New York's Adirondack Mountains.

Chance, N.A. 1966. *The Eskimo of North Alaska*. Holt, Rinehart and Winston, New York.

Chapman, R.C. 1977. The effects of human disturbance on wolves (*Canis lupus* L.). M.S. Thesis, University of Alaska.

Chapman, R.C. 1978. Rabies: Decimation of a wolf pack in Arctic Alaska. *Science* 201:365-367.

Cheney, C.D. 1978. Predator-prey interactions. In H. Markowitz and V. Stevens (eds.), *Studies of Captive Wild Animals*. Nelson-Hall, Chicago.

Cheney, C.D. 1979. A prey chamber for the experimental analysis of raptor hunting. *Behavior Research Methods and Instrumentation* 11:558-560.

Cheney, C.D. and Snyder, R.L. 1974. A chamber for separating visual from physical prey access from predators. *Behavior Research Methods and Instrumentation* 6:553-555.

Choquette, L.P.E. and Kuyt, E. 1974. Serological indication of canine distemper and of infectious canine hepatitis in wolves (*Canis lupus* L.) in northern Canada. *Journal of Wildlife Diseases* 10:321-324.

Clark, K.R.F. 1971. Food habits and behavior of the tundra wolf on central Baffin Island. Ph.D. Thesis, University of Toronto.

Couchie, D.G. and Collingwood, L. 1978. Bison concentrations and distribution on primary ranges. Wood Buffalo National Park, 21pp.

Cowan, I.M. 1947. The timber wolf in the Rocky Mountain National Parks of Canada. *Canadian Journal of Research* 25:139-174.

Cowan, I.M. and Guiguet, C.J. 1965. *The Mammals of British Columbia*. B.C. Provincial Museum, Victoria, Department of Recreation and Conservation Handbook No. 11.

Culver, W.G. 1923. Report of moose on Kenai Peninsula. Kenai National Moose Range files.

Curnow, E. 1969. The history of the eradication of the wolf in Montana. M.S. Thesis, University of Montana, Missoula.

Daniel, W.J. 1942. Cooperative problem solving in rats. *Journal of Comparative Psychology* 34:361-368.

Daniel, W.J. 1943. Higher order cooperative problem solving in rats. *Journal of Comparative Psychology* 35:297-305.

*Danilkin, A. 1979. Hunting of roe deer by wolf-dog hybrids. Okhota i okhotnich'e khozyaystov, 3.

Danilov, P.I. 1979. Novosely karel'skih lesov. Petrozavodsk, 88pp.

Danilov, P.I., Ivanter, E.V., Belkin, V.V. and Nikolaevskij, A.A. 1978. Izmeneni'a cislennosti ohotnicih zverej Karelii po materialam zimnih marsrutnyh ucetov. In E.V. Ivanter (ed.), *Fauna i Ekologi'a Ptic i Mlekopita'uscih Taeznogo Severo-Zapada SSSR*, Petrozavodsk, 128-159.

Davis, J.L. and Franzmann, A.W. 1979 (in press). Fire-moose-caribou interrelationships. *Proceedings of the 15th North American Moose Conference and Workshop*.

Dawkins, R. 1976. *The Selfish Gene*. Oxford University Press, New York.

Day, G.L. 1977. The status and distribution of the Northern Rocky Mountain Wolf (*Canis lupus irremotus*). M.S. Thesis, University of Montana, Missoula.

DeKay, J.E. 1842. Zoology. Part I in *Natural History of New York*, Appleton and Co., New York.

*Dementyev, G.P. 1933. *The Wolf*. Moscow, Leningrad.

DeVoto, B. 1953. *The Journals of Lewis and Clark*. Houghton Mifflin Company, Boston.

Dixon, W.J. (ed.). 1973. *BMD: Biomedical Computer Programs*. University of California Press, Berkeley.

Durkheim, E. 1893. *The Division of Labor in Society*. Free Press, New York, 1964.

Durkheim, E. 1895. *The Rules of the Sociological Method*. Free Press, New York, 1964.

Elder, W.H. and Hayden, C.M. 1977. Use of discriminant function in taxonomic determination of canids from Missouri. *Journal of Mammalogy* 58:17-24.

*Eliseev, N., Klokov, K. and Syroechkovskii, E. 1973. The wolf and its future. Okhota i okhotnich'e khozyaystvo, 5.

Ellerman, J.R. and Morrison-Scott, T.C.S. 1951. *Checklist of Palearctic and Indian Mammals*. British Museum of Natural History, London.

Emlen, S.T. and Demong, N.J. 1975. Adaptive significance of synchronized breeding in a colonial bird: A new hypothesis. *Science* 188:1029-1031.

Emlen, S.T. and Oring, L.W. 1977. Ecology, sexual selection, and the evolution of mating systems. *Science* 197:215-223.

Engelhart, S. and Hazard, K. 1975. Wolves in the Adirondacks. *The Conservationist*, Oct.-Nov. 1975, 9-11.

Fentress, J.C. 1967. Observations on the behavioral development of a hand-reared male timber wolf. *American Zoologist* 7:339-351.

Fentress, J.C. 1973. Specific and nonspecific factors in the causation of behavior. In P.P.G. Bateson and P.H. Klopfer (eds.), *Perspectives in Ethology*, Plenum Press, New York, 155-224.

Fentress, J.C. 1976. Dynamic boundaries of patterned behavior: Interaction and self-organization. In P.P.G. Bateson and R.A. Hinde (eds.), *Growing Points in Ethology*. Cambridge University Press, Cambridge, 135-169.

Fentress, J.C. 1978. Conflict and context in sexual behavior. In J. Hutchison (ed.), *Biological Determinants of Sexual Behavior*. Wiley, New York, 579-614.

Fentress, J.C. 1980 (in press). How can behavior be studied from a neuroethological perspective? In H. Pinsker, (ed.), *Information Processing in the Nervous System: Communication Among Neurons and Neuroscientists*. Raven Press, New York.

Fentress, J.C., Field, R. and Parr, H. 1978. Social dynamics and communication. In H. Markowitz and V. Stevens (eds.), *Behavior of Captive Wild Animals*. Nelson-Hall, Chicago, 67-106.

Fick, A. 1975. A new voice in the northeast. *Defenders of Wildlife* 50:346-347.

Field, R. 1979. A perspective on syntactics of wolf vocalizations. In E. Klinghammer (ed.), *The Behavior and Ecology of Wolves*. Garland STPM Press, New York, 182-205.

*Filimonov, A.N. 1979. On the possible selective role of the wolf during the calving period of antelopes. Ecological Foundations for Preservation and Rational Use of Predatory Animals. Moscow.

Filonov, C. 1980. Predator-prey problems in nature reserves of the European part of the R.S.F.S.R. *Journal of Wildlife Management* 44:389-396.

Firebaugh, J.E., Flath, D.L. and Knoche, K.G. 1975. *Deer-railroad relationship study final report*. Montana Department of Fish and Game, 24pp.

Firth, R. 1964. *Essays on Social Organization and Values*. London School of Economics Monographs on Social Anthropology, No. 28, Athlone Press.

Fish and Wildlife Service. 1978. Reclassification of the gray wolf in the United States and Mexico, with determination of critical habitat in Michigan and Minnesota. *Federal Register* 43:9607-9613.

Flath, D.L. 1979. The nature and extent of reported wolf activity in Montana. Joint meeting including Montana Chapter of Wildlife Society, Feb. 1, 1979, Missoula, Montana, 17pp. (and Appendices).

Floyd, T.J., Mech, L.D. and Jordan, P.A. 1978. Relating wolf scat content to prey consumed. *Journal of Wildlife Management* 42:528-532.

*Formozov, A.N. 1946. Snow cover as an environment factor and its meaning in the lives of mammals and birds in the USSR. Moscow.

Fortes, M. 1958. Introduction. In J. Goody (ed.), *The Developmental Cycle in Domestic Groups*, Cambridge University Press, Cambridge.

Fortes, M. and Evans-Pritchard, E.E. 1940. Introduction. In *African Political Systems*, International African Institute, Oxford University Press, London, 1-24.

Fowler, B. 1974. *Adirondack album.* Outdoor Association, Schenectady, New York 56-57.

Fox, M.W. 1971. *Behavior of Wolves, Dogs and Related Canids.* Harper and Row, New York.

Fox, M.W. (ed.). 1975. *The Wild Canids: Their Systematics, Behavioral Ecology and Evolution.* Van Nostrand Reinhold Company, New York.

Fox, R. 1967. *Kinship and Marriage.* Penguin, Middlesex, England.

Franzmann, A.W. and Arneson, P.D. 1973. Moose Research Center studies. Alaska Department of Fish and Game, P-R Project Report W-17-5.

Franzmann, A.W. and Arneson, P.D. 1975. Moose Research Center report. Alaska Department of Fish and Game. P-R Project Report W-17-7.

Franzmann, A.W. and Bailey, T.N. 1977. Moose Research Center report. Alaska Department of Fish and Game. P-R Project Report, W-17-9.

Franzmann, A.W. and Schwartz, C.C. 1978. Moose calf mortality study, Kenai Peninsula. Alaska Department of Fish and Game, Juneau, P-R Project Report W-17-10, Jobs. 1.24R and 17.3R.

Freuchen, P. 1915. Report on the first Thule expedition (scientific work). Medd. om Gronland 51.

Fritts, S.H. and Mech, L.D. 1981. Dynamics, movements, and feeding ecology of a newly protected wolf population in northwestern Minnesota. Wildlife Monographs. (in press)

Fuller, T.K. and Keith, L.B. 1980. Wolf population dynamics and prey relationship in northeastern Alberta. *Journal of Wildlife Management* 44:583-602.

Fuller, W.A. 1966. The biology and management of the bison of Wood Buffalo National Park. *Canadian Wildlife Service Management Bulletin*, Series 1, no. 16.

Fuller, W.A. and Novakowski, N. 1955. Wolf control operations, Wood Buffalo National Park, 1951-52. *Canadian Wildlife Service Wildlife Management Bulletin*, Series 1, No. 11.

Gates, B.R. 1968. Deer food production in certain seral stages of the coast forest. M.S. Thesis, University of British Columbia, Vancouver.

Geigy. 1959. Chemical composition of foodstuffs. In *Documenta Geigy: Scientific Tables*, Karger, New York, 240pp.

Geist, V. 1978. *Life Strategies, Human Evolution, Environmental Design.* Springer-Verlag, New York.

Gessaman, J.A. 1973. Methods for estimating the energy cost of free existence. In J.A. Gessaman (ed.), *Ecological Energetics of Homeotherms: A View Compatible with Ecological Modeling*, Utah State University Press, Monograph Series No. 20, 3-31.

Ghiselin, J. 1972. Why is ecology so near the lunatic fringe? *Bulletin of the Ecological Society of America* 53:13-14.

Gill, D. 1978. *Large Mammals of the Macmillan Pass Area Northwest Territories and Yukon.* Denver.

Gilmer, D.S., Cowardin, L.M., Duval, R.L., Mechlin, L.M., Shaiffer, C.W. and Kuechle, V.B. 1981 (in press). Procedures for the use of aircraft in wildlife biotelemetry studies. *U.S. Fish and Wildlife Service Technical Bulletin.*

Gilmer, D.S., Kuechle, V.B. and Ball, Jr., I.J. 1971. A device for monitoring radio-marked animals. *Journal of Wildlife Management* 35:829-832.

Ginsberg, B. 1979. Comparative studies of captive wolves. *Portland Wolf Symposium*, Portland, Oregon, August 1979.

Gipson, P.S., Sealander, J.A. and Dunn, J.E. 1974. The taxonomic status of wild *Canis* in Arkansas. *Systematic Zoology* 23:1-11.

Golani, I. and Keller, A. 1975. A longitudinal field study of the behavior of a pair of golden jackals. In M.W. Fox (ed.), *The Wild Canids: Their Systematics, Behavioral Ecology and Evolution.* Van Nostrand Reinhold, New York, 303-335.

Golani, I. and Mendelssohn, H. 1971. Sequences of precopulatory behaviour of the jackal (*Canis aureus* L.). *Behaviour* 38:169-192.

Goldman, E.A. 1937. The wolves of North America. *Journal of Mammalogy* 18:37-45.

Goldman, E.A. 1941. Three new wolves from North America. *Proceedings of the Biological Society of Washington* 54:109-114.

Goldman, E.A. 1944. Classification of Wolves. Part II of S.P. Young and E.A. Goldman (eds.), *The Wolves of North America.* Dover Publishers, New York.

*Golgofskaya, K.Y., Kudaktin, A.N. and Bibikov, D.I. 1979. On the study of trophic relations between predators, ungulates and their pastures in the northwestern Caucasus. Ecological Foundations for Preservation and Rational Use of Predatory Mammals, Moscow.

Gubser, N.J. 1965. *The Nunamiut Eskimos — Hunters of Caribou.* Yale University Press, New Haven.

Guthrie, R.D. 1968. Paleoecology and the large-mammal community in interior Alaska during the late Pleistocene. *American Midland Naturalist* 79:346-363.

Haber, G.C. 1977. Socio-ecological dynamics of wolves and prey in a subarctic ecosystem. Ph.D. Thesis, University of British Columbia, Vancouver. Published by the Joint Federal-State Land Use Planning Commission for Alaska, Anchorage.

Haglund, B. 1968. Winter habits of the bear and the wolf as revealed by tracking in the snow. *Swedish Wildlife* 5:213-361.

Haglund, B. 1975. The wolf in Fennoscandia. In D.H. Pimlott (ed.), *Wolves.* Proceedings of the 1st working meeting of wolf specialists and of the 1st international conference on conservation of the wolf. IUCN Publications New Series, supplementary paper, No. 43.

Haglund, B. and Nilsson, E. 1977 (mimeo). Fjallraven — en hotad djurart. Inst. f. viltekologi. Sveriges lantbruksuniversitet.

Hake, D.F. and Vukelich, R. 1972. A classification and review of cooperation procedures. *Journal of the Experimental Analysis of Behavior* 18:333-343.

Hall, E.R. 1932. Remarks on the affinities of the mammalian fauna of Vancouver Island, British Columbia, with descriptions of new subspecies. *University of California Publ. Zoology* 38:415-423.

Hamilton, W.D. 1963. The evolution of altruistic behavior. *American Naturalist* 97:354-356.

Hamilton, W.D. 1964. The genetical evolution of social behaviour. I and II. *Journal of Theoretical Biology* 7:1-52.

Hamilton, W.D. 1971. Geometry for the selfish herd. *Journal of Theoretical Biology* 31:295-311.

Hamilton, W.J. 1974. Food habits of the coyote in the Adirondacks. *New York Fish and Game Journal* 21:177-181.

Hamlin, K.L. and Schweitzer, L.L. 1979. Cooperation by coyote pairs attacking mule deer fawns. *Journal of Mammalogy* 60:849-850.

Harestad, A.S. 1979. Seasonal movements of black-tailed deer on northern Vancouver Island. B.C. Fish and Wildlife Branch, Ministry of Environment, Fish and Wildlife Report No. R-3.

Harestad, A.S. and Bunnell, F.L. 1979. Home range and body weight — a re-evaluation. *Ecology* 60:389-402.

Harrington, F.A. 1977. *A Guide to the Mammals of Iran*. Department of the Environment, Teheran.

Harrington, F.H. 1975. Response parameters of elicited wolf howling. Ph.D. Thesis, State University of New York, Stony Brook.

Harrington, F.H. 1978. Ravens attracted to wolf howling. *Condor* 80:236-237.

Harrington, F.H. 1981. Urine-marking and caching behavior in the wolf. *Behaviour* 76:280-288.

Harrington, F.H. and Mech, L.D. 1978. Howling at two Minnesota wolf pack summer homesites. *Canadian Journal of Zoology* 56:2024-2028.

Harrington, F.H. and Mech, L.D. 1979. Wolf howling and its role in territory maintenance. *Behaviour* 68:207-249.

Harrison, D.L. 1964. *The Mammals of Arabia I*. Ernest Benn, Ltd., London.

Harrison, D.L. 1968. *The Mammals of Arabia II*. Ernest Benn, Ltd., London.

Havkin, G.Z. 1977. Symmetry shifts in the development of interactive behaviour of two wolf pups (*Canis lupus*). M.A. Thesis, Dalhousie University, Halifax.

Hebert, D.M. 1979. Wildlife-forestry planning in the coastal forests of Vancouver Island. In O.C. Wallmo and J.C. Schoen (eds.), *Proceedings of the Black-tailed Deer Conference*, U.S. Forest Service, Juneau, Alaska, 133-158.

Helm, J. 1968. The nature of Dogrib socioterritorial groups. In R.B. Lee and I. Devore (eds.), *Man the Hunter*, Aldine, Chicago, 118-125.

Hemming, J.E. 1971. The distribution and movement patterns of caribou in Alaska. Alaska Department of Fish & Game, Game Technical Bulletin No. 1.

Hendrickson, J., Robinson, W.L. and Mech, L.D. 1975. Status of the wolf in Michigan, 1973. *American Midland Naturalist* 94:226-232.

Henshaw, R.E. and Stephenson, R.O. 1974. Homing in the gray wolf (*Canis lupus*). *Journal of Mammalogy* 55:234-237.

Henshaw, R.E., Lockwood, R., Schideler, R. and Stephenson, R.O. 1979. Experimental release of captive wolves. In E. Klinghammer (ed.), *The Behavior and Ecology of Wolves*, Garland STPM Press, New York, 319-395.

Hinde, R.A. 1970. *Animal Behaviour. A Synthesis of Ethology and Comparative Psychology*. (2nd Edition), McGraw-Hill, New York.

Hinde, R.A. and Stevenson-Hinde, J. 1976. Towards understanding relationships: dynamic stability. In P.P.G. Bateson and R.A. Hinde (eds.), *Growing Points in Ethology*. Cambridge University Press, Cambridge, 451-479.

Hook, R.A. 1981. A survey of public attitudes toward predators in six Michigan counties. M.A. Thesis, Northern Michigan University, Marquette.

Huot, J., Banville, D. and Jolicoeur, H. 1978. *Etude de la predation par le loup sur le cerf de virginie dans la region de l'Outaouais.* Que. Minist. Tour., Chasse et de la Peche, Dir. Rech. Faun.

Ilani, G. 1979. *Zoogeographical and Ecological Survey of Carnivores in Israel and Administered Areas.* Nature Reserves Authority, Tel Aviv (in Hebrew, English summary).

Inglis, A. 1978. *Northern Vagabond.* McClelland and Stewart Ltd., Toronto.

Ingsted, H. 1951. *Nunamiut — Among Alaska's Inland Eskimos.* Allen and Unwin, London.

Irving, L. 1953. The naming of birds by Nunamiut Eskimo. *Arctic* 6:35-43.

Irving. L. 1958. On the naming of birds by Eskimos. *Anthropological Papers of the University of Alaska* 6:61-77.

Irving, L. 1960. Birds of Anaktuvuk Pass, Kobuk, and Old Crow. *U.S. National Museum Bulletin* No. 217.

Ivanov, F.V. 1967. The history of the wolf in the Rjazan-area. In *Scientific Work at the Nature Reserve of Oka. Moskva.* Translated by Torsten Jansson, 229pp.

Jenness, D. 1957. *Dawn in Arctic Alaska.* University of Minnesota Press, Minneapolis.

Johnsen, S. 1929. Rovdyr- og rovfugistatistikken i Norge. *Bergens Museums Arbok,* 1-118.

Johnson, M.K. and Hansen, R.M. 1979. Coyote food habits on the Idaho National Engineering Laboratory. *Journal of Wildlife Management* 43:951-956.

Johnson, R.N. 1970. Spatial probability learning and brain stimulation in rats. *Psychonomic Science* 18:33.

Jolicoeur, P. 1959. Multivariate geographical variation in the wolf *Canis lupus* L. *Evolution* 13:283-299.

Jolicoeur, P. 1975. Sexual dimorphism and geographical distance as factors of skull variation in the wolf *Canis lupus* L. In M.W. Fox (ed.), *The Wild Canids,* Van Nostrand Reinhold, New York, 54-61.

Jolly, J. 1975. In defense of *Canis lupus. The Conservationist,* Oct.-Nov. 1975, 12-13.

Jordan, P.A., Shelton, P.C. and Allen, D.L. 1967. Numbers, turnover, and social structure of the Isle Royale wolf population. *American Zoologist* 7:233-252.

Joslin, P.W.B. 1966. Summer activities of two timber wolf (*Canis lupus*) packs in Algonquin Park. M.S. Thesis, University of Toronto.

Joslin, P.W.B. 1967. Movements and homesites of timber wolves in Algonquin Park. *American Zoologist* 7:279-288.

Kale, L.W. 1979. An integrated data system for wildlife management. M.S. Thesis, University of British Columbia, Vancouver.

*Kaletskaya, M.L. 1973. The wolf and its role as predator in the Darwin Preserve. *Transactions of the Darwin Preserve,* Vol. 2.

Kaley, M.R. 1976 (mimeo). Summary of wolf observations since spring of 1975. Glacier National Park files, West Glacier, Montana.

Kauri, H. 1957. Hundid Eestis. *Rev. Est. Lit. Sci.* 8:310-315.

Kellert, S.R. 1976. A study of American attitudes towards animals. Vol. 1 and 2. A report to the U.S. Fish and Wildlife Service, Yale University School of Medicine.

Kelsall, J.P. 1968. The migratory barren-ground caribou of Canada. Canadian Wildlife Service Monograph No. 3, Ottawa.

Kelsall, J.P. 1969. Structural adaptations of moose and deer for snow. *Journal of Mammalogy* 50:302-310.

Ketchledge, E.H. 1965. Changes in the forests of New York. *The Conservationist* 20:29-34.

Kleiber, M. 1961. *The Fire of Life: An Introduction to Animal Energetics.* Wiley and Sons, New York.

Kleiman, D.G. 1977. Monogamy in mammals. *Quarterly Review of Biology* 52:39-69.

Kleiman, D.G. and Brady, C.A. 1978. Coyote behavior in the context of recent canid research: problems and perspectives. In M. Bekoff (ed.), *Coyotes: Biology, Behavior, and Management.* Academic Press, New York, 163-188.

Kleiman, D.G. and Eisenberg, J.F. 1973. Comparisons of canid and felid social systems from an evolutionary perspective. *Animal Behaviour* 21:637-659.

Kleiman, D.G. and Malcolm, J.R. 1979. The evolution of male care in mammals. Animal Behavior Society Symposium, New Orleans, June 1979.

Klein, D.R. 1965. Postglacial distribution patterns of mammals in the southern coastal regions of Alaska. *Arctic* 18:7-20.

Klinghammer, E. (ed.). 1979. *The Behavior and Ecology of Wolves,* Garland STPM Press, New York.

Koenig, C. 1974. Zum Verhalten spanischer Geier an Kadavern. *Journal f. Ornithologie* 115:289-320.

Kolenosky, G.B. 1972. Wolf predation on wintering deer in eastcentral Ontario. *Journal of Wildlife Management* 36:357-369.

Kolenosky, G.B. and Johnston, D.H. 1967. Radio-tracking timber wolves in Ontario. *American Zoologist* 7:289-303.

Kolenosky, G.B. and Stanfield, R.O. 1975. Morphological and ecological variation among gray wolves (*Canis lupus*) of Ontario, Canada. In M.W. Fox (ed.), *The Wild Canids,* Van Nostrand Reinhold, New York, 62-72.

*Kostin, Y.V. 1970. Certain aspects of the problem "predator-prey" in the game management and forestry of the mountain Crimea. *Transactions of the 9th International Congress of Game Biologists.* Moscow.

Krajina, V.J. 1965. *Ecology of Western North America.* Vol. 1, University of British Columbia, Vancouver.

Krebs, J.R. and Davies, N.B. (eds.). 1978. *Behavioral Ecology: An Evolutionary Approach.* Blackwell, Oxford.

Kruuk, H. 1972. *The Spotted Hyena.* University of Chicago Press, Chicago.

Kruuk, H. 1975. Functional aspects of social hunting by carnivores. In G. Baerends, C. Beer and A. Manning (eds.), *Function and Evolution in Behaviour,* Oxford, Claredon Press, 119-141.

*Kudaktin, A.N. 1978. On the selectivity of ungulate hunting by the wolf in the Caucasus preserve. *Bulletin of the Moscow Society for Nature Research,* Biology Section 3 (see BMSNRBS).

*Kudaktin, A.N. 1979. Territorial distribution and population structure of wolves in the Caucasus preserve. BMSNRBS, 2.

Kuper, A. 1973. *Anthropologists and Anthropology.* Peregrine Books, Middlesex, England.

Kuyt, E. 1962. Movements of young wolves in the Northwest Territories of Canada. *Journal of Mammalogy* 43:270-271.

Kuyt, E. 1972. Food-habits and ecology of wolves on barren-ground caribou range in the Northwest Territories. Canadian Wildlife Service Report No. 21, Ottawa.

Langille, W.A. 1904 (microfilm). The proposed forest preserve on the Kenai Peninsula, Alaska. National Archives and Record Service, General Services Administration, Washington, DC, 1969. Correspondence relating to the Chugach National Forest, Part 1.

Lawrence, B. and Bossert, W.H. 1967. Multiple character analysis of *Canis lupus, latrans* and *familiaris,* with a discussion of the relationship of *Canis niger. American Zoologist* 7:223-232.

Leirfallom, J. 1970. Wolf management in Minnesota. In S.E. Jorgensen, C.E. Falkner and L.D. Mech (eds.), *Proceedings of Symposium on Wolf Management in Selected Areas of North America,* U.S. Fish and Wildlife Service Special Publications, Twin Cities, Minnesota, 9-15.

Lentfer, J.W. and Sanders, D.K. 1973. Notes on the captive wolf (*Canis lupus*) colony, Barrow, Alaska. *Canadian Journal of Zoology* 51:623-627.

Leopold, A. 1933. *Game Management.* Charles Scribner's Sons, New York.

LeResche, R.E., Bishop, R.H. and Coady, J.W. 1974. Distribution and habitats of moose in Alaska. *Naturaliste Canadien* 101:143-178.

LeRoux, P.A. 1975. Wolf survey-inventory report — 1973. Annual Report Survey-Inventory Act, Part III, Vol. V, Federal Aid in Wildlife Restoration Project W-17-6, Juneau, Alaska, 197pp.

Levi-Strauss, C. 1969. *The Elementary Structures of Kinship* (Les Structures elementaires de la Parente). Eyre & Spottiswoode Ltd., London.

Levi-Strauss, C. 1979. *Myth and Meaning.* Schocken Books, New York.

Lewontin, R.C. 1970. The units of selection. *Annual Review of Ecology and Systematics* 1:1-18.

Linhart, S.B. and Knowlton, F.F. 1975. Determining the relative abundance of coyotes by scent station lines. *Wildlife Society Bulletin* 3:119-124.

Lockie, J.D. 1959. The estimation of the food of foxes. *Journal of Wildlife Management* 23:224-227.

Lockwood, R. 1976. An ethnological analysis of social structure and affiliation in captive wolves (*Canis lupus*). Ph.D. Thesis, Washington University, St. Louis.

Lockwood, R. 1979. Dominance in wolves: useful construct or bad habit? In E. Klinghammer (ed.), *The Behavior and Ecology of Wolves.* Garland STPM Press, New York, 225-244.

Loether, R. 1978. An analysis of the foraging strategy of three species of wild canidae. M.S. Thesis, Utah State University.

Loizos, C. 1966. Play in mammals. *Zoological Society of London Symposium* 18:1-9.

Lopez, B.H. 1978. *Of Wolves and Men.* Charles Scribner's Sons, New York.

Ludlow, W. 1876. Report of a reconnaissance from Carroll, Montana Territory, on the upper Missouri, to the Yellowstone National Park and return, made in the summer of 1875 (cited from Curnow, 1969).

Lutz, H.J. 1960. History of the early occurrence of moose on the Kenai Peninsula and in other sections of Alaska. Misc. Pub. No. 1, Alaska Forest Research Center, U.S. Forest Service.

Macdonald, D.W. 1977. The behavioural ecology of the red fox, *Vulpes vulpes:* a study of social organisation resource exploitation. Ph.D. Thesis, Oxford University, Oxford.

Macdonald, D.W. 1979. 'Helpers' in fox society. *Nature* 282:69-71.

Mackintosh, N.J. 1969. Comparative studies of reversal and probability learning: rats, birds and fish. In R.M. Gilbert and N.S. Sutherland (eds.), *Animal Discrimination Learning.* Academic Press, London, 137-162.

Mackintosh, N.J. 1974. *The Psychology of Animal Learning.* Academic Press, New York.

Mackintosh, N.J., Lord, J. and Little, L. 1971. Visual and spatial probability learning in pigeons and goldfish. *Psychonomic Science* 24:221-223.

Macpherson, A.H. 1969. The dynamics of Canadian arctic fox populations. Canadian Wildlife Service Report Series No. 8.

Maine, H. 1861. *Ancient Law.* Reprinted by J.M. Dent & Sons Ltd., London, 1917 (last ed. 1972).

Manning, T.H. and Macpherson, A.H. 1958. *The Mammals of Banks Island.* Arctic Institute of North America, Technical Paper No. 2.

Marchinton, R.L. and Jeter, L.K. 1966. Telemetric study of deer movement-ecology in the southeast. *20th Annual Conference of the Southeastern Association of Game and Fish Commissions,* 189-206.

Markowitz, H. and Stevens, V.J. (eds.). 1978. *Behavior of Captive Wild Animals.* Nelson-Hall, Chicago.

Martin, G.C., Johnson, B.L. and Grant, U.S. 1915. Geology and mineral resources of the Kenai Peninsula, Alaska. *U.S. Geological Survey Bulletin* 587. U.S. Government Printing Office, Washington, DC.

Marvin, M.Ja. 1959. *Mlekopita'uscie Karelii.* Petrozavodsk.

Mason, W. 1979. Ontogeny of social behavior. In P. Marler and J.G. Vandenbergh (eds.), *Handbook of Behavioral Neurobiology, Volume 3: Social Behavior and Communication.* Plenum Press, New York, 1-28.

Mattson, U. and Ream, R.R. 1978. The current status of the gray wolf (*Canis lupus*) in the Rocky Mountain Front. Wolf Ecology Project, University of Montana, Missoula, 18pp.

Mauss, M. 1967. *The Gift.* W.W. Norton, New York.

Mayer, J. 1953. Caloric requirements and obesity in dogs. Gaines Veterinary Symposium, New York.

Maynard Smith, J. 1964. Group selection and kin selection. *Nature* 201:1145-1147.

Maynard Smith, J. 1977. Parental investment: a prospective analysis. *Animal Behaviour* 25:1-9.

Maynard Smith, J. 1978. *The Evolution of Sex.* Cambridge University Press, Cambridge.

Mayr, E. 1974. *Populations, Species and Evolution: An Abridgement of Animal Species and Evolution.* The Belknap Press of Harvard University Press, Cambridge, Massachusetts.

McIlroy, C. 1974. Moose survey-inventory progress report — 1972, Game Management Unit 13. In D.E. McKnight (ed.), *Annual Report of Survey-Inventory Activities. Part II. Moose, Caribou, Marine Mammals and Goat.* Alaska Federal Aid in Wildlife Restoration Report, Project W-17-5, 67-74.

McKnight, D.E. 1970 (unpublished). The history of predator control in Alaska. Alaska Department of Fish and Game.

McKnight, T. 1964. *Feral lifestock in Anglo-America.* University of California Press, Berkeley and Los Angeles.

McNab, B. 1971. On the ecological significance of Bergmann's rule. *Ecology* 52:845-854.

Mech, L.D. 1966. *The Wolves of Isle Royale.* U.S. National Park Service, Fauna Series 7, Washington, DC.

Mech, L.D. 1970. *The Wolf: The Ecology and Behavior of an Endangered Species.* Natural History Press, Garden City, New York.

Mech, L.D. 1972. Spacing and possible mechanisms of population regulation in wolves. *American Zoologist* 12:9.

Mech, L.D. 1973. *Wolf Numbers in the Superior National Forest of Minnesota.* U.S. Department of Agriculture, Forest Service Research Paper NC-97, St. Paul, Minnesota.

Mech, L.D. 1974a. *Canis lupus.* Mammalian Species, No. 37.

Mech, L.D. 1974b. Current techniques in the study of elusive wilderness carnivores. In D.H. Pimlott (ed.), *Wolves.* Proceedings of the 11th International Congress of Game Biol. 11:315-322.

Mech, L.D. 1974c. A new profile of the wolf. *Natural History* 83(4):26-31.

Mech, L.D. 1975. Disproportionate sex ratios of wolf pups. *Journal of Wildlife Management* 39:737-740.

Mech, L.D. 1977a. Population trend and winter deer consumption in a Minnesota wolf pack. In R.L. Phillips and C. Jonkel (eds.), *Proceedings of the 1975 Predator Symposium.* University of Montana, Missoula, 55-83.

Mech, L.D. 1977b. A recovery plan for the eastern timber wolf. *National Parks and Conservation Magazine,* January 1977, 17-21.

Mech, L.D. 1977c. Productivity, mortality and population trends of wolves in northeastern Minnesota. *Journal of Mammalogy* 58:559-574.

Mech, L.D. 1977d. Wolf-pack buffer zones as prey reservoirs. *Science* 198:320-321.

Mech, L.D. 1978. The wolf: an introduction to its behavior, ecology and conservation. Edinburgh Wolf Symposium.

Mech, L.D. 1979. Some considerations in re-establishing wolves in the wild. In E. Klinghammer (ed.), *The Behavior and Ecology of Wolves,* Garland STPM Press, New York.

Mech, L.D. and Frenzel, Jr., L.D. 1971. *Ecological Studies of the Timber Wolf in Northeastern Minnesota.* USDA Forest Service Research Paper, NC-52, St. Paul, Minnesota.

Mech, L.D., Frenzel, Jr., L.D. and Karns, P.D. 1971a. The effect of snow conditions on the vulnerability of white-tailed deer to wolf predation. In L.D. Mech and L.D. Frenzel (eds.), *Ecological Studies of the Timber Wolf in Northeastern Minnesota*. USDA Forest Service Research Paper, NC-52, St. Paul, Minnesota.

Mech, L.D., Frenzel, Jr., L.D., Ream, R.R. and Winship, J.W. 1971b. Movements, behaviour and ecology of timber wolves in northeastern Minnesota. In L.D. Mech and L.D. Frenzel (eds.), *Ecological Studies of the Timber Wolf in Northeastern Minnesota*. USDA Forest Service Research Paper NC-52, St. Paul, Minnesota.

Mech, L.D. and Karns, P.D. 1977. Role of the wolf in a deer decline in the Superior National Forest. USDA Forest Service Research Paper NC-148, St. Paul, Minnesota, 23pp.

Medjo, D.C. and Mech, L.D. 1976. Reproductive activity in nine- and ten-month old wolves. *Journal of Mammalogy* 57:406-408.

Mendelssohn, H. 1974. The development of the populations of gazelles in Israel and their behavioural adaptations. In V. Geist and F. Walther (eds.), *The Behaviour of Ungulates and Its Relation to Management*. International Union for Conservation of Nature and Natural Resources, Morges, Switzerland.

Mendelssohn, H., Golani, I. and Marder, U. 1971. Agricultural development and the distribution of venomous snakes and snake bite in Israel. In A. de Vries and E. Kochva (eds.), *Toxins of Animal and Plant Origin*. Gordon and Breach, London.

Mendenhall, W.C. 1900. A reconnaissance from Resurrection Bay to the Tanana River, Alaska, in 1898. In *20th Annual Report of U.S. Geological Survey*, Part 7, 271-340.

Meyer-Holzapfel, M. 1956. Das spiel bei saugetieren. *Handbook of Zoology* 8(10)5:1-36.

*Michurin, L.N. 1970. Influence of the wolf on the population of wild northern elk in the north of central Siberia. *Transactions of the 9th International Congress of Game Biologists*, Moscow.

Miller, D.R. 1975. Observations of wolf predation on barren ground caribou in winter. In J.R. Luick, P.C. Lent, D.R. Klein and R.G. White (eds.), *Proceedings of the 1st International Reindeer and Caribou Symposium*, University of Alaska, Fairbanks, 209-220.

Miller, F.L. and Broughton, E. 1974. Calf mortality during 1970 on the calving ground of the Kaminuriak caribou. Canadian Wildlife Service Report No. 26, Ottawa.

Minnesota Department of Natural Resources. 1980. Minnesota timber wolf management plan, 1980. St. Paul, Minnesota, 20pp.

Moehlman, P.D. 1979. Jackal helpers and pup survival. *Nature* 277:382-383.

Moffitt, F.H. 1906. Mineral resources of Kenai Peninsula. *U.S. Geological Survey Bulletin* 277.

Mohr, C.O. 1947. Table of equivalent populations in North American small mammals. *American Midland Naturalist* 37:223-249.

Mongeau, N. 1961. Hepatic distomatosis and infectious canine hepatitis in northern Manitoba. *Canadian Veterinarian Journal* 2:33-38.

Montgomery, G.G. 1959. Social behavior in a refuge population of white-tailed deer (*Odocoileus virginianus*) studied with specially developed marking techniques. M.S. Thesis, Pennsylvania State University, University Park.

Moore, T.D., Spence, L.E., Dugnolle, C.E. and Hepworth, W.B. 1974. Identification of dorsal guard hairs of some mammals of Wyoming. Wyoming Game and Fish Department, Bulletin No. 14, Cheyenne, Wyoming.

Moran, G. 1977. *The Structure in Supplanting Interactions in the Wolf*. Ph.D. Thesis, Dalhousie University, Halifax.

Moran, G. and Fentress, J.C. 1979. A search for order in wolf social behaviour. In E. Klinghammer (ed.), *The Behavior and Ecology of Wolves*. Garland STPM Press, New York, 245-293.

Moran, G., Fentress, J.C. and Golani, I. (in press). A description of relational patterns during "ritualized fighting" in wolves. *Animal Behaviour*.

Mueller, D. 1977. Probability learning in prey selection with a red-tailed hawk and a great horned owl. M.S. Thesis, Utah State University.

Murie, A. 1944. *The Wolves of Mt. McKinley*. U.S. National Park Service, Fauna Series 5, Washington, DC.

Murie, A. 1963. *A Naturalist in Alaska*. Doubleday, New York.

Nasimovich, A.A. 1955. *The Role of the Regime of Snow Cover in the Life of Ungulates in the USSR*. Akademiya Nauk S.S.R., Moskva (English translation, Canadian Wildlife Service, Ottawa).

National Oceanic and Atmospheric Administration. 1950-78. Climatological Data, Environmental Data Service, Asheville, Vols. 64-84.

Neiland, K.A. 1970. Rangiferine brucellosis in Alaskan canids. *Journal of Wildlife Diseases* 6:136-139.

Neiland, K.A. 1975. Further observations on rangiferine brucellosis in Alaskan carnivores. *Journal of Wildlife Diseases* 11:45-53.

Nelson, M.E. 1979. Home range location in white-tailed deer. USDA Forest Service Research Paper NC-173, St. Paul, Minnesota.

Nelson, M.M. 1971. Predator management with emphasis on the timber wolf. In M. Nelson (ed.), *Proceedings of Symposium on the White-Tailed Deer in Minnesota*. Minnesota Department of Natural Resources and Minnesota Chapter of the Wildlife Society, St. Paul, Minnesota.

Nevo, E. 1969. Mole rat *Spalax ehrenbergi:* Mating behaviour and its evolutionary significance. *Science* 163:484-486.

Nevo, E. 1973. Variation and evolution in the subterranean mole rat, *Spalax*. *Israel Journal of Zoology* 22:207-208.

Nevo, E. and Shaw, C.R. 1972. Genetic variation in a subterranean mammal, *Spalax ehrenbergi*. *Biochemical Genetics* 7:235-241.

New York Crop Reporting Service. 1972. 1969 U.S. Census of Agriculture — New York. Bureau of Statistics, New York State Department of Agriculture and Markets, New York.

Nie, N.H., Hull, C.H., Jenkins, J.G., Steinbrenner, K. and Bent, D.H. 1975. *SPSS: Statistical Package for the Social Sciences*. McGraw-Hill Book Co., 675pp.

Novikov, G.A. 1956. *Carnivorous Mammals of the Fauna of the USSR*. Israel Program for Scientific Translations, Jerusalem (1962).